SATELLITES, OCEANOGRAPHY AND SOCIETY

Elsevier Oceanography Series
Series Editor: David Halpern (1993-)

FURTHER TITLES IN THIS SERIES
Volumes 1-7, 11, 15, 16, 18, 19, 21, 23, 29 and 32 are out of print.

8 E. LISITZIN
SEA-LEVEL CHANGES
9 R.H. PARKER
THE STUDY OF BENTHIC COMMUNITIES
10 J.C.J. NIHOUL (Editor)
MODELLING OF MARINE SYSTEMS
12 E.J. FERGUSON WOOD and R.E. JOHANNES
TROPICAL MARINE POLLUTION
13 E. STEEMANN NIELSEN
MARINE PHOTOSYNTHESIS
14 N.G.JERLOV
MARINE OPTICS
17 R.A. GEYER (Editor)
SUBMERSIBLES AND THEIR USE IN OCEANOGRAPHY AND OCEAN ENGINEERING
20 P.H. LEBLOND and L.A. MYSAK
WAVES IN THE OCEAN
22 P. DEHLINGER
MARINE GRAVITY
24 F.T. BANNER, M.B. COLLINS and K.S. MASSIE (Editors)
THE NORTH-WEST EUROPEAN SHELF SEAS: THE SEA BED AND THE SEA IN MOTION
25 J.C.J. NIHOUL (Editor)
MARINE FORECASTING
26 H.G. RAMMING and Z. KOWALIK
NUMERICAL MODELLING MARINE HYDRODYNAMICS
27 R.A. GEYER (Editor)
MARINE ENVIRONMENTAL POLLUTION
28 J.C.J. NIHOUL (Editor)
MARINE TURBULENCE
30 A. VOIPIO (Editor)
THE BALTIC SEA
31 E.K. DUURSMA and R. DAWSON (Editors)
MARINE ORGANIC CHEMISTRY
33 R.HEKINIAN
PETROLOGY OF THE OCEAN FLOOR
34 J.C.J. NIHOUL (Editor)
HYDRODYNAMICS OF SEMI-ENCLOSED SEAS
35 B. JOHNS (Editor)
PHYSICAL OCEANOGRAPHY OF COASTAL AND SHELF SEAS
36 J.C.J. NIHOUL (Editor)
HYDRODYNAMICS OF THE EQUATORIAL OCEAN
37 W. LANGERAAR
SURVEYING AND CHARTING OF THE SEAS
38 J.C.J. NIHOUL (Editor)
REMOTE SENSING OF SHELF-SEA HYDRODYNAMICS
39 T.ICHIYE (Editor)
OCEAN HYDRODYNAMICS OF THE JAPAN AND EAST CHINA SEAS
40 J.C.J. NIHOUL (Editor)
COUPLED OCEAN–ATMOSPHERE MODELS
41 H. KUNZENDORF (Editor)
MARINE MINERAL EXPLORATION
42 J.C.J NIHOUL (Editor)
MARINE INTERFACES ECOHYDRODYNAMICS
43 P. LASSERRE and J.M. MARTIN (Editors)
BIOGEOCHEMICAL PROCESSES AT THE LAND-SEA BOUNDARY
44 I.P. MARTINI (Editor)
CANADIAN INLAND SEAS
45 J.C.J. NIHOUL (Editor)
THREE-DIMINSIONAL MODELS OF MARINE AND ESTUARIN DYNAMICS
46 J.C.J. NIHOUL (Editor)
SMALL-SCALE TURBULENCE AND MIXING IN THE OCEAN
47 M.R. LANDRY and B.M. HICKEY (Editors)
COASTAL OCENOGRAPHY OF WASHINGTON AND OREGON
48 S.R. MASSEL
HYDRODYNAMICS OF COASTAL ZONES
49 V.C. LAKHAN and A.S. TRENHAILE (Editors)
APPLICATIONS IN COASTAL MODELING
50 J.C.J. NIHOUL and B.M. JAMART (Editors)
MESOSCALE SYNOPTIC COHERENT STRUCTURES IN GEOPHYSICAL TURBULENCE
51 G.P. GLASBY (Editor)
ANTARCTIC SECTOR OF THE PACIFIC
52 P.W. GLYNN (Editor)
GLOBAL ECOLOGICAL CONSEQUENCES OF THE 1982–83 EL NINO–SOUTHERN OSCILLATION
53 J. DERA (Editor)
MARINE PHYSICS
54 K. TAKANO (Editor)
OCEANOGRAPHY OF ASIAN MARGINAL SEAS
55 TAN WEIYAN
SHALLOW WATER HYDRODYNAMICS
56 R. CHARLIER and J. JUSTUS
OCEAN ENERGIES, ENVIRONMENTAL, ECONOMIC AND TECHNOLOGICAL ASPECTS OF ALTERNATIVE POWER SOURCES
57 P.C. CHU and J.C. GASCARD (Editors)
DEEP CONVECTION AND DEEP WATER FORMATION IN THE OCEANS
58 P.A. PIRAZZOLI
WORLD ATLAS OF HOLOCENE SEA-LEVEL CHANGES
59 T. TERAMOTO (Editor)
DEEP OCEAN CIRCULATION-PHYSICAL AND CHEMICAL ASPECTS
60 B. KJERFVE (Editor)
COASTAL LAGOON PROCESSES
61 P. MALANOTTE-RIZZOLI (Editor)
MODERN APPROACHES TO DATA ASSIMILATION IN OCEAN MODELING
62 H.W.A. BEHRENS, J.C. BORST, L.J. DROPPERT, and J.P. VAN DER MEULEN (Editors)
OPERATIONAL OCEANOGRAPHY

Elsevier Oceanography Series, 63

SATELLITES, OCEANOGRAPHY AND SOCIETY

Edited by

David Halpern

Jet Propulsion Laboratory,
California Institute of Technology,
Pasadena, USA

2000
ELSEVIER
Amsterdam – Lausanne – New York – Oxford – Shannon – Singapore – Tokyo

ELSEVIER SCIENCE B.V.
Sara Burgerhartstraat 25
P.O. Box 211, 1000 AE Amsterdam, The Netherlands

First edition 2000

Library of Congress Cataloging in Publication Data
A catalog record from the Library of Congress has been applied for.

ISBN: 0 444 50501 6

∞ The paper used in this publication meets the requirements of ANSI/NISO Z39.48-1992 (Permanence of Paper).
Printed in The Netherlands.

Contents

Preface

Our world evolved from and increasingly depends on the waters surrounding us, making an understanding of the ocean critical to the future of this planet. Since their beginning in 1978, satellite measurements have been changing the course of oceanographic research. Twenty years later, the International Year of the Ocean and EXPO '98 provided a confluence of time and place to highlight the outstanding scientific advances made possible by satellite observations of the ocean and the benefits to all of such rapidly improving knowledge. In recognition of this unique opportunity, the International Conference on Satellites, Oceanography and Society (ICSOS) marked the first time the five space agencies involved in global observation of the ocean cosponsored an oceanographic conference. Scientists from twenty-eight nations came together to discuss forecasting weather and climate variability to mitigate natural disasters and improve the quality of life, managing fisheries for long-term conservation, preserving marine ecosystems for future generations, and creating an integrated ocean observing system. This book is a permanent legacy of the conference.

All ICSOS participants were invited to contribute a manuscript prior to 1 April 1999. Thirty-two manuscripts were submitted and underwent anonymous peer review by two reviewers, resulting in the nineteen manuscripts published here. I wish to express my sincere appreciation to all the authors.

I am truly thankful to the followings reviewers who generously gave their time and contributed substantial expertise: Mark Abbott, Oregon State University; Meinrat Andreae, Max Planck Institute for Chemistry; Des Barton, University of Wales; Amy Bower, Woods Hole Oceanographic Institution; Otis Brown, University of Miami; David Carter, Satellite Observing Systems; Ping Chang, Texas A&M University; Dudley Chelton, Oregon State University; Paula Coble, University of South Florida; David Cotton, Satellite Observing Systems; Jorge Csirke, Fisheries and Agriculture Organization; Curtiss Davis, Naval Research Laboratory; Thierry Delcroix, Centre Institut de Recherche pour le Developpement du Noumea; Paul Falkowski, Rutgers University; David Foley, National Oceanic and Atmospheric Administration; Robert Frouin, Scripps Institution of Oceanography; Michael Glantz, National Center for Atmospheric Research; David Glover, Woods Hole Oceanographic Institution; James Goerss, Naval Research Laboratory; Hans Graber, University of Miami; Nicholas Grima, Université Pierre et Marie Curie; Trevor Guymer, Southampton Oceanography Centre; David Halpern, Jet Propulsion Laboratory; Stefan Hastenrath, University of Wisconsin; Larry Hutchings, Sea Fisheries Research Institute; Kaisa Kononen, Maj and Tor Nessling Foundation; Alexei Kosarev, Moscow State University; Yochanan Kushnir, Lamont-Doherty Earth Observatory; Mojib Latif, Max-Planck-Institut für Meteorologie; William Lau, Goddard Space Flight Center; Michael Laurs, National Oceanic and Atmospheric Administration; Jean-Michel Lefevre, Meteo-France; Patrick Lehodey, Secretariat of the Pacific Community; Pierre-Yves Le Traon, Collect Localisation Satellites; Yukio Masumoto, University of Tokyo; Kendall Melville, Scripps Institution of Oceanography; Jerry Miller, Naval Research Laboratory; Ekkehard Mittelstaedt, Bundesamt für Seeschiffahrt und Hydrographie; Frank Muller-Karger, University of South Florida; Raghuran Murtugudde, University of Maryland; Neville Nicholls, Bureau of

Meteorology Research Centre; Temel Oguz, Middle East Technical University; Paulo Polito, Jet Propulsion Laboratory; Roger Pulwarty, National Oceanic and Atmospheric Administration; Keith Raney, The Johns Hopkins University; Michele Reinecker, Goddard Space Flight Center; Laurie Richardson, Florida International University; Paola Rizzoli, Massachusetts Institute of Technology; Ernesto Rodriguez, Jet Propulsion Laboratory; Edward Sarachik, University of Washington; Peter Schlüssel, European Organization for the Exploitation of Meteorological Satellites; Meric Srokosz, Southampton Oceanography Centre; Ad Stoffelen, Royal Netherlands Meteorological Institute; Dariusz Stramski, Scripps Institution of Oceanography; Halil Sur, Istanbul University; Paris Vachon, Canada Centre for Remote Sensing; Robert Weisberg, University of South Florida; Donna Witter, Lamont-Doherty Earth Observatory; Simon Yueh, Jet Propulsion Laboratory; Walter Zenk, Institut für Meereskunde.

It is with great pleasure that I acknowledge the ICSOS sponsors — Centre National d'Etudes Spatiales, European Space Agency, EXPO '98, Intergovernmental Oceanographic Commission, National Aeronautics and Space Administration, National Oceanic and Atmospheric Administration, National Space Development Agency of Japan, Scientific Committee on Oceanic Research, and the World Climate Research Programme — for their generous financial contributions, which made possible the conference and the publication of this book.

I am especially grateful to Dr. Eric Lindstrom, NASA Headquarters, for support to convene ICSOS and to edit this book. The work was performed, in part, at the Jet Propulsion Laboratory, California Institute of Technology, under contract with the National Aeronautics and Space Administration. As a statement of full disclosure, I shall not receive any royalties from this book.

David Halpern

Satellites, Oceanography and Society
edited by David Halpern

Chapter 1

Editor's note: Professor Walter Munk was called upon to make two presentations on the first day of the Conference. During Monday morning he gave a scientific lecture regarding results from the Acoustic Thermometry of Ocean Climate project. On Monday evening, at the Conference Opening Ceremony at the EXPO '98 Ocean Pavilion, Dr. Munk's entertaining keynote address is a wonderful introduction to "Satellites, Oceanography and Society."

Oceanography before, and after, the advent of satellites

Walter Munk
Scripps Institution of Oceanography, La Jolla, California

Yes, there was oceanography (Figure 1) even before the advent of satellite oceanography. More importantly, programs like the Acoustic Thermometry of Ocean Climate (ATOC) provide a strong incentive for a combined Earth- and space-based observing program.

Since the days of the Challenger expedition in the 1870s, oceanography has been traditionally conducted by sounding the oceans from a few moving ships. Accordingly, successive soundings are associated with changes in both the space dimension and the time dimension; however, the measured changes were nearly always attributed to the space dimension. The inevitable result is a climatology steady in time and increasingly complex in space. The first law of ocean research was to never waste your assets by occupying the same station twice! And when this law was violated and the results differed, the differences could be attributed to equipment malfunctioning.

This age came to an abrupt end in the 1960s with the discovery of mesoscale variability: the "ocean weather" associated with scales of 100 km and 100 days. We now know that mesoscale currents are responsible for more than 95% of the ocean's kinetic energy. For one hundred years this overwhelming mesoscale dynamics had fallen through the loose mesh of traditional sampling!

Figure 1. The Red Sea parted, allowing Moses and the Israelites to escape the pursuing soldiers of the Pharaoh (by permission of Pictures Now! Powered by Wood River Media, Inc., 1998, Wood River Media, June 1998; http://www.lycos.com/picturethis/religion/judaism/history/bible_stories/crossing_the_red_sea/310521.html). The picture suggests a tsunami with high nonlinear distortion, what is now called a soliton of depression. Tsunamis following earthquakes have been reported in the area. According to an eyewitness report in A.D. 363, "the sea...was driven back...from the land, revealing...deep valleys which nature had hidden in the unplumbed depth; then...the great mass of waters, returning when it was least expected, killed many thousands of men by drowning." A similar tsunami was experienced in the first century A.D. by John the Divine (Nur 1991; Nur and McAskill 1991).

Chronic undersampling leads to curious aberrations. Wherever and whenever you make measurements, there is unexpected activity. Fritz Fuglister once gave a paper with a title like: "Why is it that the Gulf Stream follows oceanographic research vessels?" And even today ocean climatic changes seem to be happening at the few places where long-term time series have been taken.

If I were to choose a single phrase to characterize the first century of modern oceanography, it would be "a century of undersampling." The most profound effect of satellite oceanography has not been the resulting new sensor packages (and these have been remarkable), nor the global coverage, but rather that for the first time ocean processes were adequately sampled.

In the early 1970s, when the first oceanographic satellite—SeaSat—was being planned, we were living under the axiom that what is not done from ships is not oceanography. Upon hearing that satellite altimeters would measure dynamic heights, a well-known oceanographer replied, "If you gave it to me I wouldn't know what to do with it." Satellite altimetry is now an outstanding success story. Its contribution towards understanding ocean processes goes well beyond anything that had been imagined.

But even so, there is much value added by combining the space observations with "sea truth." ATOC can serve as an illustration. Sound travels faster in a warmer ocean. Thus, the travel time of an acoustic pulse is a measure of the mean temperature of the intervening water between source and receiver. For a typical ATOC range of 5000 km, there are many tens of arrivals for a single emitted pulse. Early arrivals travel along steep rays extending from surface to bottom; therefore, their travel time is affected by the temperature profile of the entire water column. Late arrivals hug the sound "axis," typically at 1-km depth; each ray weights the water column in a different way. By combining the dataset of arrival times one can estimate the vertical temperature profile and associated heat content averaged horizontally between source and receiver.

From the changes in the acoustically derived temperature profiles between one transmission and the next one can derive the changes in the mean sea level along the acoustic paths. These derived changes can be compared to those measured by satellite altimetry. It was found that in the northeast Pacific the acoustically derived month-to-month sea level changes, and those from one year to the next, were only about half those measured by Topography Experiment (TOPEX)/Poseidon altimetry (ATOC Consortium 1998). The most plausible explanation is that thermal expansion is only one part of the process leading to seasonal changes in sea level; there must be a comparable contribution from flow divergence (as in surface tides).

Satellite altimetry is an incomplete proxy for ocean heat storage. But satellite altimetry and acoustic thermometry combined can give a very accurate measure of the heat storage in an ocean basin. The two methods are nicely complementary. Satellite altimetry has good horizontal resolution, fair time resolution, and essentially no depth resolution. Acoustic thermometry has poor horizontal resolution (there are a limited number of receiver stations), good time resolution, and fair depth resolution. The complementarity of the two sets of measurements was the theme of an early paper by Munk and Wunsch (1982) proposing the ATOC experiment. The combined measurements give more information than the sum of the two separate measurements: $1 + 1 = 3$. To the audience of satellite aficionados we plead for closer cooperation with Earth-based oceanography.

References

*ATOC Consortium, Ocean climate change: comparison of acoustic tomography, satellite altimetry, and modeling, *Science*, *281*, 1327–1332, 1998.

Munk, W., and C. Wunsch, Observing the ocean in the 1900s, *Phil. Tran. Roy. Soc. Lond. A*, *307*, 439–464, 1982.

Nur, A., And the walls came tumbling down, *New Scientist*, *6*, 45–48, 1991.

Nur, A., and C. McAskill, The walls came tumbling down—earthquakes in the Holy Land—a video documentary, Geophysics Department, Stanford University, Palo Alto, California, 1991.

* The ATOC Consortium: A. B. Baggeroer, T. G. Birdsall, C. Clark, J. A. Colosi, B. D. Cornuelle, D. Costa, B. D. Dushaw, M. Dzieciuch, A. M. G. Forbes, C. Hill, B. M. Howe, J. Marshall, D. Menemenlis, J. A. Mercer, K. Metzger, W. Munk, R. C. Spindel, D. Stammer, P. F. Worcester, and C. Wunsch.

A. B. Baggeroer is in the Department of Ocean Engineering, Massachusetts Institute of Technology (MIT), Cambridge, MA 02139; T. G. Birdsall and K. Metzger are in the Department of Electrical Engineering and Computer Sciences, University of Michigan, Ann Arbor, MI 48109; C. Clark is in the Laboratory of Ornithology, Cornell University, Ithaca, NY 14853; J. A. Colosi is in the Department of Applied Ocean Physics and Engineering, Woods Hole Oceanographic Institution, Woods Hole, MA 02543; B. D. Cornuelle, M. Dzieciuch, W. Munk, and P. F. Worcester are at Scripps Institution of Oceanography, University of California, San Diego, La Jolla, CA 92093; D. Costa is in the Biology Department, University of California, Santa Cruz, CA 95064; B. D. Dushaw, B. M. Howe, J. A. Mercer, and R. C. Spindel are at the Applied Physics Laboratory, University of Washington, Seattle, WA 98105; A. M. G. Forbes is at the Division of Oceanography, CSIRO, Hobart, Tasmania 7001, Australia; C. Hill, J. Marshall, D. Menemenlis, D. Stammer, and C. Wunsch are in the Department of Earth, Atmospheric, and Planetary Sciences, MIT, Cambridge, MA 02139.

Walter Munk, Institute of Geophysics and Planetary Physics, Scripps Institution of Oceanography, University of California at San Diego, La Jolla, CA 92093-0225, U.S.A. (email, wmunk@ucsd.edu; fax, +1-858-534-6251)

Satellites, Oceanography and Society
edited by David Halpern

Chapter 2

Development and application of satellite retrievals of ocean wave spectra

Patrick Heimbach and Klaus Hasselmann
Max-Planck-Institut für Meteorologie, Hamburg, Germany

Abstract. The launch of SeaSat in 1978 demonstrated the feasibility of measuring ocean wave heights and imaging the corresponding two-dimensional wave field from space. With the launch of the first European Remote-sensing Satellite (ERS-1) in 1991, wave researchers and operational forecasters obtained global, continuous, quasi-real-time wave data for the first time. This led to the developments of sophisticated, so-called "third-generation" wave models, such as the Wave Model (WAM), and spectral retrieval algorithms for synthetic aperture radar (SAR) data. To achieve these goals, however, significant hurdles had to be overcome. Wave modelers had to develop numerically viable parameterizations of the nonlinear wave-wave interactions. The remote-sensing challenge was to understand and resolve the strong nonlinearities besetting SAR imaging of the moving ocean wave surface. This paper reviews the progress achieved over the last twenty years and summarizes wave data assimilation methods and other current applications of ERS quasi-real-time global SAR wave spectral data or SAR wave-mode product.

Two applications are presented. A comparison of wave spectra predicted by WAM with spectra retrieved from ERS-1 on a global scale revealed that WAM overpredicted local wind-generated sea surface heights and underpredicted swell. The former can be largely attributed to wind-forcing errors, while the latter is most likely due to an overly strong swell dissipation in WAM. Assimilation of ERS-1 altimeter sea surface height data into the WAM spectra was found to not alter the qualitative conclusions of the comparison. A second application addresses the trans-ocean propagation of swell. Swell propagating from a storm in the South Pacific is traced over a period of ten days with ERS-1 SAR and compared with model predictions. Wind fields used for wave predictions are also compared with ERS-1 wind scatterometer data.

1. Introduction

Major increases in computing performance have enabled the development of comprehensive atmospheric and oceanic general circulation models (A/OGCMs). A similarly impressive expansion of global datasets available for initialization, validation, and assimilation into A/OGCMs has been enabled by a series of sophisticated Earth-observing satellite missions. Although less well known, similar efforts have been undertaken in the field of ocean wave remote sensing and modeling.

Since surface wave fields are two dimensional, a statistical description of local sea state requires the two-dimensional spectrum, $F(\mathbf{k})$, of the distribution of wave energy (or, equivalently, the variance of sea surface elevation) with respect to the propagation wavenumber, $\mathbf{k}$. Modern, state-of-the-art third-generation wave models, such as the Wave Model (WAM) (WAMDI Group 1988), solve a spectral energy balance equation for the evolution of $F(\mathbf{k})$ under the influence of wave generation by wind, nonlinear wave-wave interactions, and dissipation due to wave breaking. For a detailed overview, see Komen et al. (1994). WAM is currently operational at numerous numerical weather prediction (NWP) centers, such as the European Centre for Medium-Range Weather Forecasts (ECMWF), and is implemented at about 120 research institutions worldwide. A recent evaluation of four widely used contemporary wave models may be found in Cardone et al. (1996) in relation to two rare extreme events in the North Atlantic: the "Storm of the Century" of March 1993 and the "Halloween Storm" of October 1991, an account of which was given in the best-seller by Junger (1997).

Various techniques have been developed to assimilate satellite and in-situ wave measurements into wave models. Janssen and Viterbo (1996) have shown the important role of the moving sea surface in the transfer of momentum and, presumably, other properties across the air-sea interface. As a consequence, NWP centers have started (e.g., ECMWF) or others are planning (e.g., United Kingdom Meteorological Office, Deutsches Wetterdieust) to couple wave models to their operational atmospheric GCMs to provide a more realistic boundary condition at the air-sea interface.

The ocean-observing satellites ERS-1/2 and the follow-on Environmental Satellite (ENVISAT) motivated the development of more-sophisticated wave models than the operational models of the 1980s. Limitations of these parametric models were compiled by the SWAMP Group (1985). The ERS-1/2 missions have been able, for the first time, to provide continuous, global, near-real-time measurements of both significant wave height and the two-dimensional wave spectrum. Other satellites, such as the geodetic satellite (Geosat) and the Topography Experiment (TOPEX)/Poseidon satellite, which carry radar altimeters, have also provided accurate measurements of significant wave height, but only in an off-line mode. While these data were valuable for model validation and wave statistics, the near-real-time ERS radar altimeter wave height data, in combination with real-time measurements of the two-dimensional wave spectrum with a synthetic aperture radar (SAR) operating in a global mode, offered an exciting new prospect for operational wave forecasting and research. However, this is only possible with retrieval algorithms and wave models that fully exploit the two-dimensional wave spectral information contained in the SAR data. The focus of this review, therefore, is on methods of retrieval of two-dimensional wave spectra from SAR image spectra and application of the retrieved wave spectra. Satellite missions carrying radar sensors applicable to ocean wind and wave measurements are summarized in Table 1.

The SAR is an all-weather, day-and-night, side-looking radar that emits short microwave pulses and processes a two-dimensional image from the received backscatter electromagnetic radiation. The cross-track, or range, coordinate of the backscattered energy is inferred, as is the case with a real-aperture radar (RAR), from the travel time of the pulse from emission to reception. The along-track, or azimuth, coordinate is reconstructed from the Doppler phase history of the signal produced by the moving platform. Unfortunately, satellite SAR ocean wave imaging is usually strongly nonlinear from the motions of the backscattering wave field, which create spurious Doppler shifts in the backscattered signal and produce misplacements of the backscattering surface elements in the image plane. For a long time, the resulting image distortion and associated partial loss of information at high azimuthal wavenumbers deterred researchers from using SAR data for quantitative ocean wave studies. Thus, at the time of the launch of the first oceanographic satellite, SeaSat, in 1978, ocean wave imaging by a SAR was still being evaluated, with many open questions. Today, however, a clearer understanding of how the wave spectrum is mapped into the SAR image spectrum, leading to the derivation of a closed, nonlinear, spectral mapping relation and the development of operationally viable retrieval algorithms, together with extensive validations of satellite spectral retrievals, have clearly established the usefulness of the SAR as a quantitative wave spectral measurement system.

1.1 SeaSat

Despite having a lifetime of only three months in 1978, SeaSat clearly demonstrated that a spaceborne radar altimeter was capable of quantitatively measuring significant wave heights and that the two-dimensional ocean wave pattern could be successfully imaged with a SAR. However, analysis of SAR data also clearly demonstrated that the linear modulation transfer function relating the SAR image spectrum to ocean wave spectrum, which had been used for the airborne SAR, was in general not applicable for a spaceborne SAR. For a high-flying platform (i.e., one crossing the sky with a low angular velocity relative to a ground observer), the nonlinearity of the imaging mechanism can no longer be regarded as weak (Alpers et al. 1981). In principle, this problem could have been solved by deriving two-dimensional spectra in near-real-time from the SAR signal, without the intermediate step of first forming an image (Hasselmann 1980). This approach required excessive computer resources because of the inapplicability of fast Fourier transform (FFT) algorithms. The high volume of SAR data also precluded storing SAR data on board the satellite. As a result, the data had to be transmitted in real time to a small number of line-of-sight ground stations. Although current spaceborne SAR missions provide onboard data recorders, data storage facilities for full-swath SAR images are still limited

Table 1. A list of satellite missions for studies of ocean surface waves (Scat = scatterometer, SAR= synthetic aperture radar, ◊ = manned mission, * = planned launch, • = planned end date).

Satellite	*Country*	*From*	*Until*	*Altimeter*	*Scat*	*SAR*	*URL: http://...*
SKYLAB◊	USA	25/5/1973	8/2/1974	yes	—	—	
GEOS-3	USA	14/4/1975	1/12/1978	yes	—	—	
SEASAT	USA	27/6/1978	9/10/1978	yes	yes	yes	www.earth.nasa.gov/history/seasat/seasat.html
SIR-A◊	USA	12/11/1981	14/11/1981	—	—	yes	southport.jpl.nasa.gov/index.html
SIR-B◊	USA	05/10/1984	13/10/1984	—	—	yes	southport.jpl.nasa.gov/index.html
KOSMOS-1870	Russia	7/1987	10/1989	—	—	yes	
GEOSAT	USA	12/5/1985	1/1990	yes	—	—	
ALMAZ-1	Russia	31/3/1991	17/10/1992	—	—	yes	www.neosoft.com/Almaz/
ERS-1	ESA	16/7/1991	02/06/1996	yes	yes	yes	earth.esa.int/ersnewhome
JERS	Japan	11/2/1992	1999	—	—	yes	yyy.tksc.nasda.go.jp/Home/Earth_Obs/e/jers_e.html
TOPEX/Poseidon	USA/F	10/8/1992	2000	yes	—	—	topex-www.jpl.nasa.gov/
SIR-C◊	USA	9/4/1994	20/4/1994	—	—	yes	southport.jpl.nasa.gov/index.html
SIR-C◊	USA	30/9/1994	11/10/1994	—	—	yes	southport.jpl.nasa.gov/in7dex.html
ERS-2	ESA	21/4/1995	2000•	yes	yes	yes	earth.esa.int/ersnewhome
RADARSAT-1	Canada	4/11/1995	2000•	—	—	yes	radarsat.space.gc.ca/
PRIRODA (MIR)◊	Russia	23/04/96	2000•	yes	—	yes	www.ire.rss.ru/priroda/priroda.htm
ADEOS	USA/JP	4/8/1996	30/6/1997	—	NSCAT	—	echo.gsfc.nasa.gov/adeos/adeos.html
GFO	USA	10/2/1998	2002•	yes	—	—	gfo.bmpcoe.org/Gfo/
QUIKSCAT	USA	19/6/1999	2001•	—	SeaWinds	—	winds.jpl.nasa.gov/missions/quikscat/quikindex.html
OKEAN-O	Russia	multi-satellite series		—	—	yes	solar.rtd.utk.edu/~mwade/project/okean.htm
LACROSSE	USA	multi-satellite series		—	—	yes	solar.rtd.utk.edu/~mwade/craft/lacrosse.htm
ENVISAT-1	ESA	2000*	2004•	yes	—	ASAR	envisat.estec.esa.nl
JASON-1	USA/F	2000*	2004•	yes	—	—	topex-www.jpl.nasa.gov/jason1/
ADEOS-2	USA/JP	2000*	2004•	—	SeaWinds	—	adeos2.hq.nasda.go.jp/default_e.htm
ALMAZ-1B	Russia	2001*	2004•	yes	yes	yes	www.neosoft.com/Almaz/almaz1b/
METOP-1	EUMETSAT	2001*	2004•	—	AScat	—	earth.esa.int/METOP.html
RADARSAT-II	Canada	2002*	2005•	—	—	yes	radarsat.space.gc.ca/info/future.html
ALOS	Japan	2002*	2005•	—	—	PALSAR	yyy.tksc.nasda.go.jp/Home/Earth_Obs/e/alos_e.html

(e.g., 10 min per 100 min orbit for ERS-1/2, 28 min per 100 min orbit for RadarSat, 30 min per 100 min orbit planned for ENVISAT). In principle, the high data rate problem can be overcome by transmitting the data via relay satellites, but this is not expected to be available globally for SAR satellites in the near future. Nevertheless, limited data-relay capabilities will be provided for ENVISAT, which will be launched in 2000.

1.2 European Remote-sensing Satellite

The first European Remote-sensing Satellite (ERS-1) was launched in July 1991 from Kourou, French Guiana, into a near-polar, sun-synchronous orbit yielding coverage between 81.5°S and 81.5°N. After ERS-2 was launched in April 1995, both satellites were operated in tandem between August 1995 and May 1996. ERS-2 follows ERS-1 with an approximate 30-min time lag in the same orbital plane, so that there is a 1-day interval between ERS-1 and ERS-2 observing the same ground swath. While this simultaneous operation of two spaceborne SARs has enabled a variety of novel applications, particularly related to interferometry at time scales longer than one day, the impact for wave applications remained limited due to large temporal wind and wave variability. Nevertheless, the simultaneous operation enabled cross-calibration between both satellites' sensors. ERS-1 was switched into a dormant mode in June 1996. The launch of ENVISAT (Section 1.3) ensures continuity of SAR data into the next millennium.

With respect to surface wave measurements, the main advance of ERS relative to SeaSat is the implementation of near-real-time processing for both altimeter and SAR measurements, as well as global sampling for the SAR, enabling SAR wave mode data to be used for global studies and operational wave forecasting. Data storage limitations of the SAR are surmounted with a subframe image mode specifically designed for ocean wave measurements. This so-called SAR wave mode (SWM) is switched on every 200 km, producing a 5-km × 10-km SAR image (or "imagette"). The average data rate relative to the continuously operating standard SAR imaging mode, which has a swath width of 100 km, is reduced by a factor of 100. Imagettes in a global, locally intermittent mode are stored on board and transmitted to a ground station once per orbit. The imagettes are the amplitude averages of three successive looks and are Fourier transformed to wavenumber spectra, which are then bin-averaged to reduced 12 × 12 polar wavenumber spectra. These spectra are disseminated by the European Space Agency (ESA) as a fast delivery product (FDP) in quasi-real time to NWP centers. With the exception of occasional gaps, primarily near coasts and the ice edge, where the SAR is operated in the full-swath mode providing precision images, the SAR yields daily coverage of the global wave spectral field at an along-track resolution comparable with typical NWP models and a cross-track resolution, dependent on latitude, of 1000–2000 km, which is a lower resolution than NWP models.

The SWM is interlaced in the swath of the simultaneously operating wind scatterometer (WNS). The SWM footprint at 19.9° incident angle corresponds to the second of

19 scatterometer range nodes, from near range to far range, separated by 25 km. This enables simultaneous recording and analysis of both wind and wave data (e.g., Chapron et al. 1995; Kerbaol et al. 1998). This complements the ongoing efforts to apply ERS SAR imagery to high-resolution scatterometry (e.g., Vachon and Dobson 1996; Scoon et al. 1996; Lehner et al. 1998; Hϕgda et al. 1998). The scatterometer is comprised of three antennas measuring normalized radar cross-section (NRCS) from which the mean surface wind speed and direction over a 50-km × 50-km area can be extracted. An atlas of global wind fields produced between 1991 and 1996 has been published by Bentamy et al. (1996). The scatterometer and SAR operate at the same C-band (5.6 cm) wavelength and are combined in a single active microwave instrument (AMI), enabling the development and mutual validation of the same microwave backscatter models for both sensors (Johannessen et al. 1998).

The radar altimeter provides a third simultaneous source of wind and significant wave height data. However, it operates at Ku-band (13.8 GHz), and its nadir position is separated by about 270 km from the SWM imagette location. Wind speed is extracted from the intensity of the return echo signal; wave height is extracted from the slope of the leading edge of the return echo signal.

1.3 Environmental Satellite

The launch of ENVISAT is scheduled for late 2000. ENVISAT will carry the Advanced SAR (ASAR) instrument, which will feature a number of enhanced capabilities. ASAR can operate in five mutually exclusive modes. During a global mission, which requires a low data rate for full operationality, ASAR is switched either into the global monitoring (ScanSAR) mode or into the wave (imagette) mode, comparable to ERS SWM. During a regional mission, which requires a high data rate, ASAR is operated either in image mode or alternating polarization mode (both are 30-m × 30-m resolution with 55- to 100-km swath), or the wide-swath mode (150-m × 150-m resolution with 405-km swath). Changing the incident angle between 15° and 45° allows one of seven possible subswaths to be selected. Images may also be taken at different polarizations.

A novel feature, important for wave monitoring, is the ability to overcome the directional ambiguity problem by exploiting information on time-dependent changes in the wave field. These changes are contained in successive single-look images, which normally are simply superimposed to produce a reduced-speckle multi-look image. Pairs of successive single-look images are generated from different subbands of the full-bandwidth Doppler spectrum. A cross-spectrum is computed from these pairs, which are typically separated in time by a fraction of the dominant wave period. The wave's dominant travel direction may be determined from the cross-spectrum.

Cross-spectrum analysis enables a second problem of SAR imagery to be efficiently tackled: the reduction of speckle noise. Speckle, which refers to the grainy appearance of SAR images, arises through coherent (phase-related) contributions of differential scatterers within a resolution cell (pixel). In contrast, the point-spread functions of different pixels in the image are completely dephased. Speckle noise can be considered a random walk problem and, for Gaussian processes, reduces to a multiplicative noise contribution. Conventionally, speckle noise is reduced by means of multi-look averaging, where single-look images are added up incoherently, yielding a speckle-reduced image (e.g., Gower 1983; Vachon and West 1991; Johnsen 1992). This approach reduces the wave image contrast because the target is stationary. The cross-spectrum computed from single-look images completely removes the speckle contribution for white noise, while avoiding image contrast reduction caused by moving waves.

Ambiguity removal was first considered in the context of ship radars by Atanasov et al. (1985), Rosenthal et al. (1989), and Rosenthal and Ziemer (1991). Various studies have subsequently been performed with airborne SARs (e.g., Raney et al. 1989; Vachon and Raney 1991). The cross-spectrum was incorporated into a SAR-to-wave nonlinear spectral inversion algorithm, first for airborne SAR data (Engen and Johnsen 1995a), and then for ERS-1 SAR images (Engen and Johnsen 1995b). The cross-spectrum is planned to be part of the fast delivery wave mode product for ENVISAT (Johnsen and Desnos 1999).

1.4 Theory of synthetic aperture radar ocean wave imaging

At the time of SeaSat, there were numerous theories for the SAR imaging of ocean waves, but no consensus on the proper description of how to map a moving, random sea surface into a SAR image and the associated two-dimensional spectra. Among the main issues (e.g., Allan 1983; Ulaby et al. 1986) were questions relating to the applicability of the two-scale concept of moving point scatterers (facets), the relevance of Bragg scattering theory, the role of radar polarization, the form of linear modulation transfer functions, the relationship between scene coherence time and dynamics of scattering waves, the impact of speckle noise on signal-to-noise, calibration, image degradation due to orbital facet acceleration, the relative importance of the phase and orbital velocities of waves, and—most important of all—the quantitative description of nonlinear image distortions induced by wave motions. These nonlinearities frequently prevented the detection of ocean waves and made the interpretation of imaged waves difficult.

One of the more ambitious SAR aircraft campaigns, designed to resolve many contentious issues regarding SAR ocean wave imaging, was carried out during the Marine Remote Sensing (MARSEN) project in the summer of 1979 in the North Sea. MARSEN data reconciled different views on SAR imaging of a moving, random, ocean wave surface within the framework of a consistent, comprehensive theory (Hasselmann et al. 1985; Tucker 1985). Individual Bragg backscattering facets of large dimension compared with radar wavelength, but small compared with typical wavelengths of surface waves, are mapped individually into pixels (Wright 1968; Valenzuela 1978) in the image plane. The separability of the mapping mechanism on the facet scale is justified by the phase decorrelation, but not amplitude independence, of the separate facet return signals.

The motion of the scatterers of a facet induces a Doppler frequency in the return signal, which translates into an azimuthal displacement of the facet in the image plane. The effective velocity of the scatterers is given by the sum of the phase velocities of the backscattering wave perturbations propagating on the facets and the significantly larger eigenmotions of the facets due to the orbital motions of the long waves. To a first approximation, the scatterer velocity can be regarded as constant during the SAR illumination time, so that the Doppler spectrum of an individual facet is a single narrow line. However, the Doppler spectrum for an ensemble of facets with different effective scatterer velocities is a broad Gaussian distribution, and the corresponding distribution of the facet positions in the image plane is nonuniform. For small wave steepness, the modulation of the facet positions in the image plane by the long-wave orbital velocity ("velocity bunching") enhances the RAR imaging due to the direct modulation of the scattering cross-section by the long waves, but for higher wave steepness, the image is smeared in the azimuthal direction. For a quantitative analysis of these effects, the coherence time of the backscattered facet signals was found to be a less useful concept than the Doppler spectrum of the facet return signals, which can be expressed directly in terms of the kinematic and dynamical properties of the facet elements.

The velocity bunching mechanism has been extensively studied theoretically and verified by experiments (Alpers and Hasselmann 1978; Alpers and Rufenach 1979; Swift and Wilson 1979; Valenzuela 1980; Raney 1980; Plant and Keller 1983). The component, υ, of the wave orbital motion in the direction of the satellite induces an additional Doppler shift that leads to an additional azimuthal displacement

$$\Delta x = \frac{R}{U}\upsilon$$

of the facet in the SAR image domain, where R is the slant range and U is the platform velocity. For small displacements, Δx, compared with the wavelength, λ, of the waves being imaged, the alternate bunching and spreading of the facet positions by the orbital wave motion enhances the imaging and can be described by a linear modulation transfer function, which can be added to the analogous RAR modulation transfer function. However, when the displacements become comparable to or larger than λ, the wave structure in the image plane becomes convoluted, and the mapping of the wave spectrum into the image spectrum is no longer linear. The nonlinear image degradation is governed by the ratio:

$$\frac{\Delta x}{\lambda} = \frac{\omega}{\omega_s}\frac{h}{\lambda}$$

where ω and h denote the frequency and height of the waves, respectively, and:

$$\omega_s = \frac{U}{R}$$

is the angular frequency with which a ground observer standing in the SAR beam would see the platform crossing the sky. For high-flying satellites such as ERS, ω_s is small and the nonlinearity parameter $\Delta x/\lambda$ is large for most sea states.

The first simulations of the fully nonlinear velocity bunching mechanism for a random sea were achieved by mapping, pixel-by-pixel, the sea surface into the image plane, using a Monte Carlo simulation of the random wave height and associated orbital velocity fields (Alpers 1983, Alpers et al. 1986). This technique, however, is very expensive with respect to computing resources and does not lend itself readily to inversion, which is required to retrieve the wave spectrum from the measured SAR spectrum. It was not until the derivation by Hasselmann and Hasselmann (1991)—referred to in the following as HH—of a closed, nonlinear, spectral integral transform describing the mapping of the wave spectrum into the SAR image spectrum that the inversion problem was solved. The full integral can be expanded into a Taylor series with respect to orders of the nonlinearities in wave spectral components and velocity bunching. The individual terms represent Fourier transforms of higher order products of auto- and cross-covariance functions of RAR, hydrodynamic, and velocity bunching cross-section modulations, which may be efficiently computed with FFTs. Subsequently, Krogstad (1992) showed that the nonlinear transform could also be derived as the second-order moment of the characteristic function of a multivariate random vector that incorporates the local sea surface properties governing the SAR imaging of a moving sea surface. This framework also allowed generalizations (e.g., Krogstad et al. 1994; Engen and Johnsen 1995a).

1.5 Ocean wave spectral retrieval

Aspects of SAR imaging of ocean waves and the forward wave-to-SAR mapping relation have been validated for spaceborne SAR data recorded during Shuttle Imaging Radar (SIR) missions B and C, SIR-B/C (Alpers et al. 1986; Monaldo and Lyzenga 1988) and Russian spaceborne SAR mission ALMAZ-1 (алмаз: Russian for *diamond*) (Wilde et al. 1994), as well as in the following airborne missions and field campaigns: Labrador Ice Margin Experiment, LIMEX (Raney et al. 1989); Labrador Extreme Waves Experiment, LEWEX (Beal 1991); Norwegian Continental Shelf Experiment, NORCSEX (Johnsen et al. 1991); Synthetic Aperture Radar and X-Band Nonlinearities—Forschungsplatform Nordsee, SAXON-FPN (Plant and Alpers 1994); Surface Wave Dynamics Experiment, SWADE (Cardone et al. 1995); Hasselmann et al. (1998b). The feasibility of retrieving wave spectra from SAR image spectra was demonstrated with SeaSat data (HH), aircraft data during LEWEX (Hasselmann et al. 1991), and ERS-1 data during the Grand Banks calibration and validation campaign (Grand Banks 1994).

The first detailed evaluation of the ERS-1 SAR wave mode was carried out for a three-day dataset in the Atlantic Ocean (Brüning et al. 1994a) and yielded an improved retrieval algorithm, WASAR (Hasselmann et al. 1996; referred to as HBHH). The square deviation between the simulated and observed SAR image spectra, named cost function, is minimized iteratively. To reduce the cost function in wave spectral domain, the gradient $\Delta F^n(\mathbf{k})$ is computed with the inverse of an explicit solution for the quasi-linear wave-to-SAR mapping, M^{ql},

$$\Delta F^n(\mathbf{k}) = (M^{ql})^{-1} \bullet P^n(\mathbf{k})$$

The updated SAR image spectrum P^{n+1} is inferred from the full closed nonlinear transform M^{nl} of the updated wave spectrum,

$$F^{n+1} = F^n + \Delta F^n$$

The resolution of the 180° directional ambiguity inherent in frozen-image wave spectra is achieved using an additional term that penalizes deviations of the retrieved wave spectrum from the first-guess and favors the propagation direction corresponding to the first-guess direction. A third term penalizes deviations from the observed azimuthal cut-off wavenumber, which is a measure of the root-mean-square (rms) orbital velocity and, therefore, particularly sensitive to short waves beyond the cut-off wavenumber. In this manner it is possible to recover, at least in integral form, information on the short wavelength region of the wave spectrum that cannot be directly imaged.

Despite the explicit cut-off adjustment, the inversion method modifies the detailed form of the spectrum only in the main part of the spectrum, for which direct SAR spectral information is available. This difficulty is overcome in the HBHH algorithm by introducing a spectral partitioning scheme into the additional iteration loop that updates the input spectrum. The new input spectrum retains the continuity properties of the original input spectrum; however, the scales and propagation directions of the wave systems of the new spectrum are adjusted to the inverted spectrum.

A valuable feature of the WASAR algorithm is the availability of an internal calibration based on the level of background clutter spectrum. Thus, the retrieved spectrum can be calibrated in absolute wave height units without reference to the SAR instrument calibration or measurements of the absolute backscattering cross-section (Alpers and Hasselmann 1982; Brüning et al. 1994a). The WASAR algorithm is available from the German Climate Computing Centre in Hamburg (Hasselmann et al. 1998a).

An extensive assessment of the quality, performance, and sensitivity-related aspects of ERS-1 SWM data and the wave spectral retrieval procedure was carried out for the global 1993–1995 ERS-1 dataset (Heimbach et al. 1998; referred to as H3). Emphasis was also placed on the issue of first-guess dependence of the retrieved spectrum. A single or second iteration of the input spectrum yielded an appreciable improvement of the retrievals. Sensitivity experiments were conducted in which the first-guess was modified by changes in energy, frequency, and direction. For low and high wind speeds and for azimuth and range travelling waves, a statistical assessment of the modified retrievals showed only weak residual dependence of the retrieval on the first-guess input spectrum.

A complementary global validation of SWM-retrieved significant wave height, H_S, for 1994 was compared with independent, collocated, significant wave height data retrieved from the TOPEX and ERS-1 altimeters (Bauer and Heimbach 1999). Using the additional spectral information provided by the SWM partner of the collocated pairs, the full H_S sample was further stratified with respect to spectral properties. The SWM-retrieved swell wave heights were in particularly close agreement with altimeter-derived wave heights.

Various aspects of the inherent nonlinearities in the ERS-1 SWM product have been investigated using higher order spectral methods (e.g., Le Caillec et al. 1996; Kerbaol et al. 1998). SAR data from Canada's RadarSat have also been analyzed (Vachon et al. 1997a, 1997b).

1.6 Wave data assimilation

Several techniques for assimilation of SAR wave spectral data into wave models have been developed. In all cases, both the wave spectra and the wind field are updated; the methods differed primarily in the level of sophistication and dynamical consistency.

Optimal interpolation

This technique, also called the nudging scheme, is straightforward to implement operationally. It was developed and applied to ERS-1 SWM wave spectral retrievals by Hasselmann et al. (1997).

Green's function

This method incorporates more aspects of wave dynamics, in particular the propagation of swell, using a Green's function method developed by Bauer et al. (1996). In contrast to the optimal interpolation (OI) scheme, which yields only instantaneous wind corrections from the local wind-generated surface wave part of the spectrum, the Green's function method also derives corrections of past wind fields from the measured swell. The swell is traced to its origin by means of a two-dimensional wave age spectrum computed with an extended version of WAM. This greatly increases the proportion of wave measurements available for wind corrections. Although corrections of past wind data are of limited value for forecasting, they are useful for wind field reconstructions and statistical compilations. They also provide valuable information for wind field validation, particularly in intense wind regions that may be inadequately sampled by the normal observational network (see Section 2.2 and 3.2.3). The method also serves as a consistency check between wind corrections inferred from the local wind-generated surface waves, named windsea, and from swell.

Adjoint

The adjoint technique fully respects the model dynamics, rendering it the most dynamically consistent assimilation method. This fully variational approach was recently implemented in WAM with the Giering and Kaminski (1998) tangent linear and adjoint model compiler (TAMC). However, the adjoint method is expensive with respect to computing time and has been applied only for model tuning (Hersbach 1998), not for assimilation of satellite data.

2. Global Comparison of ERS-1 SWM and WAM Wave Spectra

The section addresses applications of ERS-1 SAR wave spectral retrievals for model validation and investigations of large-scale fields of two-dimensional wave spectral properties. We also present a validation of the WAM model using ERS-1 SWM data that differs from that presented in H3, because WAM was re-run without assimilation of ERS-1 altimeter data.

2.1 Global distribution of seasonal mean spectral properties

Monthly and seasonal mean spectral properties retrieved from 1993 to 1995 ERS-1 SWM data and from WAM, produced operationally at ECMWF, have been compared by H3. Figures 1 and 2 show examples of global distributions of seasonal mean windsea and swell wave heights, respectively, for austral winter June–August 1994. The selection criteria for swell focused on low-frequency swell. Thus, the distributions refer only to the largest swell components within each spectrum, and only swell with wavelength greater than 250 m is included; the partitioning scheme used to define the swell components is described in HBHH. The windsea distributions (Figure 1) reflect the seasonal properties of atmospheric circulation, with maxima in the Southern Hemisphere mid-latitude westerly storm belt. The influence of the trade winds is clearly seen in the tropics, while strong monsoon-driven windsea systems are found in the Arabian Sea (Figure 1). As discussed by H3, the WAM slightly overpredicts windsea wave heights. This is attributed to the strength of ECMWF winds, a contention that is supported by a comparison between ECMWF and ERS-1 scatterometer-derived wind fields (Bentamy et al. 1996).

The large-scale pattern of swell (Figure 2) differs from the windsea distribution (Figure 1). Swell radiates along great circles from the main source regions in the Southern Hemisphere mid-latitudes towards the east and the tropics. Shadowing by continents is also apparent. In contrast to the windsea, swell wave heights are systematically underpredicted by the WAM, which appears to have excessive damping (H3).

2.2 Comparison of model simulations with and without assimilation

To avoid a possible spurious bias introduced into the operational ECMWF wave spectral analysis through the assimilation of ERS-1 altimeter significant wave heights (H3),

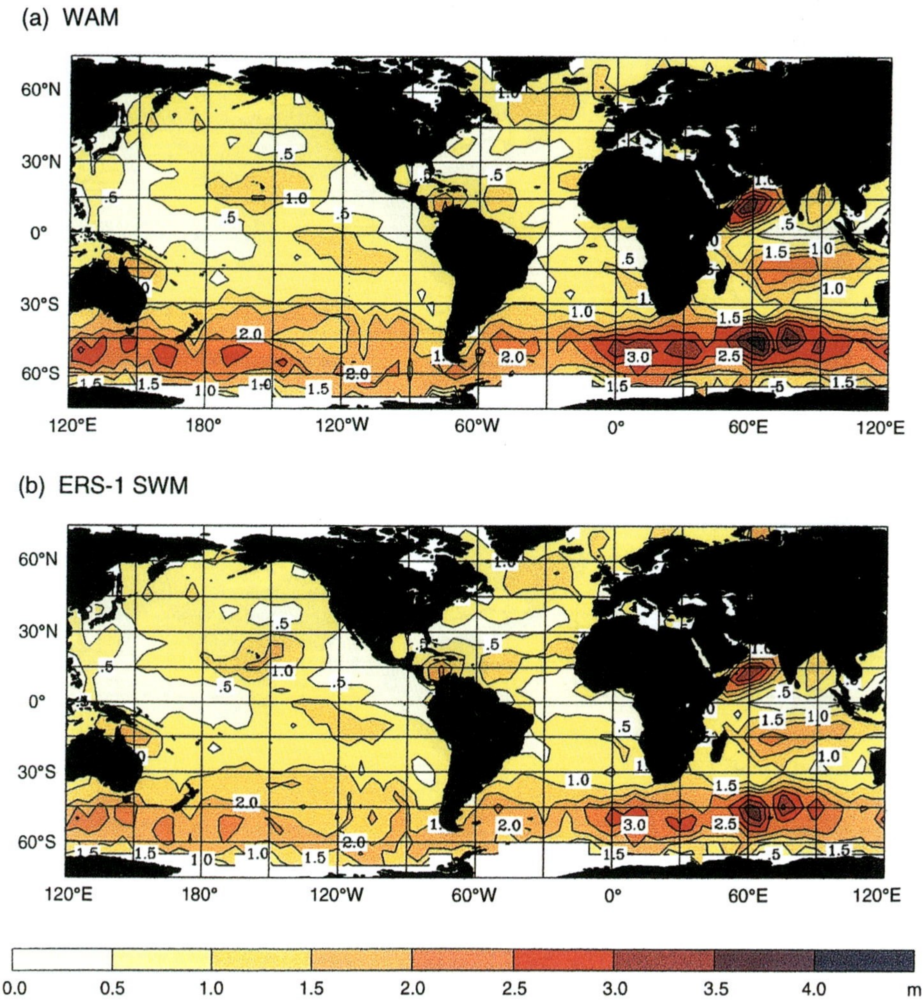

Figure 1. (a) WAM and (b) ERS-1 SWM seasonal mean wind-sea wave heights for June–August 1994.

the WAM for June–August 1995 was re-computed without assimilation of these data. The computational setup and wind forcing were identical to the operational configuration at ECMWF. A comparison of the ERS-1 SWM retrievals with the model simulations performed with and without assimilation of ERS-1 altimeter data confirms the qualitative conclusions described by H3. However, the magnitudes of the differences between the modeled and retrieved windseas and swell are affected by assimilation of the altimeter data (Figure 3).

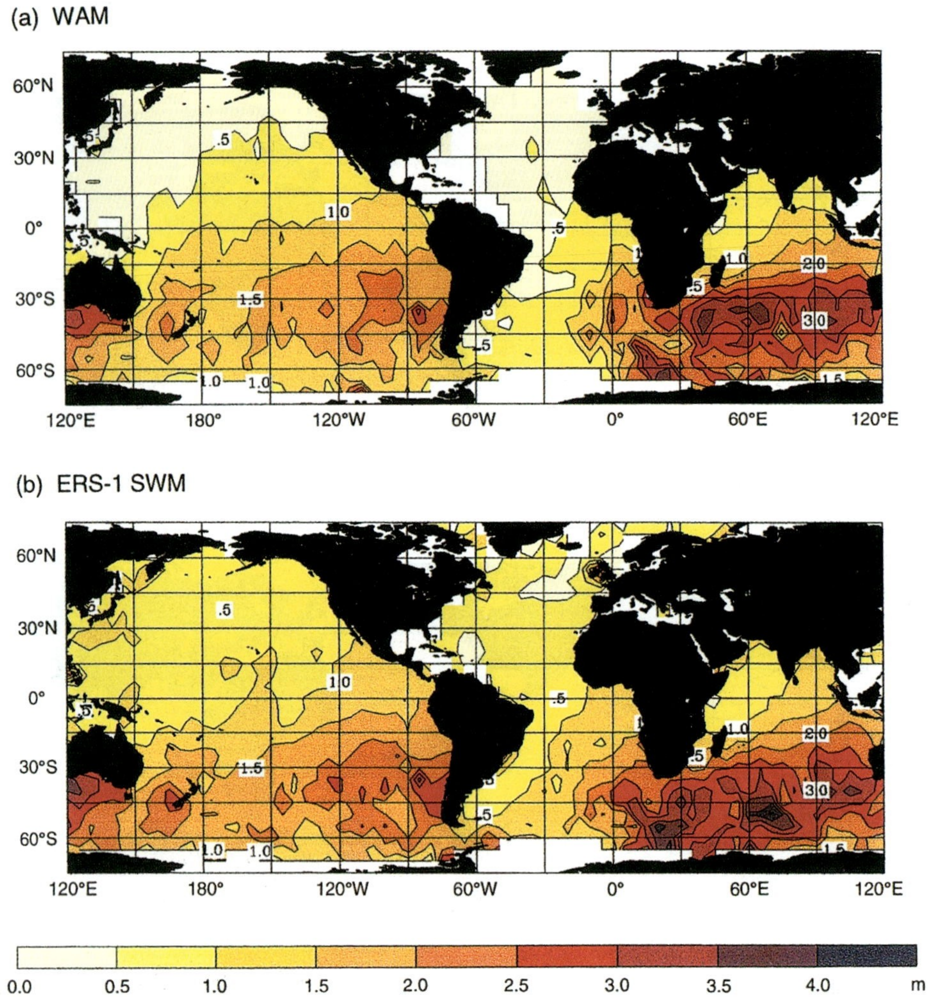

Figure 2. (a) WAM and (b) ERS-1 SWM seasonal mean wave height distributions of individual largest wavelength swell systems of wavelengths exceeding 250 m for June–August 1994.

Bauer and Heimbach (1999) reported that the ERS-1 altimeter data appear to be systematically low biased since January 1994. Thus, both mean windsea and swell wave heights for the WAM simulation without assimilation exceed the ECMWF data with assimilation. However, the effect is much smaller for windsea than for swell (Figure 3); swell covers a much larger area in the ocean than windsea and is, therefore, more frequently updated. The optimal interpolation assimilation scheme implemented at ECMWF updates windsea only when the satellite passes over the relatively limited storm area. This is a far more unlikely occurrence than updating a swell that has left the storm

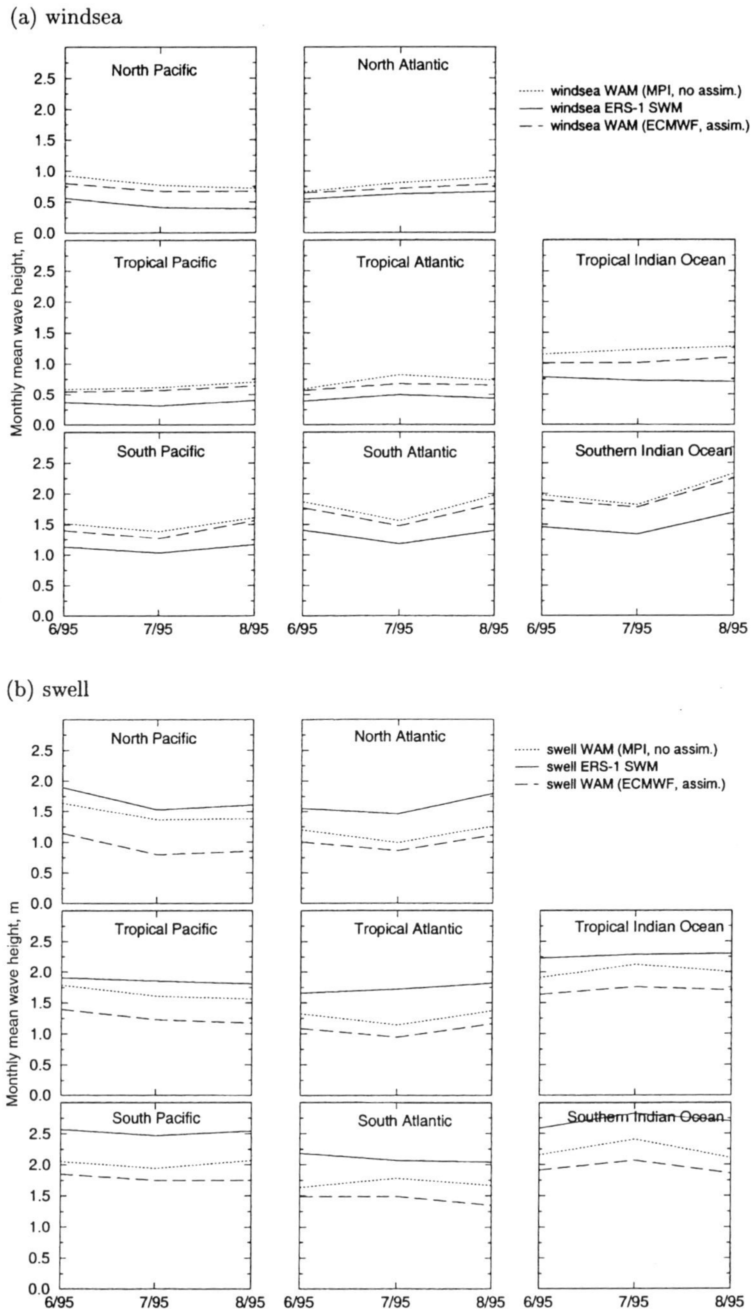

Figure 3. Monthly mean (a) windsea and (b) swell wave heights for ERS-1 SWM, WAM with assimilation (long dashed line), and WAM without assimilation (dotted line) in different ocean basins for June–August 1995.

area, and can be detected repeatedly several days later over an expanding area. This also explains the limited impact of wind corrections from the OI assimilation of significant wave height data (e.g., Breivik et al. 1998; Dunlap et al. 1998). The number of updates is small, and the updates for which windsea can be unambiguously separated from swell in the single, integral wave height provided by the altimeter is too small to make a significant impact (Sections 1.6 and 3.2). Although the sign is maintained, the bias in the swell is reduced by one-third to one-half, depending on the ocean basin, for the case without assimilation compared to the case with assimilation.

3. Trans-Ocean Propagation of Swell

3.1 Snodgrass et al. (1966) experiment

Sverdrup and Munk (1947), Barber and Ursell (1948), and Munk and Snodgrass (1957) indicated the ability of swell to travel over very large distances across ocean basins. During the austral winter of 1963, the first major ocean wave experiment was carried out to measure the propagation of swell across the Pacific Ocean (Snodgrass et al. 1966). The main goal was to determine whether, by how much, and by what mechanism swell was attenuated over a long path. Data would be used to test the Phillips (1957) and Miles (1957) theories of wave generation by wind, and the transfer of energy across the spectrum by resonant nonlinear interactions among ocean waves which generates long wavelength swell (Hasselmann 1967). In addition, the spectral action balance equation provided an elegant framework for propagation of swell wave packets along great circle rays on a sphere, viewed as a problem of Hamiltonian ray dynamics. Little was known about the dissipation of ocean swell.

To infer the source of long swell and estimate its travel time, Snodgrass et al. (1966) concentrated on a ‘great-circle’ in the Pacific, along which several wave-recording stations were installed. The ‘reference great-circle’ was chosen at an inclination of 195.5° with respect to Honolulu and connected regions of high storms east of New Zealand with the coast of Alaska at Yakutat. Snodgrass et al. (1966) were able to detect 12 major events. Their main finding was that long wavelength swell (λ >280 m), once the nonlinear wave-wave interactions had become negligible, propagated without detectable attenuation beyond the immediate vicinity of the generation region. A weak attenuation was marginally detected for large wave height swell events with wavelengths between 240 and 280 m. For wavelengths below 240 m, individual events could no longer be identified above the background swell radiating out continually from the high wind belt of the Southern Ocean. For all events, wave shadowing by intervening small islands was an important consideration that complicated the computed attenuation. Contributing to scatter in the observed attenuation were differences in the geometry and intensity of the wind fields of individual storm events.

In the next section, we examine swell propagation over very large distances, making use of the greatly expanded database provided by ERS-1 SAR wave mode data. Although the satellite cannot provide continuous time series at specific locations, it does yield global, quasi-continuous coverage, enabling various stages of the travelling swell to be observed at many locations and times.

3.2 The 4–6 June 1995 South Pacific storm

An extreme storm event began southeast of Tasmania (60°S, 145°E) on 4 June 1995. According to the ECMWF analysis, a low pressure system at 55°S, 165°E on 6 June 1995 slowly moved eastward, producing southwesterly wind speeds up to 20 m s^{-1}. The wave field radiating from the storm was analyzed using ERS-1 SAR wave-mode data and collocated WAM spectral values of wave energy and wave age at all frequency-direction spectral bins (Bauer et al. 1996).

Wave field

Figure 4a shows the vector field of significant wave height in the Pacific at 0000 Universal Time (UT) 6 June 1995. The dominant feature is the pronounced storm center, with wave heights of 12 m. At that time, no significant wave propagation seems to have occurred across the tropics from the Southern into the Northern Hemisphere: wave heights in the tropics remained below 2 m, and wave vectors followed more or less the local wind pattern. The situation changed considerably nine days later (Figure 4b). A northwest-to-southeast oriented ridge of 2.8- to 3.5-m wave heights with wave direction toward the northeast occurred 1000–2000 km off North America, extending into the tropical region. To select a collocated sample of WAM and ERS-1 SWM spectra associated with this storm, all available WAM and SWM-retrievals were decomposed into their principal wave systems using the HBHH spectral partitioning algorithm. A fan grid of 'great-circle' rays with origin at 45°S, 180°E was constructed, and all swell partitionings with directions aligned within ±30° with the nearest 'great-circle' direction were selected.

Wavelength-propagation diagram

To investigate variations in wavelength with position and time for the swell, a travel-time and distance diagram or Hovmüller diagram was constructed for WAM and SWM swell partitionings. Both ERS-1 SWM and WAM display (Figure 5) the propagation of a long wavelength signal along a ridge of maximum wavelength. The wavelength along the ridge increases with increasing distance and travel time. Moreover, a broadening and slight positive curvature of the ridge, with increasing distance from the generating storm, occurred. Both the increase in wavelength with increasing distance from the source and the positive curvature are consistent with the selective attenuation of shorter swell, leaving behind longer swell components with higher speed. This is also consistent with the effect of wave dispersion, in which long waves travel faster than short waves. Thus, the ridge

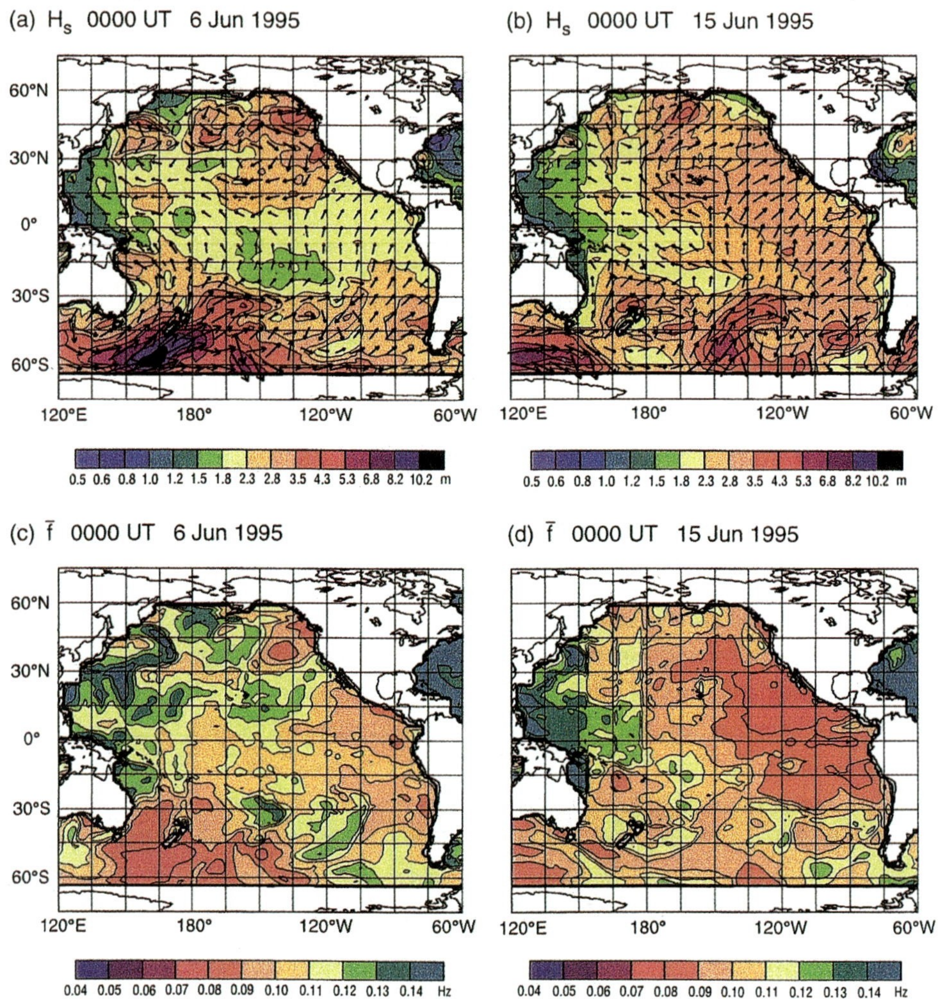

Figure 4. Significant wave height (m) and direction (top panels) and mean frequency (Hz) (bottom panels) on (a, c) 6 June 1995 and (b, d) 15 June 1995, both at 0000 UT.

bends upwards from the constant slope line, which would correspond to a constant mean frequency of the swell system. Faster waves with larger wavelengths appear first at a specific location. WAM had a broader ridge compared with SWM because of the numerical dispersion of the first-order upwind propagation scheme used in WAM. Although the propagation scheme conserves energy, barely impacting the overall wave statistics, for the coarse 30° directional resolution used by the ECMWF operational model, this characteristic must be taken into account when studying individual propagation events.

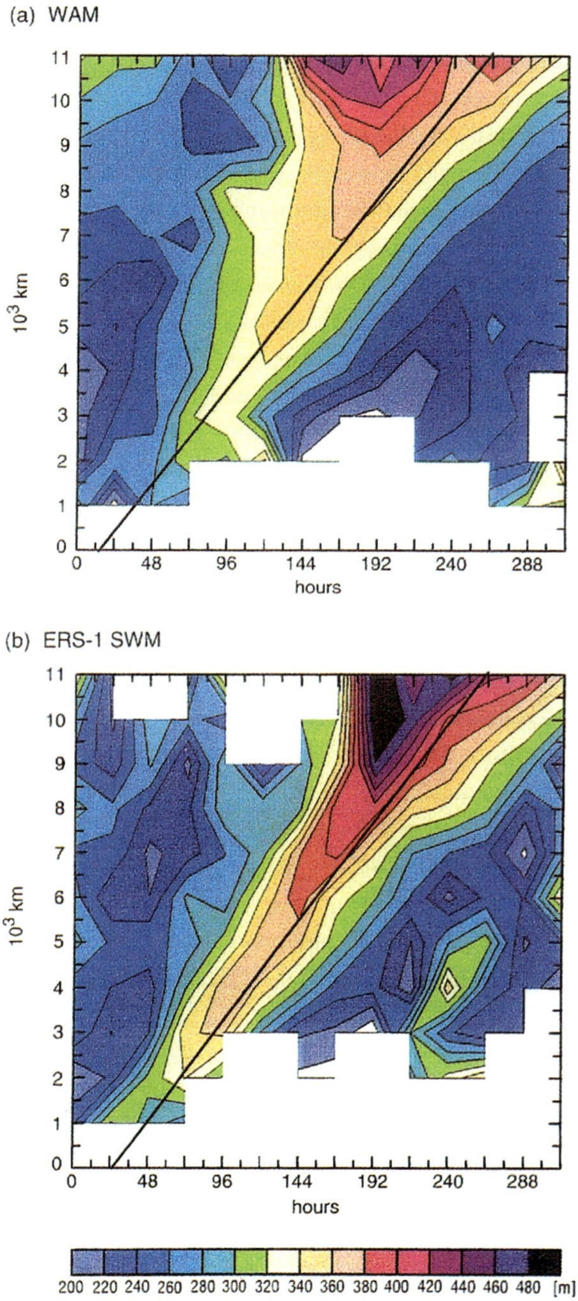

Figure 5. Travel time and distance diagram for (a) WAM and (b) ERS-1 SWM data showing wavelengths with respect to a reference time of 0000 UT 6 June 1995 and a reference location of 55°S, 165°E.

The slope of the maximum wavelength ridge (Figure 5) is the group velocity of the dominant swell. This allows the associated wavelength to be inferred from the dispersion relation, yielding an independent wavelength estimate that can be compared with the modeled or measured local wavelength. The slope may be computed locally at various points along the ridge, idealized as a curve, and compared to the observed wavelength. The only requirement for the slope to be detectable is for the local wavelength to be larger than the wavelength of the background swell (i.e., the ridge is significant relevant to the background). The WAM and ERS-1 SWM group velocities (Table 2) were in good agreement. However, the WAM significantly underestimated the wavelength compared to ERS-1 SWM (Table 2), which is discussed below.

Ray back-tracing using spectral wave age

To focus on long wavelength swell, we restrict the collocated swell dataset to waves with wavelengths above 300 m. We then trace the path of the swell from observed positions to position and time of origin using the locally determined spectral wave age, wavelength, and propagation direction to reconstruct each 'great-circle' propagation path. WAM results (Figure 6c) show the west-to-east shift of the generation area, in accord with the movement of the storm center. The spectral wave age is clearly a useful variable for classifying swell history and should be incorporated into wave models, such as the dynamically consistent data assimilation scheme proposed by Bauer et al. (1996).

ERS-1 SAR wave mode data along individual 'great-circles' are very scattered (Figure 6b), perhaps by island shadowing (Snodgrass et al. 1966), numerical dispersion that result in inaccuracies along individual rays, and contamination by background swell not associated with the storm. It is unlikely that inaccurate SWM data and the partitioning scheme to determine the swell wavelengths and wave heights are significant sources of scatter, because SWM data were successfully validated against altimeter wave heights (e.g., Bauer and Heimbach 1999).

Wind field in the storm region

WAM swell wavelengths are considerably lower than the wavelengths of ERS-1 retrievals (Table 2). This suggests that ECMWF wind forcing is too weak compared with actual winds. In an analysis of South Pacific wave data, H3 conjectured that ECMWF underestimated wind speed south of 50°S. We also tested the hypothesis of underesti-

Table 2. Group velocity and wavelength inferred from slope and main ridges in Figure 6.

	Slope (= group velocity)	*Wavelength (inferred from slope)*	*Wavelength (inferred from ridge)*
WAM	12.1 m s^{-1}	375 m	340–380 m
ERS-1 SWM	13.4 m s^{-1}	460 m	420–460 m

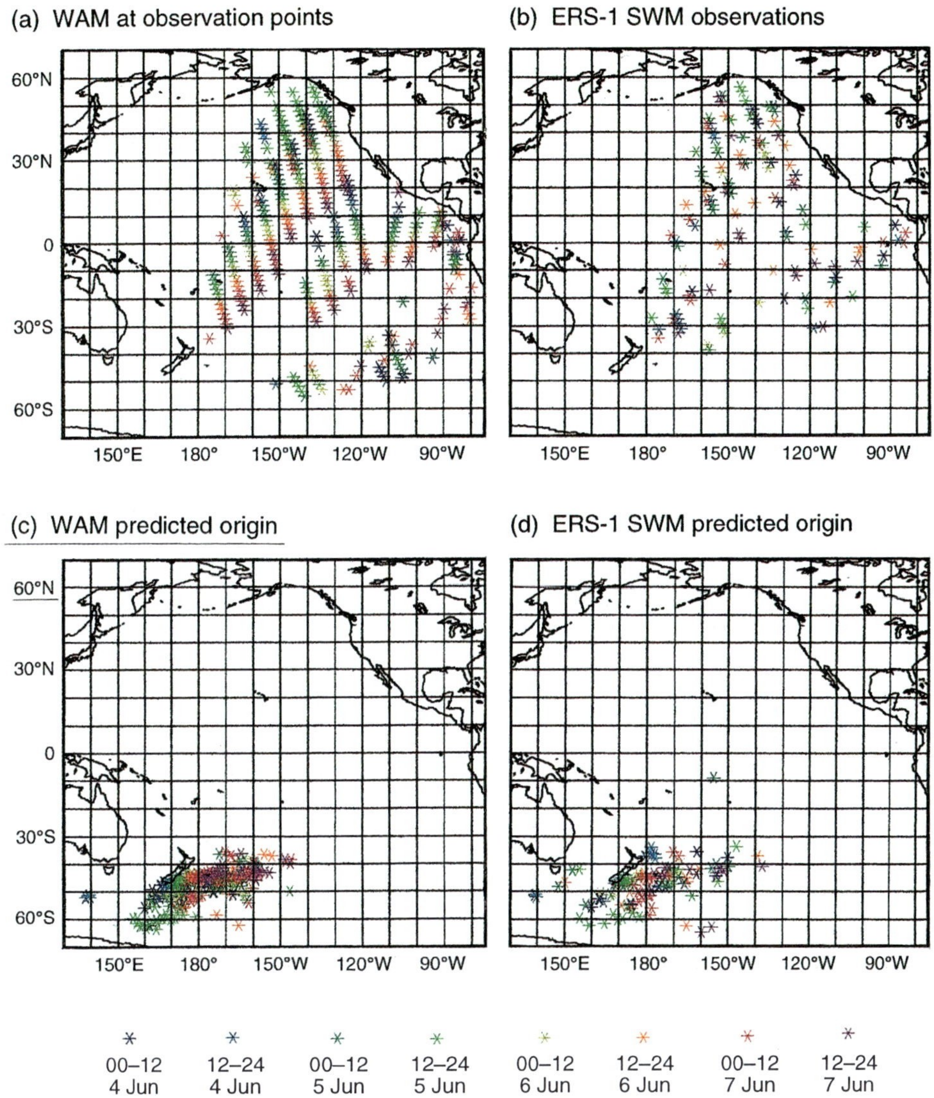

Figure 6. (a) WAM and (b) ERS-1 SWM location of swell. Predicted locations of origin of swell from (c) WAM and (d) ERS-1 SWM data, within ±2 days of 00 UT 6 June 1995.

mated ECMWF winds in the intense storm region by a direct comparison with wind vectors retrieved from the ERS-1 scatterometer. The storm area is about 45°–70°S and 140°–180°E. Between 0000 UT 5 June and 1800 UT 6 June 1995, the area was overflown six times by ERS-1. Only overflights 3 and 5 had footprints passing close to the storm center (Table 3) because wind directions were about 225° (Figure 7a), i.e., normal to the mean windsea direction (045°) inferred from the swell propagating away from the storm region. Other orbits exhibit higher differences in wind directions. The scatterometer and ECMWF wind speeds (Figure 7b) for overflight 3 are in reasonable agreement. However, orbit 5 exhibits large deviations between ECMWF and ERS-1 wind speeds. ERS-1 wind speeds reach a maximum of 29 m s^{-1}, one-third larger than ECMWF wind speeds (Figure 7b). These data provide evidence that ECMWF analysis did not resolve peak wind speeds of the storm. For wind speeds below about 20 m s^{-1} encountered during overflights 1, 2, 4, and 6, there is no systematic bias between ECMWF and ERS-1 winds. The lower wind speeds relative to the peak values observed on orbit 5 confirm that these overflights passed close to, but not through, the high wind center of the storm.

4. Conclusions and Perspectives

Research and modeling of ocean waves and the development and application of satellite remote-sensing methods for ocean waves have made remarkable progress since the launch of the first ocean satellite, SeaSat, more than 20 years ago. Before SeaSat, the prospect of the availability of global wave height and two-dimensional spectral data, and the challenge of retrieving that data from complex—and as yet inadequately validated—microwave sensor systems, provided a major stimulus for research. Sophisticated third-generation wave models and complex SAR wave-spectral retrieval algorithms were developed to meet this challenge. By the time ERS-1 was launched 13 years later in 1991, providing continuous, global, near-real-time wave data for the first time, most of the techniques for the application of satellite wave data for research and operational wave forecasting had been developed. The following years have seen a validation of the basic techniques and models, and a series of interesting applications for research, operational wave forecasting, and ship routing optimization (Lehner et al. 1996). ERS data have also been used to detect ocean wave refraction in the marginal ice zone, and to quantify sea ice thickness from the damping rate (Schulz-Stellenfleth et al. 1999; Schulz-Stellenfleth and Lehner 1999).

Table 3. ERS-1 overflights over the storm area. Orbits were either (a) ascending or (d) descending. Individual orbit footprints were estimated to be situated on the forward face, in the center, or on the backward face of the storm.

Number	*Day*	*Time interval*	*Orbit*	*Longitude*	*Position relative to storm center*
1	95/06/05	12:14–12:20	a	165°E–180°	forward
2	95/06/05	13:56–14:01	a	140°E–149°E	back
3	95/06/05	21:59–22:04	d	164°E–174°E	center
4	95/06/05	23:41–23:45	d	140°E–149°E	back
5	95/06/06	11:45–11:49	a	172°E–180°	center
6	95/06/06	13:25–13:29	a	148°E–158°E	back

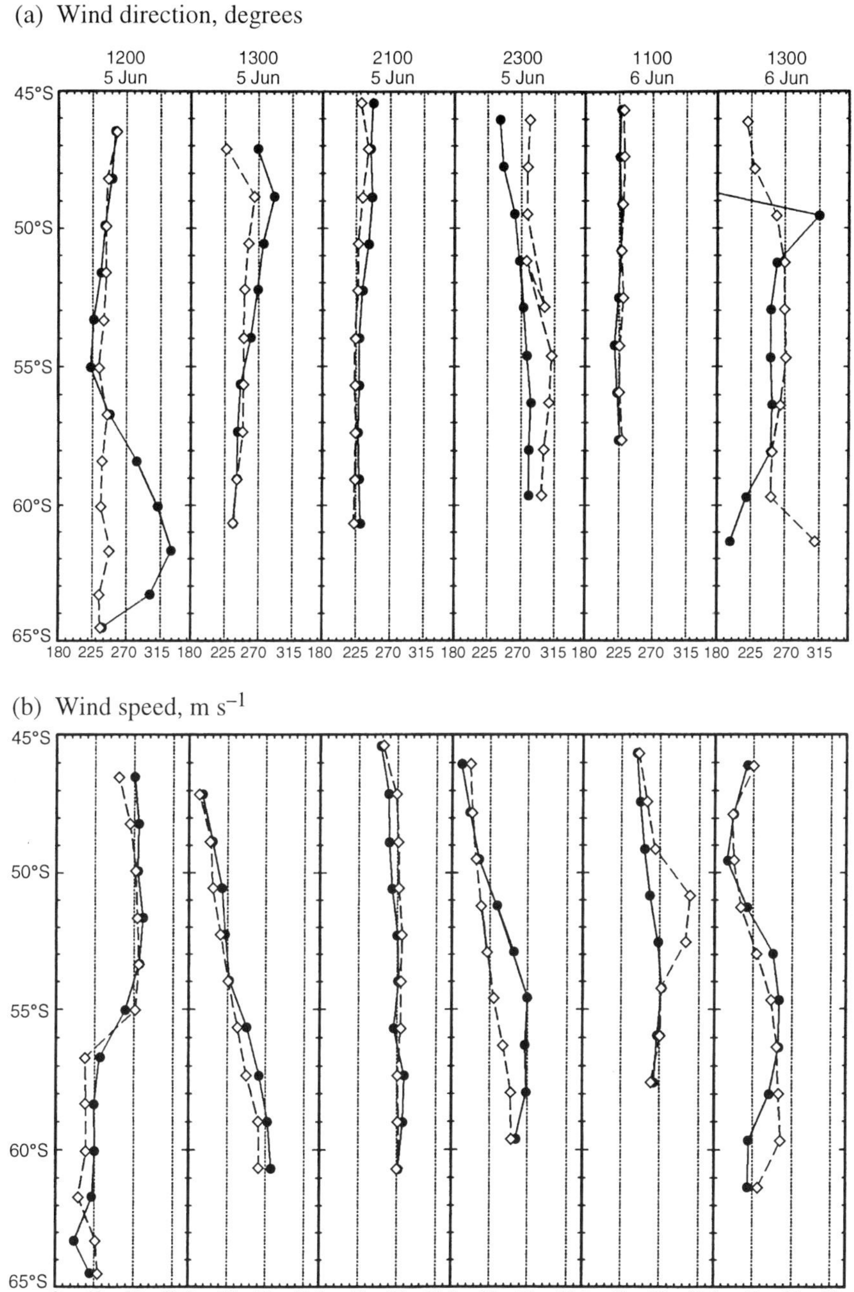

Figure 7. (a) Wind direction and (b) wind speed from ERS-1 scatterometer (dashed line) and ECMWF analysis (solid line) along ERS-1 ground tracks over the storm area.

However, much remains to be done. Interactions between operational wave forecasting, wave research, and satellite ocean wave remote sensing should be intensified. While operational ocean wave forecasts are being used successfully to provide first-guess inputs for the operational retrieval of SAR wave spectral data, the assimilation of these retrievals for operational wave forecasting has yet to be implemented. Existing operational data assimilation techniques are limited to relatively simple optimal interpolation methods for altimeter wave height data, which, as single integral values with undefined partitioning between windsea or swell, can have only a limited, imprecise impact on wave and wind updates. Assimilation schemes for SAR wave spectral retrievals have been validated and should now be tested in operational forecasts.

Another important area of development is the construction of coupled ocean-atmosphere models with a wave model, such as the European Coupled Atmosphere Wave Ocean Model (Weisse and Alvarez 1997). A coupled ocean-atmosphere model with a dynamical interface would be valuable for weather, wave, and storm surge forecasting (Mastenbroek et al. 1993; Weber et al. 1993), for seasonal and interannual climate forecasts, and for scenario computations of anthropogenic climate change. This type of model would be particularly relevant for predicting the statistics of extreme events, which are becoming increasingly important in the context of natural climate variability and anthropogenic climate change (WASA Group 1998). Comprehensive coupled models will also be needed for advanced data assimilation schemes that strive to achieve a dynamically consistent, simultaneous update of all relevant coupled fields, using all available data.

There continues to be room for progress in the area of sensors and algorithms. An example is the development and operational implementation in ENVISAT of an improved SAR wave cross-spectral retrieval system that makes use of the additional information contained in the individual looks of a multi-look SAR image to remove the 180° ambiguity of current frozen-image spectra (Johnsen and Desnos 1999). Improvements are also to be expected in the formulation of the hydrodynamic modulation transfer function. Research is directed towards a better description of shortwave-to-longwave modulation (Kudryasvtsev et al. 1997), including the wind dependence (Plant 1982; Feindt et al. 1986; Hara and Plant 1994; Brüning et al. 1994b).

Finally, essential to the success of a satellite ocean wave remote-sensing program is the maintenance of continuous, long-term global observations. The achievements to date, beginning with SeaSat and continuing with ERS-1/2, must still be regarded as a prolonged proof-of-concept, while development of the requisite sophisticated methodology is very much a work in progress. The real value of satellite ocean wave remote sensing will be realized when the techniques have been fully implemented in operational wave forecasting, the data are routinely used in research, and the observational time series have become sufficiently long to be used in global change studies.

Acknowledgments. In addition to new results, this paper reviews work carried out by the authors with Susanne Hasselmann and Eva Bauer, whose contributions are gratefully acknowledged. The work was supported in part by grants N00014-92-J-1840 and N00014-1-0541 from the U.S. Office of Naval Research (ONR), through the SFB-318 project funded by the Deutsche Forschungsgemeinschaft (DFG), and through the European Space Agency (ESA) pilot project PP2.D1.

References

Alpers, W., and K. Hasselmann, The two-frequency microwave technique for measuring ocean wave spectra from an airplane or satellite, *Bound. Layer Meteorol.*, *13*, 215–230, 1978.

Alpers, W., and C. L. Rufenach, The effect of orbital motions on synthetic aperture radar imaging of ocean waves, *IEEE Trans. Antennas Propag.*, *27*, 685–690, 1979.

Alpers, W., and K. Hasselmann, Spectral signal-to-clutter and thermal noise properties of ocean wave imaging synthetic aperture radars, *Int. J. Remote Sensing*, *3*, 423–446, 1982.

Alpers, W., D. B. Ross, and C. L. Rufenach, On the detectability of ocean surface waves by real and synthetic aperture radar, *J. Geophys. Res.*, *86*, 6481–6498, 1981.

Alpers, W., C. Brüning, and K. Richter, Comparison of simulated and measured synthetic aperture radar image spectra with buoy-derived ocean wave spectra during the Shuttle Imaging Radar-B mission, *IEEE Trans. Geosci. Remote Sensing*, *24*, 559–566, 1986.

Atanasov, V., W. Rosenthal, and F. Ziemer, Removal of ambiguity of two-dimensional power spectra obtained by processing ship radar images of ocean waves, *J. Geophys. Res.*, *90*, 1061–1067, 1985.

Barber, B. F., and F. Ursell, The generation and propagation of ocean waves and swell: I. Wave periods and velocities, *Phil. Trans. Roy. Soc. Lond. A*, *240*, 527–560, 1948.

Bauer, E., and P. Heimbach, Annual validation of significant wave heights of ERS-1 synthetic aperture radar wave mode spectra using TOPEX/Poseidon and ERS-1 altimeter data, *J. Geophys. Res.*, *104*, 13345–13357, 1999.

Bauer, E., K. Hasselman, I.R. Young, and S. Hasselmann, Assimilation of wave data into the wave model WAM using an impulse response function, *J. Geophys. Res.*, *101*, 3801–3816, 1996.

Beal, R. C., editor, *Directional Ocean Wave Spectra,* The Johns Hopkins University Press, Baltimore, 1991.

Bentamy, A., N. Grima, Y. Quilfen, V. Harscoat, C. Maroni, and S. Pouliquen, An atlas of surface wind from ERS-1 scatterometer measurements 1991–1996, Technical Report, IFREMER/CERSAT, Plouzané, France, 1996.

Breivik, L.-A., M. Reistad, H. Schyberg, J. Sunde, H. E. Krogstad, and H. Johnsen, Assimilation of ERS SAR wave spectra in an operational wave model, *J. Geophys. Res.*, *103*, 7887–7900, 1998.

Brüning, C., W. Alpers, and J. Schröter, On the focusing issue of synthetic aperture radar imaging of ocean waves, *IEEE Trans. Geosci. Remote Sensing*, *29*, 120–128, 1991.

Brüning, C., S. Hasselmann, K. Hasselmann, S. Lehner, and T. W. Gerling, A first evaluation of ERS-1 synthetic aperture radar wave mode data, *Global Atmos. Ocean System*, *2*, 61–98, 1994a.

Brüning, C., R. Schmidt, and W. Alpers, Estimation of the ocean wave—radar modulation transfer function from synthetic aperture radar imagery, *J. Geophys. Res.*, *99*, 9803–9816, 1994b.

Cardone, V. J., H. C. Graber, R. E. Jensen, S. Hasselmann, and M. J. Caruso, In search of the true surface wind field in SWADE IOP-1: Ocean wave modeling perspective, *Global Atmos. Ocean System*, *3*, 107–150, 1995.

Cardone, V. J., R. E. Jensen, D. T. Resio, V. R. Swail, and A. T. Cox, Evaluation of contemporary ocean wave models in rare extreme events: The "Halloween Storm" of October 1991 and the "Storm of the Century" of March 1993, *J. Atmos. Oceanic Tech.*, *13*, 198–230, 1996.

Chapron, B., T. Elfouhaily, and V. Kerbaol, Calibration and validation of ERS Wave Mode products, DRO/OS/95-02, IFREMER/CERSAT, Plouzané, France, 1995.

Dunlap, E. M., R. B. Olsen, L. Wilson, S. De Margerie, and R. Lalbeharry, The effect of assimilating ERS-1 fast delivery wave data into the North Atlantic WAM model, *J. Geophys. Res.*, *103*, 7901–7915, 1998.

Engen, G., and H. Johnsen, SAR ocean wave inversion using image cross-spectra, *IEEE Trans. Geosci. Remote Sensing*, *33*, 1047–1056, 1995a.

Engen, G., and H. Johnsen, Analysis and inversion of ERS-1 image cross-spectra, In Proc. IGARSS'95, IEEE Press, Piscataway, NJ, 1863–1865, 1995b.

Feindt, F., J. Schröter, and W. Alpers, Measurement of the ocean wave—radar modulation transfer function at 35 GHz from a sea-based platform in the North Sea, *J. Geophys. Res.*, *91*, 9701–9708, 1986.

Giering, R., and T. Kaminski, Recipes for adjoint code construction, *ACM Trans. Math. Software*, *24*, 437–474, 1998.

Gower, J. F. R., "Layover" in satellite radar images of ocean waves, *J. Geophys. Res.*, *88*, 7719–7720, 1983.

Grand Banks, The Grand Banks ERS-1 SAR Wave Spectra Validation Experiment, *Atmos.-Ocean*, *32*, 3–256, 1994.

Hara, T., and W. R. Plant, Hydrodynamic modulation of short wind-wave spectra by long waves and its measurement using microwave backscatter, *J. Geophys. Res.*, *99*, 9767–9784, 1994.

Hasselmann, K., Nonlinear interactions treated by the methods of theoretical physics (with application to the generation of waves by the wind), *Phil. Trans. Roy. Soc. Lond. A*, *299*, 77–100, 1967.

Hasselmann, K., A simple algorithm for the direct extraction of the two-dimensional surface image spectrum from the return signal of a synthetic aperture radar, *IEEE Trans. Geosci. Remote Sensing*, *1*, 219–240, 1980.

Hasselmann, K., and Hasselmann, S., On the nonlinear mapping of an ocean wave spectrum into a synthetic aperture radar image spectrum and its inversion, *J. Geophys. Res.*, *96*, 10713–10729, 1991.

Hasselmann, K., R. K. Raney, W. J. Plant, W. Alpers, R. A. Shuchman, D. R. Lyzenga, C. L. Rufenach, and M. J. Tucker (MARSEN Group), Theory of synthetic aperture radar ocean imaging: A MARSEN view, *J. Geophys. Res.*, *90*, 4659–4686, 1985.

Hasselmann, K., S. Hasselmann, C. Brüning, and A. Speidel, Interpretation and application of SAR wave image spectra in wave models, In *Directional Ocean Wave Spectra*, edited by R. Beal, The Johns Hopkins University Press, Baltimore, 117–124, 1991.

Hasselmann, S., P. Heimbach, and C. Bennefeld, The WASAR algorithm for retrieving ocean wave spectra from SAR image spectra, Technical Report 14, Deutsches Klimarechenzentrum (DKRZ), Hamburg, 1998a.

Hasselmann, S., P. Lionello, and K. Hasselmann, An optimal interpolation assimilation scheme for wave data, *J. Geophys. Res.*, *101*, 16615–16629, 1997.

Hasselmann, S., C. Brüning, K. Hasselmann, and P. Heimbach, An improved algorithm for the retrieval of ocean wave spectra from SAR image spectra, *J. Geophys. Res.*, *101*, 16615–16629, 1996.

Hasselmann, S., C. Bennefeld, H. Graber, D. Hauser, F. Jackson, P. Vachon, E. J. Walsh, K. Hasselmann, and R. B. Long, Intercomparison of two-dimensional wave spectra obtained from microwave instruments, buoys and WAModel simulations during the Surface Wave Dynamics Experiment, Report 258, MPI für Meteorologie, Hamburg, 1998b.

Heimbach, P., S. Hasselmann, and K. Hasselmann, Statistical analysis and intercomparison of WAM model data with global ERS-1 SAR Wave Mode spectral retrievals over three years, *J. Geophys. Res.*, *103*, 7931–7978, 1998.

Hersbach, H., Application of the adjoint of the WAM model to inverse wave modeling, *J. Geophys. Res.*, *103*, 10469–10487, 1998.

Hϕgda, K. A., G. Engen and H. Johnsen, Wind field estimation from SAR ocean images, In Proc. IGARSS'98, IEEE Press, Piscataway, NJ, 1998.

Jain, A., Focusing effects in the synthetic aperture radar imaging of ocean waves, *App. Phys.*, *15*, 323–333, 1978.

Janssen, P. A. E. M., and P. Viterbo, Ocean waves and the atmospheric climate, *J. Climate*, *9*, 1269–1287, 1996.

Johannessen, J. A., E. Attema, and Y.-L. Desnos, Wind field retrieval from SAR, Earth Observation Quarterly EOQ No. 59, European Space Agency, ESA Publications Division, ESTEC, Noordwijk (NL), 1998.

Johnsen, H., Multi-look versus single-look processing of synthetic aperture radar images with respect to ocean wave spectra estimation, *Int. J. Remote Sensing*, *13*, 1627–1643, 1992.

Johnsen, H., and Y.-L. Desnos, Expected performance of ENVISAT ASAR wave mode product, In Proc. IGARSS'99, IEEE Press, Piscataway, NJ, 1999.

Johnsen, H., K. A. Hϕgda, T. Guneriussen, and J. P. Pedersen, Azimuth smearing in synthetic aperture radar ocean image spectra from the Norwegian Continental Shelf Experiment of 1988, *J. Geophys. Res.*, *96*, 10443–10452, 1991.

Junger, S., *The Perfect Storm,* W. W. Norton and Company, New York, 227 pp, 1997.

Kasilingam, D. P., and O. H. Shemdin, Theory for synthetic aperture radar imaging of the ocean surface: With application to the tower ocean wave and radar dependence experiment on focus, resolution, and wave height spectra, *J. Geophys. Res.*, *93*, 13837–13848, 1988.

Kerbaol, V., B. Chapron, and P. W. Vachon, Analysis of ERS-1/2 synthetic aperture radar wave mode imagettes, *J. Geophys. Res.*, *103*, 7833–7846, 1998.

Komen, G. J., L. Cavaleri, M. Donelan, K. Hasselmann, S. Hasselmann, and P. A. E. M. Janssen, *Dynamics and Modeling of Ocean Waves,* Cambridge University Press, Cambridge, 560 pp, 1994.

Krogstad, H. E., A simple derivation of Hasselmann's nonlinear ocean-synthetic aperture radar transform, *J. Geophys. Res.*, *97*, 2421–2425, 1992.

Krogstad, H. E., O. Samset, and P. W. Vachon, Generalization of the nonlinear ocean-SAR transform and a simplified SAR inversion algorithm, *Atmos.-Ocean*, *32*, 61–82, 1994.

Kudryavtsev, V. N., C. Mastenbroek, and V. K. Makin, Modulation of wind ripples by long surface waves via the air flow: a feedback mechanism, *Bound. Layer Meteorol.*, *83*, 99–116, 1997.

Le Caillec, J. M., R. Garello, and B. Chapron, Two dimensional bispectral estimates from ocean SAR images, *Nonlin. Proc. Geophys.*, *3*, 196–215, 1996.

Lehner, S., T. Bruns, and K. Hasselmann, Test of a new onboard ship routing system, In Proc. Second ERS Applications Workshop, ESA SP-383, ESA Publications Division, ESTEC, Noordwijk (NL), 297–301, 1996.

Lehner, S., J. Horstmann, W. Koch, and W. Rosenthal, Mesoscale wind measurements using recalibrated ERS SAR images, *J. Geophys. Res.*, *103*, 7847–7856, 1998.

Lyzenga, D. R., An analytic representation of the synthetic aperture radar image spectrum for ocean waves, *J. Geophys. Res.*, *93*, 13859–13865, 1988.

Mastenbroek, C., G. Burgers, and P. A. E. M. Janssen, The dynamical coupling of a wave model and a storm surge model through the atmospheric boundary layer, *J. Phys. Oceanogr.*, *23*, 1856–1866, 1993.

Miles, J., On the generation of surface waves by shear flows, *J. Fluid Mech.*, *3*, 185–204, 1957.

Monaldo, F. M., and D. R. Lyzenga, Comparison of Shuttle Imaging Radar-B ocean wave spectra with linear model predictions based on aircraft measurements, *J. Geophys. Res.*, *93*, 374–388, 1988.

Munk, W. H., and F. E. Snodgrass, Measurements of southern swell at the Guadalupe Islands, *Deep-Sea Res.*, *4*, 272–286, 1957.

Phillips, O. M., On the generation of waves by turbulent wind, *J. Fluid Mech.*, *2*, 417–445, 1957.

Plant, W. J., A relationship between wind stress and wave slope, *J. Geophys. Res.*, *87*, 1961–1967, 1982.

Plant, W. J., Reconciliation of theories of synthetic aperture radar imagery of ocean waves, *J. Geophys. Res.*, *97*, 7493–7501, 1992.

Plant, W. J. and W. C. Keller, The two-scale radar wave probe and SAR imagery of the ocean, *J. Geophys. Res.*, *88*, 9776–9784, 1983.

Plant, W. J. and W. Alpers, An introduction to SAXON-FPN, *J. Geophys. Res.*, *99*, 9699–9703, 1994.

Raney, R. K., and P. W. Vachon, Synthetic aperture radar imaging of ocean waves from an airborne platform: focus and tracking issues, *J. Geophys. Res.*, *93*, 12475–12486, 1988.

Raney, R. K., P. W. Vachon, R. A De Abreu, and A. S. Bhogal, Airborne SAR obersvations of ocean surface waves penetrating floating ice, *IEEE Trans. Geosci. Remote Sensing*, *27*, 492–500, 1989.

Rosenthal, W., F. Ziemer, R.K. Raney, and P. Vachon, Removal of 180° ambiguity in SAR images of ocean waves, In Proc. IGARSS'89, IEEE Press, Piscataway, NJ, 1989.

Schulz-Stellenfleth, J., and S. Lehner, A new SAR inversion scheme for ocean waves traveling into sea ice, In Proc. IGARSS'99, IEEE Press, Piscataway, NJ, 1999.

Schulz-Stellenfleth, J., S. Lehner, and K. Hasselmann, ERS SAR observations of ocean waves traveling into sea ice, *J. Geophys. Res.*, submitted, 1999.

Scoon, A., I. S. Robinson, and P. J. Meadows, Demonstration of an improved calibration scheme for ERS-1 SAR imagery using a scatterometer wind model, *Int. J. Remote Sensing*, *12*, 413–418, 1996.

Snodgrass, F. E., G. W. Groves, K. Hasselmann, G. R. Miller, W. H. Munk and W. H. Powers, Propagation of ocean swell across the Pacific, *Phil. Trans. Roy. Soc. Lond. A*, *249*, 431–497, 1966.

Sverdrup, H. U., and W. H. Munk, Wind, sea, and swell: Theory of relations for forecasting, Scripps Institution of Oceanography New Series, No. 303 H.O. Pub. No. 601/ Technical Report No. 1, U.S. Navy, U.S. Hydrographic Office Publication, La Jolla, California, 1947.

SWAMP Group, *Ocean Wave Modeling,* Plenum Press, New York, 256 pp, 1985.

Swift, C. F., and L. R. Wilson, Synthetic aperture radar imaging of moving ocean waves, *IEEE Trans. Antennas Propag.*, *27*, 725–729, 1979.

Tucker, M. J., The imaging of waves by satellite synthetic aperture radar: the effect of surface motion, *Int. J. Remote Sensing*, *6*, 1059–1074, 1985.

Ulaby, F. T., R. K. Moore, and A. K. Fung, *Microwave Remote Sensing*, Artech House, Dedham, Massachusetts, 3 Vol., 1986.

Vachon, P. W., and R. K. Raney, Resolution of the ocean wave propagation direction in SAR imagery, *IEEE Trans. Geosci. Remote Sensing*, *29*, 105–112, 1991.

Vachon, P. W., and J. C. West, Spectral estimation techniques for multilook SAR images of ocean waves, *IEEE Trans. Geosci. Remote Sensing*, *30*, 568–577, 1991.

Vachon, P. W. and F. W. Dobson, Validation of wind vector retrieval from ERS-1 SAR images over the ocean, *Global Atmos. Ocean Sys.*, *5*, 177–187, 1996.

Vachon, P. W., J. W. M. Campbell, and F. W. Dobson, Comparison of ERS and RADARSAT SAR's for wind and wave measurements, In Proc. of the Third ERS Symposium, ESA SP-414, Vol. 3, ESA Publications Division, ESTEC, Noordwijk (NL), 1997a.

Vachon, P. W., J. W. M. Campbell, and F. W. Dobson, ERS and RADARSAT SAR images for wind and wave measurement, In Proc. of CEOS Wind and Wave Validation Workshop, ESTEC, ESA WPP-147, ESA Publications Division, Noordwijk (NL), 57–60, 1997b.

Valenzuela, G. R., Theories for the interaction of electromagnetic and ocean waves: A review, *Bound. Layer Meteorol.*, *13*, 61–85, 1978.

Valenzuela, G. R., An asymptotic formulation for SAR images of the dynamical ocean surface, *Radio Sci.*, *15*, 105–114, 1980.

WAMDI Group, The WAM model—a third generation ocean wave prediction model, *J. Phys. Oceanogr.*, *18*, 1775–1810, 1988.

WASA Group, Changing waves and storms in the Northeast Atlantic?, *Bull. Amer. Meteorol. Soc.*, *79*, 741–760, 1998.

Weber, S. L., H. von Storch, P. Viterbo, and L. Zambresky, Coupling an ocean wave model to an atmospheric general circulation model, *Climate Dyn.*, *9*, 63–69, 1993.

Weisse, R., and E. F. Alvarez, The European Coupled Atmosphere Wave Ocean Model: ECAWOM, MPI-Report No. 116, Max-Planck-Institut Meteorologie, Hamburg, 1997.

Wilde, A., C. Brüning, W. Alpers, V. Etkin, K. Litovchenko, A. Ivanov, and V. Zajtsev, Comparison of ocean wave imaging by ERS-1 and ALMAZ-1 synthetic aperture radar, In Proc. of the Second ERS-1 Symposium, ESA SP-361, ESA Publications Division, ESTEC, Noordwijk (NL), 239–245, 1994.

Wright, J. W., A new model for sea clutter, *IEEE Trans. Antennas Propag.*, *16*, 217–223, 1968.

Zurk, L. M. and W. J. Plant, Comparison of actual and simulated synthetic aperture radar image spectra of ocean waves, *J. Geophys. Res.*, *101*, 8913–8931, 1996.

Patrick Heimbach, Department of Earth and Planetary Sciences, Room 54-1518, Massachusetts Institute of Technology, 77 Massachusetts Avenue, Cambridge, MA 02139, U.S.A.
(email, heimbach@mit.edu; fax, +1-617-253-4464)

Satellites, Oceanography and Society
edited by David Halpern

Chapter 3

ECMWF wave modeling and satellite altimeter wave data

Peter Janssen
European Centre for Medium-Range Weather Forecasts, Reading, United Kingdom

Abstract. Satellite altimeter wave height data have important benefits to society: ship routing, fisheries, coastal protection, oil exploration, specification of initial sea state for ocean wave forecasting, and validation of wave forecast. This presentation briefly describes the role of altimeter data in modern ocean wave forecasting. Also, some assumptions to obtain wave height from the radar backscattered pulse are discussed. Comparison of European Remote-sensing Satellite (ERS-2) altimeter wave height data with buoy observations revealed that the ERS-2 wave height is too low by about 7%. The underestimation is discussed and, by using model wave spectra, a method is proposed to remove this problem.

1. Introduction

Sea state forecasting started more than fifty years ago when there was a need for knowing the wave state during sea-land operations in the Second World War. The past five decades have seen ocean wave forecasting develop from simple manual techniques to sophisticated numerical wave models based on physical principles. In the 1990s, development was rapid because of availability of wave data from satellites such as geodetic satellite (Geosat), European Remote-sensing Satellite (ERS-1 and ERS-2), and Topography Experiment (TOPEX)/Poseidon, named T/P. In the mid-1980s, there was a convergence of the need to improve wave modeling, availability of powerful computers, and prospects for remote sensing techniques to provide sea state data on a global scale (SWAMP 1985). As a consequence, a group of mainly European wave modelers, who called themselves the Wave Model (WAM) group, started to develop a surface wave model from first principles, i.e., a model that solves the energy balance equation for surface gravity waves. The source functions in the energy balance included an explicit representation of wind input, nonlinear interactions, and dissipation by white capping. WAMDI (1988) describes the first version of this new wave model, called WAM.

The quality of the initial WAM was evaluated with SeaSat (Janssen et al. 1989; Bauer et al. 1992) and Geosat (Romeiser 1993) altimeter wave height data. Also, WAM has been validated against buoy data (Zambresky 1989; Wittman et al. 1995; Khandekar and Lalbeharry 1996; Janssen et al. 1997b). Modeled wave heights obtained by forcing WAM with European Centre for Medium-Range Weather Forecasts (ECMWF) winds showed good agreement with altimeter wave heights, but there were also considerable regional and seasonal differences. During the Southern Hemisphere winter, WAM underestimated wave height by about 20% in large parts of the Southern Hemisphere and in the tropical regions. The discrepancies could be ascribed to shortcomings in WAM physics and ECMWF wind fields, which at the end of the 1980s were too low in the Southern Hemisphere because of a fairly low-resolution (T106) atmospheric general circulation model. WAM contained too much dissipation of swell and a weak wind input source function.

In November 1991, the next version of WAM, named WAM cy4 (Janssen 1991; Komen et al. 1994), became part of the ECMWF wave prediction system. In addition, in September 1991 the horizontal and vertical resolutions of the ECMWF atmospheric general circulation model were doubled to produce a better representation of surface winds, in particular for the Southern Ocean. Therefore, in late 1991 there was sufficient confidence in the quality of the ECMWF wind-wave forecasting system that it could be used for the validation of ERS-1 altimeter wind and wave products. ERS-1 was launched in July 1991. Comparison of ERS altimeter wind and wave products with corresponding ECMWF fields identified problems in the ERS wind speed and wave height retrieval algorithms (Hansen and Guenther 1992; Janssen et al. 1997a).

In August 1993, assimilation of ERS-1 altimeter wave heights was introduced into the ECMWF wave forecasting system (Janssen et al. 1989; Lionello et al. 1992), which led to an improved wave analysis (Bauer and Staabs 1998). However, the ECMWF wave height analysis was too low by about 25 cm compared to buoy data because the ERS altimeter underestimated wave height by 15% (Janssen et al. 1997a).

This paper shows how comparisons of satellite altimeter wave height and wind speed data with corresponding data products computed from the ECMWF wave forecasting system have benefited both satellite observations and numerical model simulations of surface waves.

2. Surface Wave Modeling and Prediction

2.1 Brief history

Interest in surface wave prediction started during the Second World War because of the practical need for knowledge of the sea state during amphibious operations. The first predictions were based on the work of Sverdrup and Munk (1947), who used empirical relations to predict windsea and swell. An important step forwards was the introduction of the concept of a wave spectrum (Pierson et al. 1955), but a dynamical equation describing the evolution of the spectrum was not known until Gelci et al. (1957) introduced the spectral transport equation. However, Gelci et al. (1957) used an empirically-derived net source function to describe the rate of change of the wave spectrum. The Phillips (1957) and Miles (1957) new theories of wave generation by wind and Hasselmann's (1962) development of the source function for nonlinear transfer of energy between waves provided the ingredients for the source function analytical model, consisting of input from wind, nonlinear transfer, and dissipation by white capping. For deep-water waves, the mathematical form is still used today.

None of the wave models developed in the 1960s and 1970s computed the wave spectrum from the full energy balance equation. Additional ad-hoc assumptions were introduced to ensure that the wave spectrum complies with preconceived notions of wave development that were in some cases not consistent with the source functions. Reasons for introducing simplifications in the energy balance equation were twofold: the important role of wave-wave interactions in wave evolution was not recognized; and limited computer power precluded the use of nonlinear transfer in the energy balance equation.

The relative importance of nonlinear transfer and wind input became evident from experiments on wave growth (Mitsuyasu 1968, 1969; Hasselmann et al. 1973) and from direct measurements of wind input to waves (Snyder et al. 1981). Eventually, this led in the 1980s, with availability of powerful computers, to the development of a wave prediction model that yielded the wave spectrum by integration of the energy balance equation without any prior restriction on spectral shape. Denoting the two-dimensional frequency (f)-direction (θ) wave variance spectrum by $F(f,\theta)$, the time rate of change of the wave spectrum is derived from the energy balance equation for deep-water surface gravity waves,

$$\frac{\partial}{\partial t}F + \boldsymbol{v}_g \bullet \nabla F = S_{in} + S_{nl} + S_{ds} \tag{1}$$

where $\boldsymbol{v}_g$ is the group velocity, and the source functions on the right side of equation (1) are the rates of change of the wave spectrum by wind input (S_{in}), nonlinear four wave interactions (S_{nl}), and dissipation by white capping (S_{ds}).

In the present version of WAM, the wind input is based on a parameterization of the Miles (1957) instability, including feedback of growing waves on the wind profile (Janssen 1989, 1991). As a result, the airflow drag over the ocean is sea-state dependent, in agreement with observations (Donelan 1982; Smith et al. 1992; Donelan et al. 1993; Johnson et al. 1998). A sea-state dependent drag may have consequences for the atmospheric climate (Janssen and Viterbo 1996). However, whether sea state has a significant influence on the drag coefficient remains an ongoing debate. For a pure windsea, Donelan et al. (1993) find a relation between the enhancement of the drag coefficient and a measure for the sea state, namely the wave age; however, alternatives to the wave age parameter exist (Monbaliu 1994; Anctil and Donelan 1996; Janssen 1997). In general, the sea state is confused, consisting of a mixture of windsea and swell and, therefore, a characterization of sea state in terms of wave age is not a viable option. For example, Yelland et al. (1998) could not detect a wave age dependence of the drag coefficient for the open ocean. However, Hare et al. (1999) did find indications of a sea-state dependence of the drag coefficient because the Charnock parameter increased with increasing wind speed.

Phillips (1960) and Hasselmann (1962) showed that resonant energy transfer among four surface waves, or nonlinear wave-wave interactions, is an important component to determine the shape of the wave spectrum. Nonlinear transfer of energy also plays a vital role in shifting the spectrum towards lower frequency (Hasselmann et al. 1973). Even with present-day computing capabilities, a wave model based on the exact representation of nonlinear transfer is not feasible. Therefore, some form of parameterization of S_{nl} is needed. In WAM the Hasselmann et al. (1985) approximation is utilized.

The least-known source function is energy dissipation due to white capping. Hasselmann (1974) obtained some general constraints on the form of the dissipation source term, but a few parameters remained undetermined until Komen et al. (1984) insisted that for large time the wave spectrum would evolve towards the Pierson and Moskowitz (1964) spectrum. Felizardo and Melville (1995) found good agreement between dissipation rates of waves observed at sea and rates computed from the Komen et al. (1984) expression for S_{ds}.

WAM results are highly sensitive to the quality of the wind field. Manually analyzed winds have much lower errors compared to numerical weather prediction wind data products and, as a consequence, yield dramatically improved wave forecasts (Cardone et al. 1995). The sensitivity of modeled waves to the quality of the winds was confirmed by Janssen (1998), who showed that random wind speed errors dominate the forecast wave height error after day two in the forecast. It is shown in this paper that in the tropics, where sea state is dominated by swell, WAM depends on the quality of the wind field in the extratropics where swells are generated. Of course, this does not imply that there are no WAM errors; it means that WAM errors are smaller than the ones associated with the wind field. WAM errors can presumably be exposed only when the error in the wind field is reduced sufficiently, i.e., a reduction in wind speed error to 0.8 m s^{-1} (Janssen et al. 1997b).

Recent studies suggest that WAM may have a problem with swell energy. Sterl et al. (1998) found that WAM overpredicted swell wave height by about 20 cm. In contrast, Heimbach et al. (1998) found that WAM swell wave heights were lower than swell wave heights retrieved from synthetic aperture radar (SAR) data. However, Heimbach et al. (1998) used an operational ECMWF-WAM analysis that assimilated ERS-1 altimeter data, which underestimates wave height by 10–15% (Queffeulou 1996). In this paper, analyzed wave heights are shown to be sensitive to the quality of the altimeter wave height data used to produce the wave analysis.

The next section shows that relatively subtle changes in the wind may produce fairly considerable changes in systematic wave height forecast error.

2.2 ECMWF wave forecasting

Experimental wave forecasting with the initial version of WAM began at ECMWF in early 1987. Operational global sea state forecasting started at ECMWF in June 1992 with a 3°-latitude × 3°-longitude WAM. Shortly afterwards, a limited-area 0.5° × 0.5° WAM for the Mediterranean Sea was introduced. In August 1993, assimilation of ERS-1 altimeter wave height data commenced. Presently, global and limited-area versions of WAM simulations are computed at ECMWF.

The limited-area model, now called the European shelf model, covers the North Atlantic, North Sea, Baltic Sea, Mediterranean Sea, and Black Sea, and uses an irregular latitude-longitude grid with an approximately constant 28-km × 28-km resolution. The wave spectrum has 25 frequencies and 24 directions. Shallow-water effects, in particular bottom friction, are included.

The global WAM also has an irregular latitude-longitude grid with a 55-km grid spacing. The wave spectrum has 25 frequencies and 12 directions and shallow-water effects are included. In accord with Janssen (1989, 1991), WAM is now a component of the ECMWF operational atmospheric forecast-analysis system, with surface winds from the atmospheric general circulation model provided frequently to WAM. In addition, the sea-state dependent drag coefficient is determined with the stress induced by the ocean waves on the airflow. This two-way interaction of wind and waves yields a more consistent momentum budget at the ocean surface, producing a better balance between wind and waves. Presently, the operational atmospheric general circulation model has a T319 horizontal resolution and 31 layers in the vertical.

A sea-state drag coefficient has substantial impact on the forecast of a rapid developing, fast-moving atmospheric low pressure system. For example, the maximum difference in the minimum mean sea level pressures between a two-day forecast made with and without the sea-state dependent drag coefficient is 9 hPa for a North Pacific storm (Figure 1). Also, there is some impact on the 500-hPa geopotential height, and even at 200 hPa (Janssen and Viterbo 1996), indicating that ocean waves modify the momentum budget to produce a barotropic variation in the atmosphere. This example is exceptional because it shows a large-scale impact. Normally, as expected of physical processes near the surface, the impact on the atmosphere of two-way interaction is relatively small scale. In addition, extreme events are relatively rare.

Two-way interaction between wind and wave has considerable impact on forecasting surface wave height. For example, in the tropics, the mean forecast wave height error computed with (without) a sea-state dependent drag coefficient, decreased (increased)

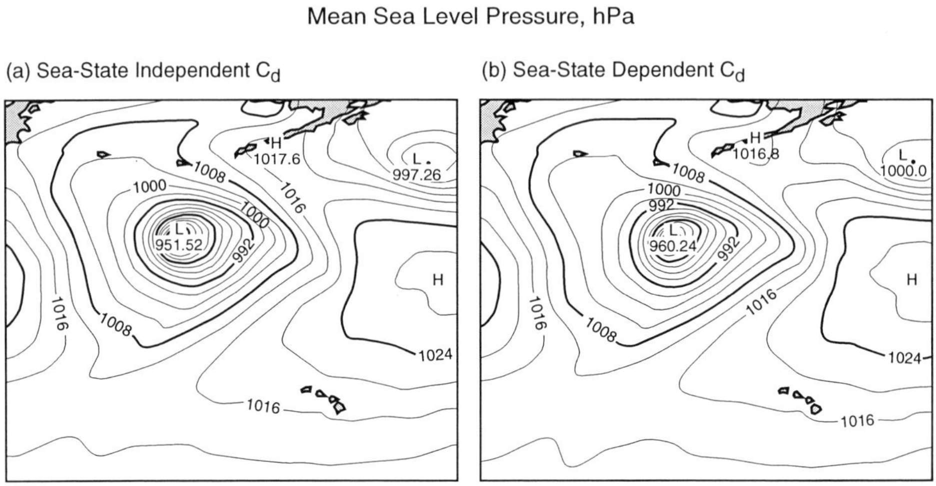

Figure 1. Two-day forecasts of mean sea level pressure made from initial conditions on 12 UT 12 December 1997 for (a) without sea-state dependent drag coefficient (C_d) i.e., without two-way interaction between wind and wave, and (b) with sea-state dependent C_d.

with forecast time (Figure 2). Having a sea-state dependent drag coefficient removes a long-standing problem of systematic forecast error growth in the ECMWF wave forecasting system. In 1994, systematic wave height errors in 5- to 10-day forecasts in the global 20°S–20°N tropical region were about 20% of the mean wave height. However, changes in the ECMWF atmospheric general circulation model in April 1995 (and continuing), and changes in the assimilation method for altimeter data in May 1996 reduced systematic errors to 5–10% of mean wave height. With introduction of an operational coupled atmosphere-ocean wave model at ECMWF on 29 June 1998, the systematic forecast error of wave height is 2–3% and has virtually disappeared.

The reduction of systematic forecast error of wave height in the tropics is an interesting problem because of the combination of local wind-generated waves, windsea, and remotely-forced wind-generated waves that have propagated long distances from the extratropics and are known as swell. In the tropics, swell is the main component of the sea state. In an atmospheric general circulation model, the momentum loss at the ocean surface is described by a drag coefficient. For a logarithmic wind profile, the drag coefficient, C_d, at height $z = L$ is

$$C_d = \left(\kappa / \ln \frac{L}{z_0}\right)^2 \tag{2}$$

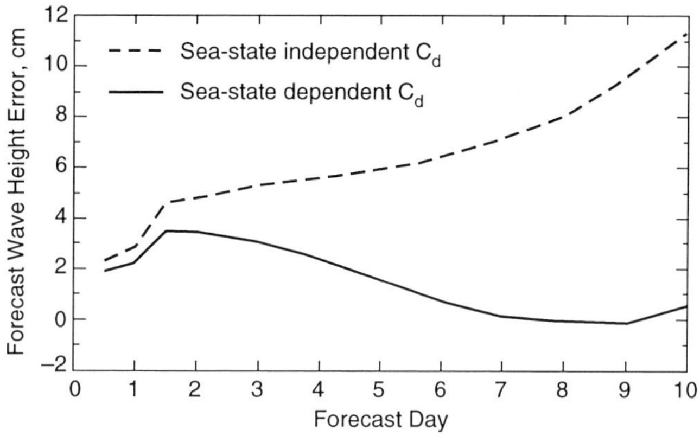

Figure 2. Ten-day forecasts of surface wave height error in the tropics (20°S–20°N, 360° longitudes), relative to the ECMWF verifying analysis, computed at 0.5-day intervals for 74 forecasts (16 April–28 June 1998) made with and without a sea-state dependent drag coefficient.

where κ is the von Karman constant and the roughness length z_0 is given by the Charnock (1955) relation,

$$z_0 = \frac{\alpha u_*^2}{g} \tag{3}$$

where u_* is the friction velocity, g is acceleration of gravity, and α is the Charnock parameter. In pre-June 1998 versions of the ECMWF atmospheric model with sea-state independent C_d, α has the constant value of 0.0185. With two-way interaction between wind and waves, the Charnock parameter is not constant but depends on sea state (Janssen 1991),

$$\alpha = 0.01\left(1 - \frac{\tau_w}{\tau}\right)^{-1/2} \tag{4}$$

where $\tau = \rho_a u_*^2$ is the total wind stress, ρ_a is the density of air at $z = L$, and τ_w is the wave-induced part of the total stress, which can be determined when the wind input source function S_{in} of the energy balance equation is known.

Young windseas, which are ocean waves just generated by wind, are usually steeper than old windseas (Hasselmann et al. 1973). For a young windsea, the contribution of the wave-induced stress to the total stress is larger than that for an old windsea, and the Charnock parameter will be larger than the nominal value of 0.0185 used in the pre-June 1998 ECMWF model. Therefore, a young windsea reduces the strength of the surface wind.

Consequently, wave heights computed from the extratropical wind field with a sea-state dependent C_d will be reduced and systematic forecast wave height error will be lower compared to those computed with the constant Charnock parameter (Janssen and Viterbo 1996). Improved forecasting of extratropical wave heights produces better estimates of swell propagating through the tropics, which reduces the forecast error of wave height in the tropics. However, the treatment of swell is not fully solved as Sterl et al. (1998) suggested that the propagation of wave energy is in error. But the quality of the ECMWF wind field improved as well with the implementation of two-way wind-wave interaction in the ECMWF forecast-analysis system on 29 June 1998. The root-mean-square (rms) difference computed between the ECMWF 6-hour and ERS-2 scatterometer winds shows that a 20 cm s^{-1} (about 10% of total error) reduction occurred on 29 June 1998 (Figure 3). The bias is reduced by about the same amount, although not as clearly visible in Figure 3.

2.3 Future developments

Beginning in 1993, ECMWF started ensemble forecasting to obtain information on the uncertainty of the deterministic forecast, and ensemble prediction of ocean waves began 29 June 1998. The present ensemble prediction system consists of the coupling of the ECMWF T159 atmosphere model and the 1.5° × 1.5° WAM. The ensemble consists of 50 members which are generated by perturbing the deterministic atmospheric analysis by the most unstable singular vectors. Preliminary results (not shown) indicate a promising future for probabilistic forecasting of waves.

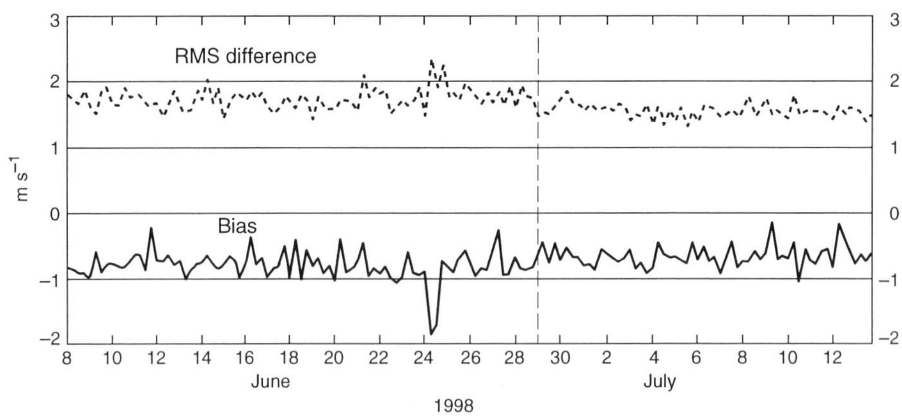

Figure 3. Bias (ERS-2 minus ECMWF) and rms difference between 6-h ECMWF surface wind data product and ERS-2 scatterometer wind measurements.

An important aspect of wave forecasting is the ability to predict extreme events associated with hurricanes and extratropical storms. However, numerical weather prediction models have difficulty simulating the wind field because of lack of horizontal resolution and inaccurate representation of physical processes. In recent years, considerable progress occurred on increasing model resolution and model parameterization of sub-grid-size-scale atmospheric dynamical processes. For example, the operational 36-h forecast of Hurricane Luis (initial conditions on 9 September 1995) in its extratropical phase south of Nova Scotia, Canada, is compared with a 36-h forecast made with the current system. In the current system, the experimental forecast was generated with the four-dimensional variational system, while the operational forecast was based on optimum interpolation. Operational (T213 resolution) and experimental (T639 resolution) forecasts of minimum sea level pressures were 977 and 963 hPa, respectively. According to the National Oceanic and Atmospheric Administration (NOAA) National Hurricane Center, the observed minimum sea level pressure was 965 hPa, which was only 2 hPa higher than that computed with the recent ECMWF system but 12 hPa lower than the value predicted with the operational system in use at ECMWF in September 1995. The newer version of the ECMWF forecast-analysis system not only gives a much deeper atmospheric low pressure, but also the location of the low is in better agreement with observations. The consequences for wave prediction are remarkable (Figure 4). Maximum wave height recorded at a NOAA buoy was 17 m, which was about 30 cm off the prediction from the newer ECMWF system but nearly 7 m different than predicted with the old operational system. An attempt was made to examine reasons for the large differences in the simulation of Hurricane Luis. The change of data assimilation method from optimal interpolation to

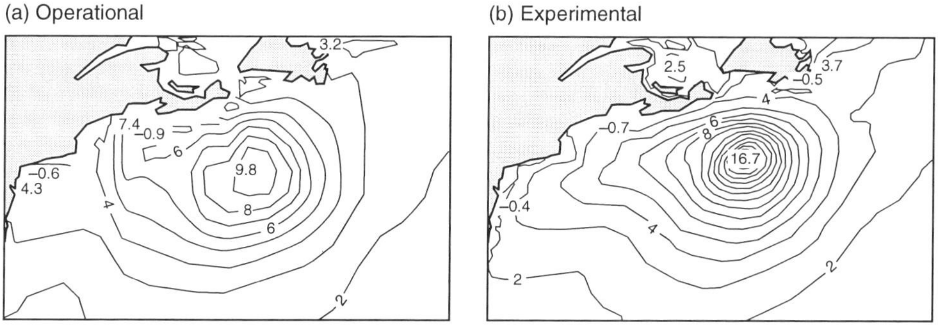

Figure 4. Thirty-six-hour wave height forecasts made with two different ECMWF forecast-analysis systems: (a) T213 operational system of 9 September 1995, named Operational, and (b) T639 system, named Experimental. Initial conditions were at 12 UT 9 September 1995.

four-dimensional variation resulted in a relatively minor improvement of forecast wave height. The relative insensitivity of forecast maximum wave height to the change of assimilation method is not typical, and is probably related to the particular circumstance that the atmospheric low was small scale, while the present version of the variational assimilation method affects relatively large scales. Improvements in wave forecasts due to changes within the atmospheric general circulation model, including horizontal and vertical resolutions, are more dramatic. The most likely candidate for the better forecast is the improved representation of convection (A. Beljaars, private communication 1999), which was introduced into the operational ECMWF system at the end of 1997. Recent changes in the semi-Lagrangian scheme may also have contributed to the improvement. The experimental simulation of Hurricane Luis (not shown) suggests that the present T319 resolution is adequate for small-scale extreme events. The future appears even more promising because in a couple of years a further increase is expected in the horizontal resolution of the ECMWF operational atmospheric general circulation model.

3. Altimeter Wave Height

3.1 ERS-2 data

A numerical weather prediction center, such as ECMWF, requires satellite observations within three hours after the remote-sensed observations have been made in order to assimilate the data into the operational forecast-analysis system. For ERS-1/2, quasi-real-time data products are Fast Delivery products.

Numerical weather prediction data products are quite useful to validate and calibrate satellite data products. For example, just after the launch of ERS-1, the altimeter global mean wave height was about 1-m higher than that simulated with the ECMWF model. Investigation of the detected bias led to the discovery of a small offset in the pre-launch instrument characterization data. When the processing algorithm was updated at all ground stations, the ERS-1 altimeter wave height was found to be satisfactory. The ERS-2 altimeter wave heights showed, from the first day onwards, a remarkably good agreement with the ECMWF 6-h wave height, except at low wave height where ERS-2 had a higher cut-off value than ERS-1. This higher cut-off value is caused by the somewhat different instrumental specifications of the ERS-2 altimeter. Because ERS-2 was launched while ERS-1 was still operational, a comparison of ERS-1 and ERS-2 wave heights revealed that ERS-2 altimeter wave heights were 8% higher than those from ERS-1. This difference was regarded as favorable because ERS-1 wave heights underestimated buoy data (Janssen et al. 1997a). Since the altimeters on ERS-1 and ERS-2 use the same wave height algorithm, the improved performance of the ERS-2 altimeter (Janssen et al. 1997a) is probably related to a different data processing procedure. Indeed, the ERS-2 data processor uses a more accurate procedure to obtain the wave form, which results in better estimation of wave height.

In-situ (NOAA buoys) and ERS altimeter wave heights are used to validate the ECMWF wave forecasting system. The monthly mean bias between ECMWF and buoy wave heights in the Northern Hemisphere was about 25 cm during ERS-1 (Figure 5), became virtually zero during the summer months of 1996 (at the beginning of ERS-2), and then was 15 cm during autumn 1996 (Figure 5). With the introduction of ERS-2 wave height data, the bias during the Northern Hemisphere summer, in which the sea state is characterized by swell and windsea of low steepness, is removed, but the ECMWF wave product still underestimates buoy wave height during the following winter (Figure 5), when the sea state has a considerable fraction of windsea with large steepness. It could be argued that the ECMWF wave forecasting system is underestimating windsea wave height. However, the ERS-2 altimeter wave height was less than the buoy data (Figure 6) by about 22 cm or 7%, with an rms difference of 47 cm.

The satellite altimeter wave height retrieval depends in a sensitive manner on the procedure how to obtain the slope of the wave form, and this could, at least to some extent, explain the discrepancy between altimeter wave height and buoy wave height. However, it does not explain why in cases of swell the altimeter performs better. This led us to suspect that perhaps there are problems with the retrieval of altimeter wave height in cases of large steepness, because most wave height algorithms assume that wave height and steepness distributions are Gaussian, which for large steepness, when nonlinear effects become important, may not be valid. It would thus be natural to study the dependence of altimeter wave height error on ocean wave steepness. However, if buoy observations are used as truth, a few years of collocated data are needed to obtain statistically significant results. Using an ECMWF data product as truth requires, however, only a month of collocations, since there are typically about 40,000 collocations between altimeter and ECMWF-modeled wave heights during one month. Hence, we used the ECMWF 6-h wave heights as truth. For relatively small slopes when swells dominate the sea state, the ERS-2 altimeter

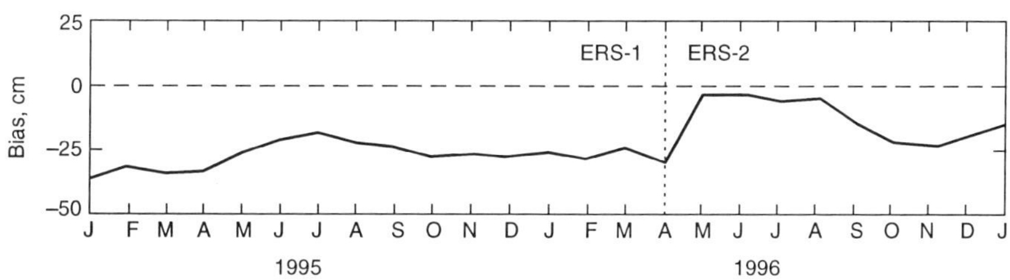

Figure 5. Bias (ECMWF minus buoy) between ECMWF modeled wave height and moored-buoy wave height measurements in the North Atlantic and North Pacific.

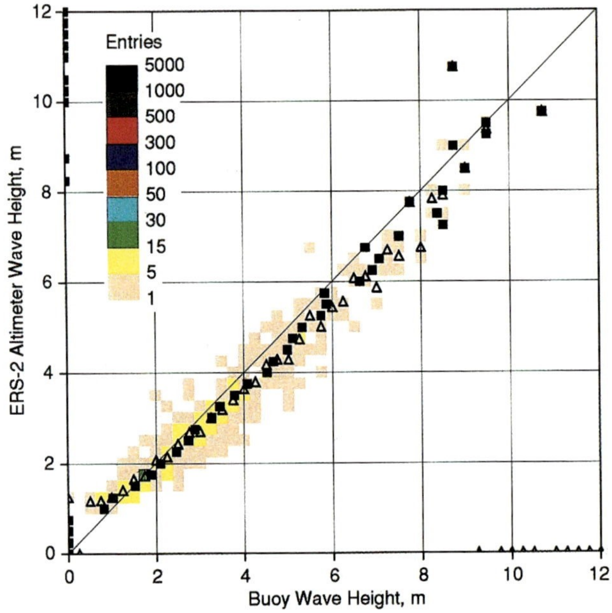

Figure 6. Scatter diagram of mean values of 0.25-m binned moored-buoy and ERS-2 altimeter wave heights for February 1997 for buoy data in the North Atlantic and North Pacific. Solid squares (open triangles) denote mean buoy (altimeter) wave height versus binned altimeter (buoy) wave height. Color code denotes the number of collocations.

wave height error is small; for large slopes, the altimeter underestimates wave height by as much as 50 cm (Figure 7). The next section describes the role of nonlinearity in the retrieval of wave height from the altimeter wave form.

3.2 Electromagnetic bias and altimeter retrieval algorithm

The average radar cross section for backscatter of randomly distributed specular points in the rough surface approximation (Barrick 1972; Barrick and Lipa 1985) is a function of time because contributions from ocean wave crests arrive at the altimeter before those from wave troughs. The time-dependent cross section is called the wave form. For a nadir-scanning radar, the wave form depends on the joint probability distribution (jpd) of surface elevation and surface slope under the condition of zero mean slope. In order to obtain a practical altimeter retrieval algorithm, the probability distributions of the surface elevation and slope are each assumed to have a Gaussian shape, which is reasonable for weakly nonlinear ocean waves (Longuet-Higgins 1963). Although this yields

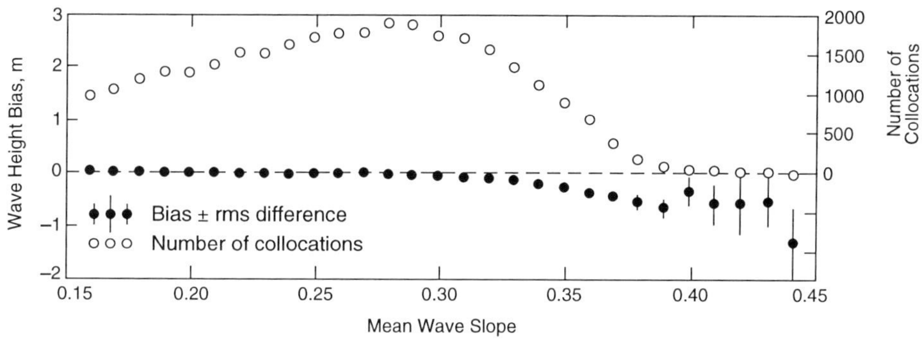

Figure 7. Wave height bias (altimeter minus ECMWF 6 h) computed at 0.01 increments of wave slope for February 1997.

good estimates for altimeter wave height, the Gaussian assumption does not include a weak, nonlinear wave process that affects the altimeter range measurement, known as electromagnetic bias (EMB), which is caused by waves having a sharp crest and a wide trough. An altimeter retrieval algorithm without consideration of EMB would emphasize the part of the sea surface below mean sea level, and, therefore, the altimeter range measurement would estimate a somewhat longer distance between satellite and ocean surface. For linear waves with small steepness the Gaussian probability distribution is valid and EMB vanishes. Also, waves may distort the altimeter pulse to produce an additional error in the altimeter range measurement of the height of the satellite above the sea surface; this error is called the instrumental error. The sum of instrumental error and electromagnetic bias is called the sea state bias (SSB).

Deviations from the Gaussian distribution may occur for a number of reasons, and we explore the consequences of a nonlinear wave surface with sharp crests and wide troughs, i.e., EMB, on altimeter retrieval of wave height. For a radar pulse with Gaussian shape and width, ν, Srokosz (1986) showed that the wave form, *W*, is

$$W \sim \frac{1}{2}\left[1 + erf(T) + \frac{e^{-\frac{T^2}{2}}}{\sqrt{\pi}}\left\{\frac{\lambda\sqrt{2}}{3}(1+T^2) - \frac{1}{\sqrt{2}}(\lambda+\delta)\right\}\right] \tag{5}$$

where $erf(T)$ is the error function, c is the speed of light, H_s is the significant wave height, T is the normalized time $\frac{t}{t_p}$, and

$$t_p = \frac{2}{c}\sqrt{v^2 + \frac{1}{8}H_s^2} \tag{6}$$

Deviations from a Gaussian distribution are measured by the skewness factor, λ, and the elevation-slope correlation, δ, which depend in a complicated way on the wave spectrum (Longuet-Higgins 1963; Jackson 1979). Using the Phillips (1958) spectrum

$$F(k) = \frac{1}{2}\alpha_p k^{-3} \tag{7}$$

then

$$\delta = 2\lambda \qquad \lambda = 2\sqrt{\alpha_p} \tag{8}$$

where α_p is the Phillips parameter, and, as noted by Srokosz (1986), λ is corrected by a factor two compared to that obtained by Longuet-Higgins (1963). Swell has typically a smaller Phillips parameter by a factor of 4–10 than windsea; therefore, swell has in practice a Gaussian distribution, while windsea, with α_p approximately equal to 0.01, may show considerable deviations from a Gaussian distribution.

The half-power of a radar pulse wave form, W, with a Gaussian jpd occurs at the origin, $T = 0$ (Figure 8), which corresponds to mean sea level. For a Gaussian jpd, EMB vanishes, and only the first two terms in equation (5) remain. H_s is determined from the half-power wave form slope, s,

$$H_s = 4\left(\frac{\kappa_1}{s^2} - \kappa_2\right)^{1/2} \tag{9}$$

where κ_1 and κ_2 depend on the power and width of the transmitted pulse and on the speed of light.

For a non-Gaussian jpd, the half-power point of the wave form (equation (5)) is shifted towards positive time (Figure 8) because in the presence of weakly nonlinear waves the radar altimeter range measurement overestimates the distance between mean surface and satellite. H_s is also determined by the half-power slope, which, however, does not coincide with $T = 0$. Assuming small deviations from a Gaussian distribution, i.e., λ and δ are small, an approximate expression for the observed H_s is

$$H_s = 4\left(\frac{\kappa_3}{s^2} - \kappa_2\right)^{1/2} \tag{10}$$

and

$$\kappa_3 = \kappa_1\left(1 + 2\left(\frac{\lambda}{3} + \delta\right)\left(\frac{5}{24}\lambda + \frac{1}{8}\delta\right)\right) \tag{11}$$

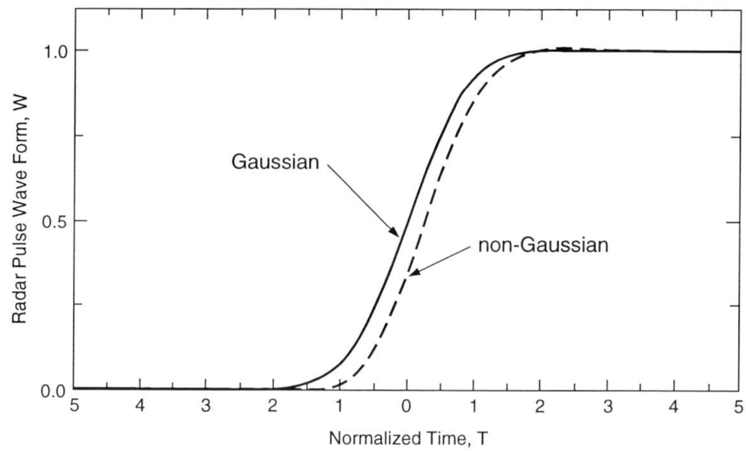

Figure 8. Distribution of radar pulse wave form, *W*, with normalized time, *T*, for Gaussian and non-Gaussian joint probability distributions of wave height and slope.

The EMB correction to the altimeter range measurement is

$$\mathrm{EMB} = -\frac{1}{8}\left(\frac{\lambda}{3} + \delta\right)H_S \tag{12}$$

For the Phillips spectrum defined by equation (7), then

$$\kappa_3 = \kappa_1\left(1 + \frac{154}{18}\alpha_p\right) \tag{13}$$

and

$$\frac{\mathrm{EMB}}{H_s} = -\frac{7}{12}\sqrt{\alpha_p} \tag{14}$$

Deviations from a Gaussian distribution produce a modest impact on altimeter wave height retrieval because the correction depends on α_p, which is, in the extreme condition of young windsea, at most 0.025; thus, at the maximum, the correction to wave height would be 10%. EMB may vary by a factor of two to three, depending on sea state, being small for swell and large for a young windsea (Minster et al. 1992).

WAM spectra are used to determine λ and δ, and, consequently, EMB. The chosen period was February 1997, when a number of extreme events occurred in the North Atlantic. EMB corrections were applied to the ERS-2 Fast Delivery altimeter data. The corrected altimeter wave heights are in slightly better agreement with buoy data. The bias has been reduced from 22 cm (Figure 6) to 14 cm (Figure 9), i.e., the mean corrected wave height is about 3% larger. The rms difference between corrected ERS-2 wave heights and buoy data is 46 cm and is the same as that computed with uncorrected ERS-2 data. This is a disappointing result, but it should be realized that in deriving the correction for wave height, we have assumed that wave height was obtained from the half-power slope of the wave form. This is not the ESA procedure to obtain the Fast Delivery wave height (R. Francis, private communication 1997). Further tests are warranted because the impact of the nonlinear sea state on wave height retrieval is sensitive to the procedure of how to obtain the slope of the wave form.

Comparison of the EMB correction computed from WAM spectra and the SSB correction computed from ERS-2 data by the Gaspar and Ogor (1996) method showed fair agreement and remarkable differences (Figure 10). For low values of EMB and SSB, EMB is too low compared to SSB, while for large corrections, the opposite is found. Realizing that to first approximation, EMB depends linearly on wave height, it is concluded that for low wave height, i.e., swell conditions, the WAM approach underestimates

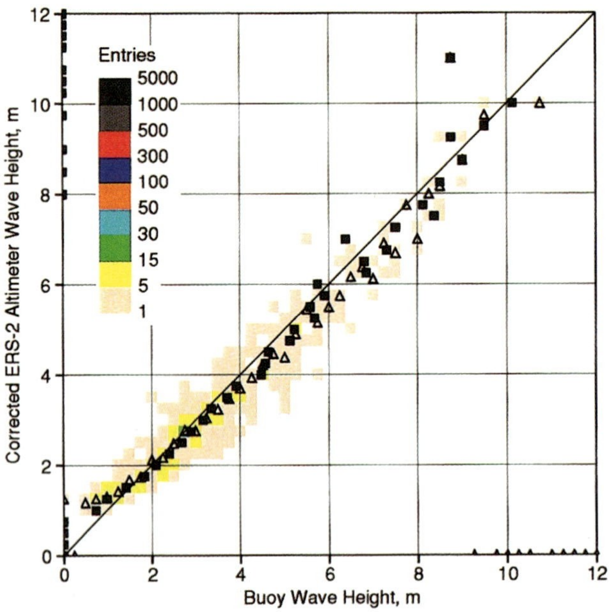

Figure 9. Scatter diagram of mean values of 0.25-m binned moored-buoy and nonlinear-corrected ERS-2 altimeter wave heights for February 1997 for buoy data in the North Atlantic and North Pacific. Solid squares (open triangles) denote mean buoy (altimeter) wave height versus binned altimeter (buoy) wave height. Color code denotes the number of collocations.

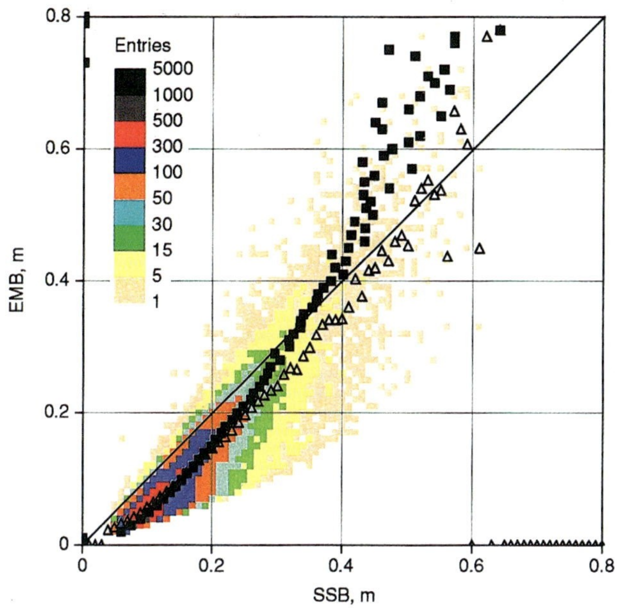

Figure 10. Scatter diagram of WAM-derived electromagnetic bias (EMB), according to the Srokosz (1986) method, and ERS-2 sea state bias (SSB), according to the Gaspar and Ogor (1996) method. Global data for February 1997 were used. Color code denotes the number of collocations.

SSB, but for windseas, the WAM method seems to give reasonable estimates of SSB. According to equation (14), EMB is lower for swell than for windsea.

The question, why the ERS-derived SSB has little sea-state dependency, is examined with reference to the dimensionless wave height, gH_s/U_{10}^2 (where U_{10} is wind speed at 10-m height), which is a measure of wave development. For an old windsea, the dimensionless wave height is about 0.25; a young windsea has smaller values and swell has larger values. The WAM-derived EMB has a much greater sensitivity to the dimensionless wave height compared to SSB computed from ERS-2 data with the Gaspar and Ogor (1996) method (Figure 11). A direct estimate of EMB is found in Melville et al. (1991), which shows a clear sea-state dependence with dimensionless wave height (Figure 11). There is a fair agreement with the WAM-derived EMB. Perhaps, the absence of sea-state dependency of the Gaspar and Ogor (1996) SSB is caused by the instrumental error having an effect opposite to that created by the EMB, but it is evident that more research is needed to clarify this issue.

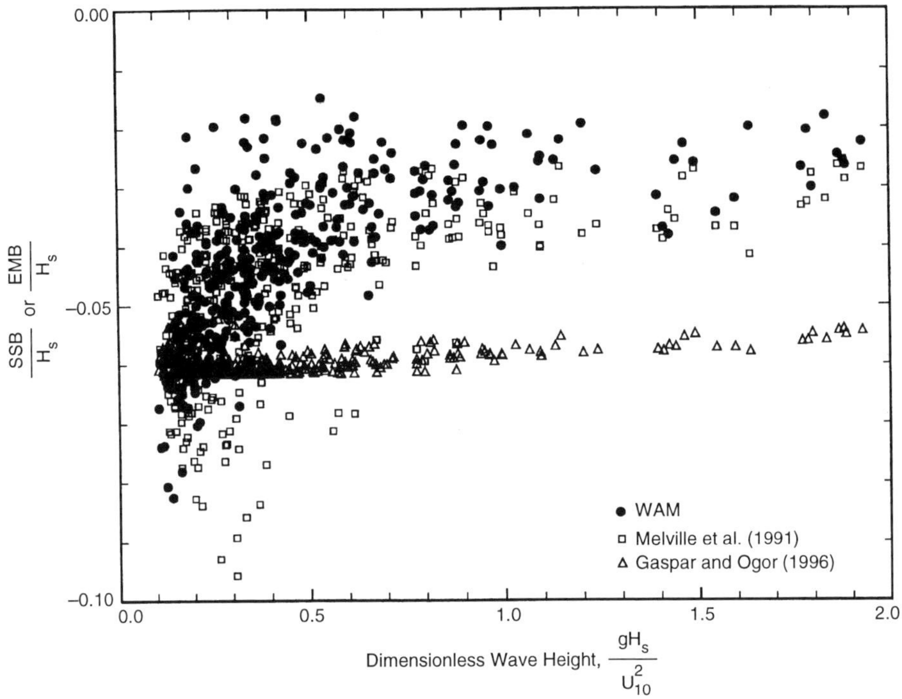

Figure 11. Scatter diagram between the dimensionless wave height and WAM-derived EMB, EMB computed from the Melville et al. (1991) formulation, and SSB computed in accord with Gaspar and Ogor (1996). Each bias is relative to ERS-2 measurements of H_s during February 1997. The 10-m height wind speed was computed from ERS-2 altimeter data.

4. Conclusions

Considerable progress has been made in the past twenty years in wave modeling and in satellite wind and wave products. The combined use of satellite and wave model products has also revealed problems. There may a problem in how WAM treats swell, although presently it is not clear whether there is too much swell or not enough. There may be a problem with the altimeter wave height retrieval for young, nonlinear sea states. Finally, two-dimensional WAM spectra may provide information on EMB. Further studies are planned to resolve some of the issues mentioned in this paper.

Acknowledgments. The author acknowledges the support by the members of the ECMWF wave group Jean Bidlot, Björn Hansen and Martin Hoffschildt. Furthermore, useful discussions with members of ESA's Altimeter Scientific Advisory Group Johnny Johannessen, Richard Francis, Remko Scharroo, Seymoor Laxon and Monica Roca are very much appreciated. I thank Lars Isaksen for providing Figure 3, whilst stimulating discussions with Martin Miller are gratefully acknowledged as well. Two reviewers are thanked for their comments, which improved the paper.

References

Anctil, F., and M. A. Donelan, Air-water momentum flux observations over shoaling waves, *J. Phys. Oceanogr.*, *26*, 1344–1353, 1996.

Barrick, D. E., Remote sensing of sea state by radar, In Remote Sensing of the Troposphere, edited by V. E. Derr, U. S. Govt. Printing Office, Washington, D.C., 12-1 to 12-46, 1972.

Barrick, D. E., and B. Lipa, Analysis and interpretation of altimeter sea echo, *Adv. Geophys.*, *27*, 60–99, 1985.

Bauer, E., S. Hasselmann, K. Hasselmann, and H. C. Graber, Validation and assimilation of Seasat altimeter wave heights using the WAM wave model, *J. Geophys. Res.*, 12671–12682, 1992.

Bauer, E., and C. Staabs, Statistical properties of global significant wave heights and their use for validation, *J. Geophys. Res.*, *103*, 1153–1166, 1998.

Cardone, V. J., H. C. Graber, R. E. Jensen, S. Hasselmann, and M. J. Caroso, In search of the true surface wind field in SWADE IOP-1: Ocean wave modeling perspective, *Global Atmos. Ocean Sys.*, *3*, 107–150, 1995.

Charnock, H., Wind stress on a water surface, *Quart. J. Roy. Meteorol. Soc.*, *81*, 639–640, 1955.

Donelan, M. A., The dependence of the aerodynamic drag coefficient on wave parameters, In Proc. First International Conference on Meteorological and Air/Sea Interaction of the Coastal Zone, Amer. Meteorol. Soc., Boston, 381–387, 1982.

Donelan, M. A., F. W. Dobson, S. D. Smith, and R. J. Anderson, On the dependence of sea surface roughness on wave development, *J. Phys. Oceanogr.*, *23*, 2143, 1993.

Felizardo, F. C., and W. K. Melville, Correlations between ambient noise and the ocean surface wave field, *J. Phys. Oceanogr.*, *25*, 513–532, 1995.

Gaspar, P., and F. Ogor, Estimation and analysis of the sea state bias of the new ERS-1 and ERS-2 altimetric data (OPR version 6), Tech. Rep. CLS/DOS/NT/96.041, Collect. Localisation, Satell. Agne, Toulouse, France, 1996.

Gelci, R., H. Cazale, and J. Vassal, Prevision de la houle: La methode des densites spectroangulaires, *Bull. Inform. Comité Central Oceanogr. Etudes Côtes*, *9*, 416–435, 1957.

Hansen, B., and H. Guenther, ERS-1 radar altimeter validation with the WAM model, In Proc. ERS-1 Geophysical Validation Workshop, European Space Agency, Paris, 157–161, 1992.

Hare, J. E., P. O. G. Persson, C. W. Fairall, and J. B. Edson, Behaviour of Charnock's relation for high wind conditions, In Proc. 13th AMS Conference on Boundary Layers and Turbulence, Amer. Meteorol. Soc., Boston, 252–255, 1999.

Hasselmann, K., On the non-linear energy transfer in a gravity-wave spectrum: General theory, *J. Fluid Mech.*, *12*, 481–500, 1962.

Hasselmann, K., On the spectral dissipation of ocean waves due to white capping, *Bound. Layer. Meteorol.*, *6*, 107–127, 1974.

Hasselmann, K., T. P. Barnett, E. Bouws, H. Carlson, D. E. Cartwright, K. Enke, J. A. Ewing, H. Gienapp, D. E. Hasselmann, P. Kruseman, A. Meerburg, P. Mueller, D. J. Olbers, K. Richter, W. Sell, and H. Walden, Measurements of wind-wave growth and swell decay during the Joint North Sea Wave Project (JONSWAP), *Deuts. Hydrogr. Z. Suppl.*, *A8*, 1–95, 1973.

Hasselmann, S., and K. Hasselmann, Computations and parameterisations of the nonlinear energy transfer in a gravity-wave spectrum: A new method for efficient computations of the exact nonlinear transfer integral, *J. Phys. Oceanogr.*, *15*, 1369–1377, 1985.

Heimbach, P., S. Hasselmann, and K. Hasselmann, Statistical analysis and intercomparison of WAM model data with global ERS-1 SAR wave mode spectral retrievals over 3 years, *J. Geophys. Res.*, *103*, 7931–7977, 1998.

Jackson, F. C., The reflection of impulses from a nonlinear random sea, *J. Geophys. Res.*, *84*, 4939–4943, 1979.

Janssen, J. A. M., Does wind stress depend on sea-state or not? A statistical error analysis of HEXMAX data, *Bound. Layer. Meteorol.*, *83*, 479–503, 1997.

Janssen, P. A. E. M., Wave-induced stress and the drag of air flow over sea waves, *J. Phys. Oceanogr.*, *19*, 745–754, 1989.

Janssen, P. A. E. M., Quasi-linear theory of wind-wave generation applied to wave forecasting, *J. Phys. Oceanogr.*, *21*, 1631–1642, 1991.

Janssen, P. A. E. M., On error growth in wave models, ECMWF Tech. Memo 249, European Centre for Medium-Range Weather Forecasts, Reading, United Kingdom, 12 pp, 1998.

Janssen, P. A. E. M., P. Lionello, M. Reistad, and A. Hollingsworth, Hindcasts and data assimilation studies with the WAM model during the Seasat period, *J. Geophys. Res.*, *94*, 973–993, 1989.

Janssen, P. A. E. M., and P. Viterbo, Ocean waves and the atmospheric climate, *J. Climate*, *9*, 1269–1287, 1996.

Janssen, P. A. E. M., B. Hansen, and J. Bidlot, Validation of ERS satellite wave products with the WAM model, In CEOS Wind and Wave Validation Workshop Report, ESA WPP147, ESTEC, The Netherlands, 101–108, 1997a.

Janssen, P. A. E. M., B. Hansen, and J.-R. Bidlot, Verification of the ECMWF wave forecasting system against buoy and altimeter data, *Wea. Forecasting*, *12*, 763–784, 1997b.

Johnson, H. K., J. Hojstrup, H. J. Vested, and S. Larson, On the dependence of sea surface roughness on wind waves, *J. Phys. Oceanogr.*, *28*, 1702–1716, 1998.

Khandekar, M. L., and R. Lalbeharry, An evaluation of Environment Canada's operational wave model based on moored buoy data, *Wea. Forecasting*, *11*, 139–152, 1996.

Komen, G. J., L. Cavaleri, M. Donelan, K. Hasselmann, S. Hasselmann, and P. A. E. M. Janssen, editors, *Dynamics and Modelling of Ocean Waves*, Cambridge University Press, Cambridge, 532 pp, 1994

Komen, G. J., K. Hasselmann, and S. Hasselmann, On the existence of a fully developed windsea spectrum, *J. Phys. Oceanogr.*, *14*, 1271–1285, 1984.

Lionello, P., H. Günther, and P. A. E. M. Janssen, Assimilation of altimeter data in a global ocean wave model, *J. Geophys. Res.*, *97*, 14453–14474, 1992.

Lipa, B., and D. E. Barrick, Ocean surface height-slope probability density function from SEASAT altimeter echo, *J. Geophys. Res.*, *86*, 10921–10930, 1981.
Longuet-Higgins, M. S., The effect of nonlinearities on statistical distributions in the theory of sea waves, *J. Fluid Mech.*, *17*, 459–480 1963.
Melville, W. K., R. H. Stewart, W. C. Keller, J. A. Kong, D. V. Arnold, A. T. Jessup, M. R. Loewen, and A. M. Slinn, Measurements of electromagnetic bias in radar altimetry, *J. Geophys. Res.*, *96*, 4915–4924, 1991.
Miles, J. W., On the generation of surface waves by shear flows, *J. Fluid Mech.*, *3*, 185–204, 1957.
Minster, J. F., D. Jourdan, Ch. Boissier, and P. Midol-Monnet, Estimation of the sea-state bias in radar altimeter GEOSAT data from examination of frontal systems, *J. Atmos. Oceanic Tech.*, *9*, 174–187, 1992.
Mitsuyasu, H., On the growth of the spectrum of wind-generated waves: 1, Rep. Res. Inst. Appl. Mech., Kyushu Univ., *16*, 251–264, 1968.
Mitsuyasu, H., On the growth of the spectrum of wind-generated waves: 2, Rep. Res. Inst. Appl. Mech., Kyushu Univ., *17*, 235–243, 1969.
Monbaliu, J., On the use of the Donelan wave spectral parameter as a measure for the roughness of wind waves, *Bound. Layer. Meteorol.*, *67*, 277–291, 1994.
Pierson, W. J., G. Neumann, and R. W. James, Practical Methods for Observing and Forecasting Ocean Waves by Means of Wave Spectra and Statistics, H.O. Pub 603, U.S. Navy Hydrographic Office, Washington, D.C., 284 pp, 1955.
Pierson, W. J., Jr,. and L. Moskowitz, A proposed spectral form for fully developed wind-seas based on the similarity theory of S. A. Kitaigorodskii, *J. Geophys. Res.*, *69*, 5181, 1964.
Phillips, O. M., On the generation of waves by turbulent wind, *J. Fluid Mech.*, *2*, 417–445, 1957.
Phillips, O. M., The equilibrium range in the spectrum of wind-generated waves, *J. Fluid Mech.*, *4*, 426–434, 1958.
Phillips, O. M., The dynamics of unsteady gravity waves of finite amplitude: 1, *J. Fluid Mech.*, *9*, 193–217, 1960.
Queffelou, P., Significant wave height and backscatter coefficient at ERS-1/2 and Topex/Poseidon ground track crossing points, FDP, IFREMER contribution to the ERS-2 radar altimeter commissioning phase, IFREMER Tech Rep., IFREMER, DRP/OS, BP 70, Plouzane, France, 25 pp, 1996.
Romeiser, R., Global validation of the wave model WAM over a one year period using Geosat wave height data, *J. Geophys. Res.*, *98*, 4713–4726, 1993.
Smith, S. D., R. J. Anderson, W. A. Oost, C. Kraan, N. Maat, J. DeCosmo, K. B. Katsaros, K. L. Davidson, K. Bumke, L. Hasse, and H. M. Chadwick, Sea surface wind stress and drag coefficients: The HEXOS results, *Bound. Layer. Meteorol.*, *60*, 109–142, 1992.
Snyder, R. L., F. W. Dobson, J. A. Elliot, and R. B. Long, Array measurements of atmospheric pressure fluctuations above surface gravity waves, *J. Fluid Mech.*, *102*, 1–59, 1981.
Srokosz, M. A., On the joint distribution of surface elevation and slope for a nonlinear random sea, with application to radar altimetry, *J. Geophys. Res.*, *91*, 995–1006, 1986.
Sterl, A. G., G. J. Komen, and P. D. Cotton, Fifteen years of global wave hindcasts using ERA winds: Validating the reanalysed winds and assessing the wave climate, *J. Geophys. Res.*, *103*, 5477–5492, 1998.

Sverdrup, H. U., and W. H. Munk, Wind Sea and Swell: Theory of Relations for Forecasting, H.O. Pub. 601, U.S. Navy Hydrographic Office, Washington, D.C., 44 pp, 1947.

SWAMP Group: J. H. Allender, T. P. Barnett, L. Bertotti, J. Bruinsma, V. J. Cardone, L. Cavaleri, J. Ephraums, B. Golding, A. Greenwood, J. Guddal, H. Gunther, K. Hasselmann, S. Hasselmann, P. Joseph, S. Kawai, G. J. Komen, L. Lawson, H. Linne, R. B. Long, M. Lybanon, E. Maeland, W. Rosenthal, Y. Toba, T. Uji, and W. J. P. de Voogt, *Sea Wave Modeling Project (SWAMP): An Intercomparison Study of Wind Wave Prediction Models, Part 1: Principal Results and Conclusions*, Plenum, New York, 256 pp, 1985.

WAMDI Group: S. Hasselmann, K. Hasselmann, E. Bauer, P. A. E. M. Janssen, G. J. Komen, L. Bertotti, P. Lionello, A. Guillaume, V. C. Cardone, J. A. Greenwood, M. Reistad, L. Zambresky, and J. A. Ewing, The WAM model: A third generation ocean wave prediction model, *J. Phys. Oceanogr.*, *18*, 1775–1810, 1988.

Wittmann, P. A., R. M. Clancy, and T. Mettlach, Operational wave forecasting at Fleet Numerical Meteorology and Oceanography Center, Monterey, CA., In Fourth Int. Workshop on Wave Hindcasting and Forecasting, Atmospheric Environment Service, Ottawa, 335–342, 1995.

Yelland, M. J., B. I. Moat, P. K Taylor, R. W. Pascal, J. Hutchings, and V. C. Cornell, 1998. Wind stress measurements from the open ocean corrected for airflow distortion by the ship, *J. Phys. Oceanogr.*, *28*, 1511–1526, 1998.

Zambresky, L., A verification study of the global WAM model, December 1987–November 1988, ECMWF Tech. Rep. 63, European Centre for Medium-Range Weather Forecasts, Reading, United Kingdom, 86 pp, 1989.

Peter Janssen, European Centre for Medium-Range Weather Forecasts (ECMWF), Shinfield Park, Reading RG2 9AX, United Kingdom. (email, peter.janssen@ecmwf.int;fax, +44-118-986-9450)

Satellites, Oceanography and Society
edited by David Halpern

Chapter 4

The use of satellite surface wind data to improve weather analysis and forecasting at the NASA Data Assimilation Office

R. Atlas
Goddard Space Flight Center, National Aeronautics and Space Administration, Greenbelt, Maryland

R. N. Hoffman
Atmospheric and Environmental Research, Inc., Cambridge, Massachusetts

Abstract. One important application of satellite surface wind observations is to improve the accuracy of weather analyses and forecasts. The first satellite to measure surface wind over the ocean was SeaSat in 1978. The initial impact of satellite surface wind data on weather analysis and forecasting was very small, but extensive research has been conducted since SeaSat to improve data accuracy and utilization of these data in atmospheric models. Satellite surface wind data are now used to detect intense storms over the ocean as well as to improve the overall representation of the wind field in numerical weather prediction models. Satellite wind data contribute to improved warnings for ships at sea and to more accurate global weather forecasts.

Experiments with the Goddard Earth Observing System atmospheric general circulation model and data assimilation system indicate that the impact of satellite wind data measured by the National Aeronautics and Space Administration Scatterometer was approximately twice as large as the impact of Special Sensor Microwave Imager or European Remote-sensing Satellite wind data. Locations of cyclones over the ocean were up to 500 km more accurate, and the useful forecast skill in the Southern Hemisphere extratropics was extended by 24 hours.

1. Introduction

Accurate observations of surface wind velocity over the oceans are required for a wide range of meteorological and oceanographic applications. Surface winds are needed to drive ocean models and surface wave models, to calculate surface fluxes of heat, moisture and momentum, and to construct surface climatologies. In addition, surface wind data are essential for nowcasting weather and wave conditions at sea, and to provide initial conditions and verification data for numerical weather prediction (NWP) models.

Prior to the launch of satellites capable of determining surface wind, observations of surface wind velocity were provided primarily by ships and buoys. Such conventional observations are important components of the global observing system, but are limited in coverage and accuracy. For example, reports of surface wind by ships cover only very limited regions of the global ocean, occur at irregular intervals of time and space, tend to avoid the worst (and therefore most interesting) weather, and are at times of poor accuracy. Buoys, while of higher accuracy, have even sparser coverage. As a result, analyses based only on these in-situ observations can misrepresent surface wind over large regions and are generally not adequate for weather forecasting.

Satellites offer an effective way to fill data voids as well as to provide higher resolution data. The European Remote-sensing Satellite (ERS) scatterometer provides coverage over 90% of the ocean within 96 hours. The National Aeronautics and Space Administration (NASA) Scatterometer (NSCAT) provides coverage over 90% of the ocean within 48 hours. The QuikScat SeaWinds scatterometer provides over 90% coverage within 24 hours. If the wind direction ambiguity is properly resolved, scatterometer data are very accurate. Reliably resolving the 180° ambiguity using spatial filters is feasible if a priori information may be used.

Moore and Pierson (1971) were the first to propose a space-based scatterometer, which led to a demonstration aboard SkyLab in 1973 and to the launch of the first satellite scatterometer on SeaSat in 1978. SeaSat failed after three months of operation and the subsequent data impact studies on weather forecasts showed a negligible impact. Vertical extension of surface wind observations increases the impact on weather forecasts, but an extension that does not account for the synoptic situation gives negative as well as positive effects. Stability-dependent vertical correlation functions yield a positive impact (Bloom and Atlas 1990, 1991).

After SeaSat, the next scatterometer was launched in 1991 on ERS-1, which was followed by ERS-2, NSCAT, and SeaWinds (Table 1). Initial experiments to show an impact of ERS-1 data on weather forecasting were inconclusive. Since then, substantial progress has been made. ERS-2 data are now used operationally at several NWP centers, and NSCAT data impact experiments showed large positive impacts. In brief, simplifying assumptions made in the initial simulation studies were incorrect, and substantial refinements were required to cope with the special characteristics of scatterometer data. This paper describes the current situation.

Table 1. Ocean surface wind observations from space. Instruments include microwave radiometers (MR) and Ku-band and C-band scatterometers. Altimeter instruments are not included.

Spacecraft/Instrument	*Dates*	*Type*	*Information*
NIMBUS-5/ESMR	1972–1976	MR	Wilheit (1979)
SeaSat/SMMR	1978	MR	Gloersen and Barath (1977)
SeaSat/SASS	1978	Ku-band	Grantham et al. (1977)
NIMBUS-7/SMMR	1978–1987	MR	Gloersen and Barath (1977)
DMSP/SSMI	1987–Present	MR	Hollinger et al. (1990)
MOS/MSR	1987–1996	MR	http://yyy.tksc.nasda.go.jp/ Home/Earth_Obs/e/mos_e.html
ERS-1/AMI	1991–1996	C-band	Attema (1991)
ERS-2/AMI	1996–Present	C-band	http://earth.esa.int/ERS/
ADEOS-1/NSCAT	1996–1997	Ku-band	Naderi et al. (1991)
QuikScat/SeaWinds	1999–Present	Ku-band	http://winds.jpl.nasa.gov/
DMSP/SSMIS	2000 Launch	MR	http://www.aerojet.com/program/ detail/about_ssmis.htm
ADEOS-2/AMSR	2001 Launch	MR	http://wwwghcc.msfc.nasa.gov/ AMSR/
ADEOS-2/SeaWinds	2001 Launch	Ku-band	http://winds.jpl.nasa.gov/
METOP/ASCAT	2003 Launch	C-band	http://www.esrin.esa.it/esa/progs/ METOP.html

2. Measurement of Surface Winds from Space

2.1 Active microwave sensors

Over the ocean, scatterometer surface winds are estimated from multiple backscatter measurements made from several directions. At moderate incidence angles (20°–70°), the major scattering mechanism is Bragg scattering from centimeter-scale waves, which are, in most conditions, in equilibrium with the local wind. Backscatter depends very nonlinearly on wind speed and direction. Most scatterometer winds are derived from empirical relationships, called model functions, which relate backscatter to geophysical parameters, and which are derived from collocated observations (Jones et al. 1977; Stoffelen and Anderson 1997; Wentz and Smith 1999). Several scatterometer measurements are made of the same location, and winds are obtained by optimally fitting these data to a model function. Although scatterometer winds are usually provided as neutral winds at some reference height, the measurement is most closely connected with surface stress (Brown 1986).

Nonlinearity of the model function allows several wind vectors consistent with the backscatter observations (Price 1976). These multiple wind vectors are called aliases in the early literature and are now generally referred to as ambiguities. The ambiguities are the minima of a cost function, which measures the differences between the observed backscatter and those calculated for the given wind speed and direction. Each ambiguity is assigned a probability of being closest to the true wind vector. Since the cost function approximates (or is) the negative of the likelihood function, ambiguities associated with smaller values of the cost function are more probable. The highest probability ambiguity is termed the rank 1 solution. For the SeaSat scatterometer, with only two antennas, all four ambiguities have approximately the same likelihood of being correct. For ERS with three antennas on one side and NSCAT with three antennas on both sides, usually only the first two probabilities are large and the associated ambiguous wind vectors point in nearly opposite directions (Stoffelen and Anderson 1997). Various filtering approaches (called dealiasing or ambiguity removal algorithms) include subjective (Wurtele et al. 1982), median filter (Schultz 1990; Shaffer et al. 1991), variational (Hoffman 1982, 1984), other horizontal filtering methods (Stoffelen and Anderson 1997, and references therein), neural net (Badran et al. 1991), and simply choosing the ambiguity closest to some reference field (Endlich et al. 1981; Baker et al. 1984). Once the ambiguity is removed, the wind vector chosen is called the unique wind.

Schroeder et al. (1985) show that three antennas increase the instrument skill to about 0.6, i.e., the rank 1 solution is the ambiguity closest to the true wind vector 60% of the time. Actual instrument skill is less than 0.60 and, as shown by Schultz (1990), as the instrument skill decreases from 0.60 to 0.50, the ambiguity removal skill decays from nearly perfect to nearly useless. In practice, purely autonomous ambiguity removal schemes have not worked well. Fortunately, these techniques do work well if initialized with a good first guess, usually based on a recent NWP forecast.

Satellite surface wind data are listed in Table 1. The first space-based scatterometer was on SkyLab during June 1973–February 1974. Based on this experience, SeaSat carried a scatterometer in 1978 (Grantham et al. 1977). Although the SeaSat mission failed after approximately 100 days, the scatterometer data were of sufficient quality and interest (Stewart 1988, Katsaros and Brown 1991) that plans for a follow-on mission were quickly formulated (O'Brien et al. 1982). The SeaSat scatterometer follow-on instrument (NSCAT) was launched in 1996 aboard the Japanese Advanced Earth Observation Satellite (ADEOS-1) (Naderi et al. 1991). Sadly, ADEOS-1 failed after nine months. The follow-on to NSCAT, called SeaWinds, was launched in June 1999 aboard QuikScat, with SeaWinds-2 to be launched in 2001 aboard ADEOS-2. SeaSat, NSCAT, and SeaWinds are NASA instruments and all operate in the Ku-band. In the interim between SeaSat and NSCAT, the European Space Agency (ESA) launched ERS-1 in July 1991 and ERS-2 in April 1995, each with an active microwave instrument (AMI) operating as a C-band scatterometer for most of each orbit (Francis et al. 1991).

Scatterometers have similar orbit characteristics: sun-synchronous, near-polar at roughly 800-km altitude, and an approximately 100-min. period. SeaSat and NSCAT have antennas on both sides of the spacecraft to produce two simultaneous swaths (each

500 km wide for SeaSat and 600 km wide for NSCAT) separated by a nadir gap (450 km wide for SeaSat and 350 km wide for NSCAT). SeaSat and NSCAT had two and three antennas on each side of the spacecraft, respectively. The NSCAT fore and aft antennas were vertically polarized (V-pol) and the mid antenna was vertically and horizontally polarized. The ERS-1 scatterometer has three V-pol antennas only on the right side of the spacecraft. Due to the geometry of fan beam observations, the time difference between backscatter values observed by the fore and aft antennas at a single location on the surface varies from approximately 70 to 200 seconds, increasing with incidence angle. SeaWinds has a radically new design, using a 1-m diameter rotating-dish antenna to illuminate two circles on the ocean surface, eliminating the nadir gap.

2.2 Passive microwave sensors

In the 1978–1991 period between SeaSat and ERS-1, there are no scatterometer wind data and the satellite surface wind speed are only measured from the Scanning Multichannel Microwave Radiometer (SMMR) and Special Sensor Microwave Imager (SSMI) (Table 1). The lack of wind direction limited their initial utility in scientific studies. SMMR data have lower resolution and less accuracy compared to SSMI data. In the 1990s, more than one SSMI was operational. Different schemes, ranging from a simple direction assignment method to a variational analysis method, have been used to convert SSMI wind speeds into vector winds (Atlas et al. 1991, 1993, 1996). Also, SSMI wind speeds are assimilated at several operational centers (e.g., Goerss and Phoebus 1992).

3. Impact of Scatterometer Data on Numerical Weather Prediction

To evaluate the importance of a particular type of data for NWP, a "Control" numerical simulation is performed. Then, experimental numerical simulations are made in which data are either withheld or added to the conditions associated with the Control simulation. Forecasts are generated every few days (to achieve relative independence of the forecast sample) for both the Control and Experiment conditions. Analyses and forecasts from each Experiment simulation are then verified to determine the data impact. Further details are described by Atlas (1997).

In the earliest study, Cane et al. (1981) showed a substantial positive impact of SeaSat scatterometer data could be expected in the surface pressure (Figure 1). However, several simplifying assumptions limited the practicality of this study. First, the same model was used to generate both "nature" and forecasts, which yield unrealistically accurate predictions. Second, the simulated SeaSat wind observations were defined for the lowest model level (nominally 945 hPa), not at the surface. Third, no errors, including ambiguity removal errors, were used. SeaSat scatterometer data impact studies performed with global models (Baker et al. 1984; Duffy et al. 1984; Yu and McPherson 1984; Anderson et

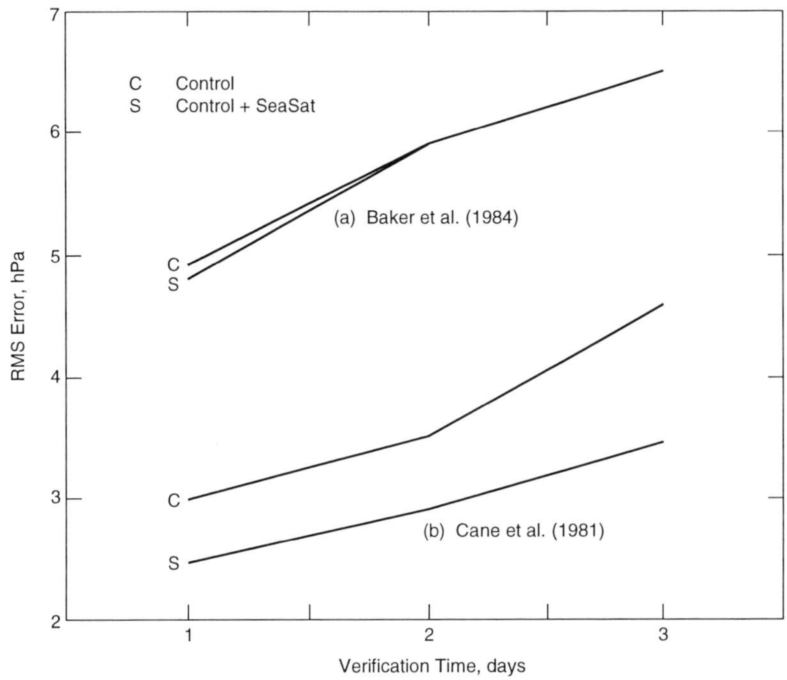

Figure 1. Root-mean-square (rms) error for North Pacific (30°–86°N, 120°E–120°W) sea level pressure computed by Cane et al. (1981) using simulated SeaSat data and by Baker et al. (1984) using real SeaSat data.

al. 1991; Ingleby and Bromley 1991) demonstrated potential for SeaSat scatterometer winds to significantly affect surface analyses, but failed to show a meaningful improvement in NWP forecasts. For example, Baker et al. (1984) showed a negligible effect of the SeaSat scatterometer data in the Northern Hemisphere extratropics (Figure 1). The following factors appear to have limited the impact: coarse resolution of the models; failure to explicitly resolve the planetary boundary layer; ambiguity and other errors in the SeaSat scatterometer winds; treatment of SeaSat scatterometer winds as synoptic; failure to account for statistical characteristics of the data; lack of or inappropriate coupling of surface winds to higher levels; and data redundancy.

In the Southern Hemisphere extratropics, SeaSat scatterometer data had a positive effect on the analyses and forecasts, but the effect was smaller than that produced with Vertical Temperature Profile Radiometer (VTPR) data. The SeaSat scatterometer data impact was reduced by VTPR data, indicating redundancy between the two datasets. Another limiting factor, the lack of or inappropriate coupling of surface winds to higher levels, was explored by Atlas and Pursch (1983) and Atlas et al. (1985). Their results suggested that the impact of surface wind data on the analyses and forecasts could be significantly enhanced by extending the influence of surface winds through the planetary boundary layer (Figure 2). How to vertically extend the influence of surface winds is a challenging problem.

Duffy and Atlas (1986) used characteristics of the synoptic event to extend the vertical influence of SeaSat scatterometer winds to produce a significant improvement in the prediction of an intense storm. Similar conclusions have been obtained by Stoffelen and

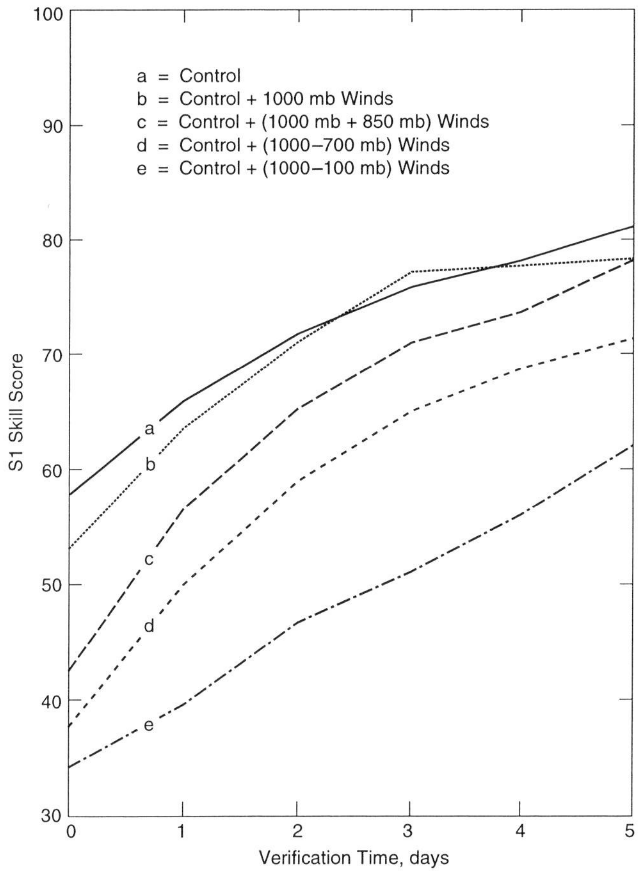

Figure 2. The forecast S1 skill score (Tweles and Wobus 1954) for sea level pressure over the Southern Hemisphere (86°–2°S, 360° longitudes) for different amounts of vertical extension of surface wind data. The S1 skill score is defined by the gradient of sea level pressure; the lower the score, the more accurate is the forecast.

Cats (1991), Atlas (1988), and Lenzen et al. (1993). In regard to the limiting factor created by the coarse resolution of the model, Stoffelen and Cats (1991) showed that scatterometer data provide useful small-scale information that is otherwise unobserved.

Satellite surface winds differ from conventional surface wind measurements and, therefore, require a specialized data processing procedure. Directional ambiguity inherent in every scatterometer wind measurement must be removed. The resolution of satellite wind observations is virtually instantaneous in time and tens of kilometers in space. An anemometer measures the wind at a single location (Pierson 1983). Satellite observations are asynoptic, requiring four-dimensional data assimilation, but the optimal updating time interval may be smaller than that used in some NWP centers.

3.1 Goddard Earth Observing System atmospheric model and data assimilation

Several features of the Goddard Earth Observing System (GEOS) atmospheric general circulation model and data assimilation system (DAS) are designed to enhance the impact of satellite surface wind data. A first-guess surface wind field, which is the 6-hour previous forecast and which is consistent with the model planetary boundary layer, is generated. Scatterometer winds are combined with the first-guess surface winds to yield an adjusted surface wind field, from which the adjusted surface pressure is computed, using the model planetary boundary layer. Based on the hydrostatic relationship, the first-guess 1000-hPa geopotential height field is modified in accordance with the adjusted sea level pressure field. Incremental modifications of the 1000-hPa geopotential height are calculated wherever satellite surface wind observations are used. All geopotential heights below 850 hPa are similarly adjusted to influence the model upper-air analysis. In addition, stringent quality control tests are employed to eliminate scatterometer observations contaminated by precipitation, boundary layer stability, atmospheric attenuation, large-scale waves, oil slicks, and transient wind fields (Atlas et al. 1999).

Impact results are expected to vary with different atmospheric general circulation models and DAS methods. Therefore, experiments have been conducted to assess the impact of assimilation of ERS-1, SSMI, and NSCAT data in GEOS-1 DAS (Schubert et al. 1993), GEOS-2 DAS (Atlas et al. 1999), and the 1995 T62-truncation version of the National Centers for Environmental Prediction (NCEP95) DAS (Parrish et al. 1997). In general, impacts tend to be smaller with the more advanced model and DAS. However, in the experiments to be reported here, more dramatic improvements were obtained with GEOS-2 than with GEOS-1. Further, some of our experiments with the T62-resolution NCEP DAS were also made at T126 resolution. Skill scores at T62 are lower than at T126. T62 and T126 impacts were essentially the same, and the T126 control forecast reproduced exactly the operational forecast. Current NWP operational systems, e.g., at the European Center for Medium-Range Weather Forecasts (ECMWF) and NCEP, and the higher-resolution GEOS-3 DAS, are more advanced than the systems used here, and impacts on the newer systems would probably be different from those reported here.

Anomaly correlations were computed with all wave numbers, including wave numbers higher than 20. This would lower the skill score compared to anomaly correlations computed with wave numbers less than 20. In addition, in most instances, we verify our forecasts against independent analyses generated by another model and DAS, while most operational centers verify forecasts against their own analyses. The operational ECMWF analyses are used for verification, and anomaly correlations for 500-hPa geopotential heights are computed between ECMWF and our forecasts. During the first 48 hours of the forecast, this also lowers the score, but leaves impacts (i.e., differences between scores) relatively unaffected. For brevity we do not show any impacts on the surface wind analysis, which are reported in Atlas et al. (1999).

3.2 Impact of ERS-l scatterometer data

Hoffman (1993) showed that an early version of ERS-1 scatterometer winds were substantially different than the first-guess ECMWF surface winds, but the forecast impacts were neutral, with no consistent improvement or degradation. Similarly, Stoffelen and Anderson (1997) found no significant improvement in ECMWF forecast accuracy beyond 12 hours with high-quality ERS-1 wind vectors. In contrast, the results reported below and those reported by Andrews and Bell (1998), who used the United Kingdom Meteorological Office (UKMO) model and DAS, show substantial improvement in forecast accuracy in the Southern Hemisphere extratropics with assimilation of ERS-1 wind vectors.

The impacts of five different ERS-1 scatterometer datasets were evaluated: “ESA”, operational ERS-1 wind vectors; “JPL”, ERS-1 wind vectors created at the Jet Propulsion Laboratory (JPL) with the Freilich and Dunbar (1993) method; “NCEP”, ERS-1 winds generated at the Goddard Space Flight Center (GSFC) using modified UKMO wind retrieval methodology and the operational NCEP analysis as the background; “GLA”, ERS-1 winds generated at GSFC, using modified UKMO wind retrieval methodology and the GEOS-1 control analysis as the background; and “VAR”, ERS-1 winds generated at GSFC by direct utilization of ERS-1 backscatter measurements in a two-dimensional variational analysis using the GEOS-1 control analysis as the background. The GEOS-1 control analysis, named “Control”, used all conventional data plus satellite temperature soundings and cloud-tracked winds. In all cases, the scatterometer data were thinned to approximately 75-km resolution. Each observation retained was then treated in the same manner as a buoy observation.

In the initial experiment, five simulations with four forecasts each were generated: “Control”; “ESA”, ESA surface wind vectors added to the Control; “JPL”, JPL surface wind vectors added to the Control; “Alias”, ambiguous JPL wind vectors, choosing the ambiguity closest to the model first guess, added to the Control; and "Speed", JPL wind speeds added to the Control. The 4° × 5° GEOS-1 model was used, with spin-up time from 12 UT 25 February to 03 UT 1 March 1993. Assimilation time period was 03 UT 1 March to 03 UT 24 March 1993. Forecasts were made on 6, 11, 16, and 21 March 1993. JPL and ESA winds show a substantial positive impact on forecasts in the Southern Hemisphere extratropics (Figure 3), although, in general, the simulations with ESA winds yield higher forecast accuracy. Comparison of the JPL, Alias, and Speed forecasts (Figure 3) shows that both the ERS-1 direction and speed improve the GEOS-1 forecasts. In the Northern Hemisphere extratropics and tropics (not shown), the impact of ERS-1 winds on GEOS-1 forecasts was negligible.

In Experiment 2, the 2° × 2.5° GEOS-1 model was used. Spin-up, assimilation, and forecast times were the same as those in the initial experiment. In addition to Control, four forecasts each were generated with ESA, JPL, NCEP, or GLA versions of ERS-1 scatterometer wind vectors added to Control. Each of the ERS-1 datasets yields a significant and equal improvement in forecast accuracy of the 500-hPa geopotential height in the Southern Hemisphere extratropics (Figure 4), and, in agreement with the 4° × 5° GEOS-1 results of Experiment 1, no improvement in forecast accuracy occurs in the tropics or Northern Hemisphere extratropics (not shown).

A synoptic evaluation of Experiment 2 forecasts showed that, compared to the Control, ERS-1 data created substantial modifications to ocean surface winds and to the baroclinic structures above the planetary boundary layer. In the Southern Hemisphere extratropics, cyclone displacement and development were improved significantly by assimilation of ERS-1 winds. However, occasional examples of significant negative impact were also

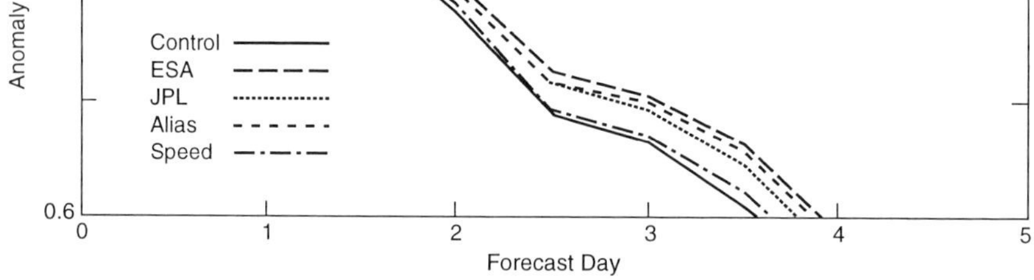

Figure 3. Anomaly correlation of the 4° × 5° GEOS-1 500-hPa geopotential height forecasts averaged over four cases for the Southern Hemisphere (86°–30°S, 360° longitudes) extratropics.

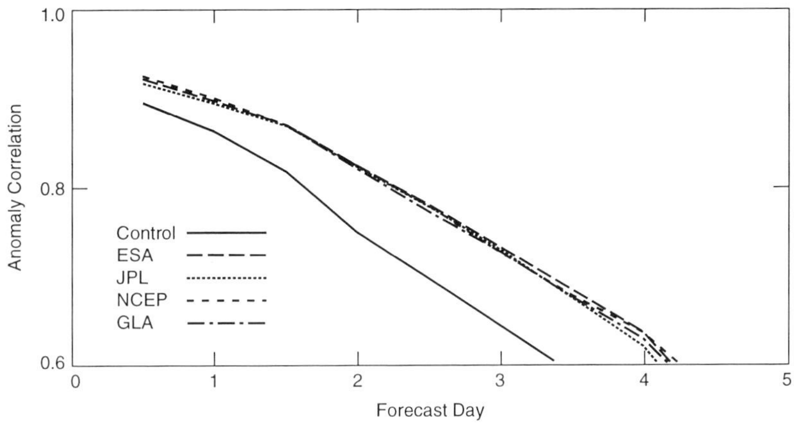

Figure 4. Anomaly correlation of the 2° × 2.5° GEOS-1 500-hPa geopotential height forecasts averaged over four cases for the Southern Hemisphere (86°–30°S, 360° longitudes) extratropics.

observed. In an effort to reduce the occurrence of negative impacts and improve the accuracy and spatial coherence of ERS-1 winds, VAR ERS-1 winds were added to Control (Experiment 3). Spin-up time was the same as in Experiment 2 and the assimilation time period was 03 UT 1 March–03 UT 6 March 1993. Forecasts were generated on 6 March. The VAR forecast was clearly improved relative to that made with the GLA simulation (Figure 5). The advantage of the variational approach is that it embeds the ambiguity in a large data-fitting problem which includes other observations, a background constraint based on balanced error covariances, and the model dynamics. The last two factors lead to a dynamically consistent use of the data. Comparison of VAR and GLA winds (not shown), which used the same background field, reveals examples of VAR having improved wind directions.

In Experiment 4, no satellite surface winds were added to Control, but JPL, ESA, and NCEP versions of ERS-1 winds were assimilated into NCEP95. The spin-up, forecasts, and assimilation periods were identical to those in Experiment 2. Both JPL and ESA ERS-1 winds produce a positive and equal impact on NCEP95 T62 forecasts (Figure 6). The NCEP winds give slightly better forecasts than either the JPL or ESA winds (Figure 6). Comparison of these results with those for GEOS-1 (Figure 4) indicates that the impact of ERS-1 data is of similar magnitude in both models. In addition, evaluation of synoptic events in NCEP forecasts revealed an impact on cyclone prediction comparable to that found with GEOS-1.

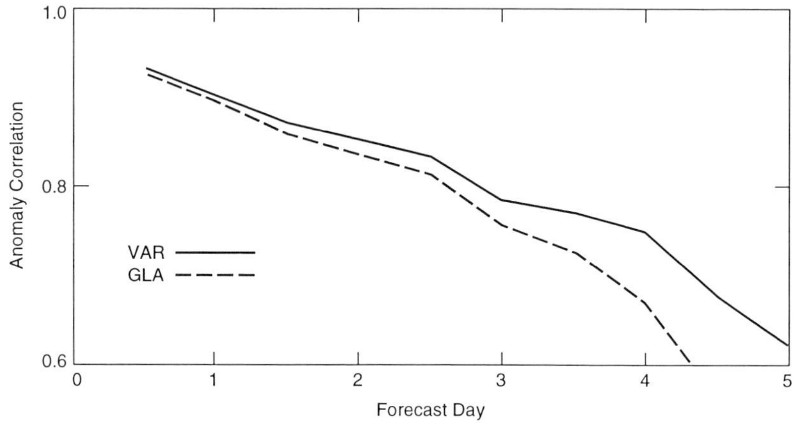

Figure 5. Anomaly correlation of the 2° × 2.5° GEOS-1 500-hPa geopotential height forecasts for the Southern Hemisphere (86°–30°S, 360° longitudes) extratropics.

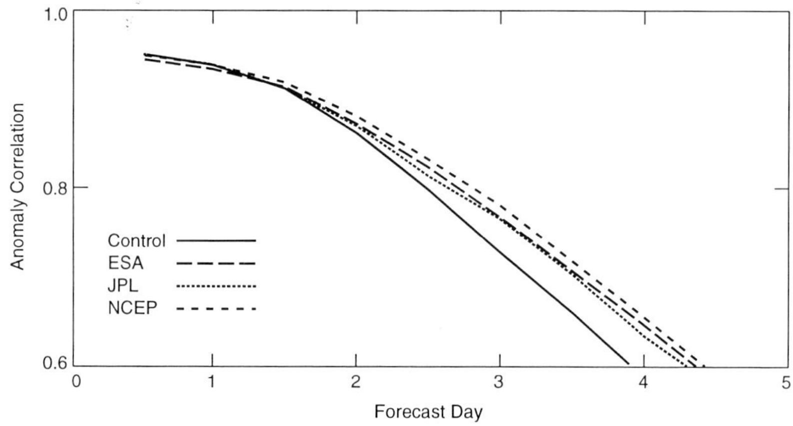

Figure 6. Anomaly correlation of the NCEP95 T62 500-hPa geopotential height forecasts averaged over four cases for the Southern Hemisphere (86°–30°S, 360° longitudes) extratropics.

3.3 Impact of NSCAT data

Control simulations were generated with the GEOS-1 and GEOS-2 models, both with 2° × 2.5° resolution, using all available data (conventional surface data, rawinsondes, aircraft observations, satellite temperature soundings, and cloud-drift winds), with the exception of satellite surface winds. Then, simulations were made with either SSMI wind speeds ("SSMI"), NSCAT wind speeds ("NSCAT-S"), or NSCAT wind vectors ("NSCAT-N") added to Control.

In Experiment 1, the GEOS-1 model was used; the spin-up time was 03 UT 10 September to 03 UT 15 September 1996; the assimilation period was 03 UT 15 September to 03 UT 12 November 1996; and eight forecasts were generated from initial states obtained at approximate five-day intervals within the two-month assimilation period. In the Northern Hemisphere extratropics, there is virtually no difference between any of the average forecast scores (Figure 7a). However, a very significant positive impact of NSCAT data is evident in the Southern Hemisphere extratropics (Figure 7b). Assimilation of NSCAT-N wind vectors produced a large increase in anomaly correlation relative to Control for days 2–5 of the forecast. In addition, the NSCAT-N forecast at five days is more accurate than the Control forecast at four days. The impact of NSCAT-N data on ocean surface wind

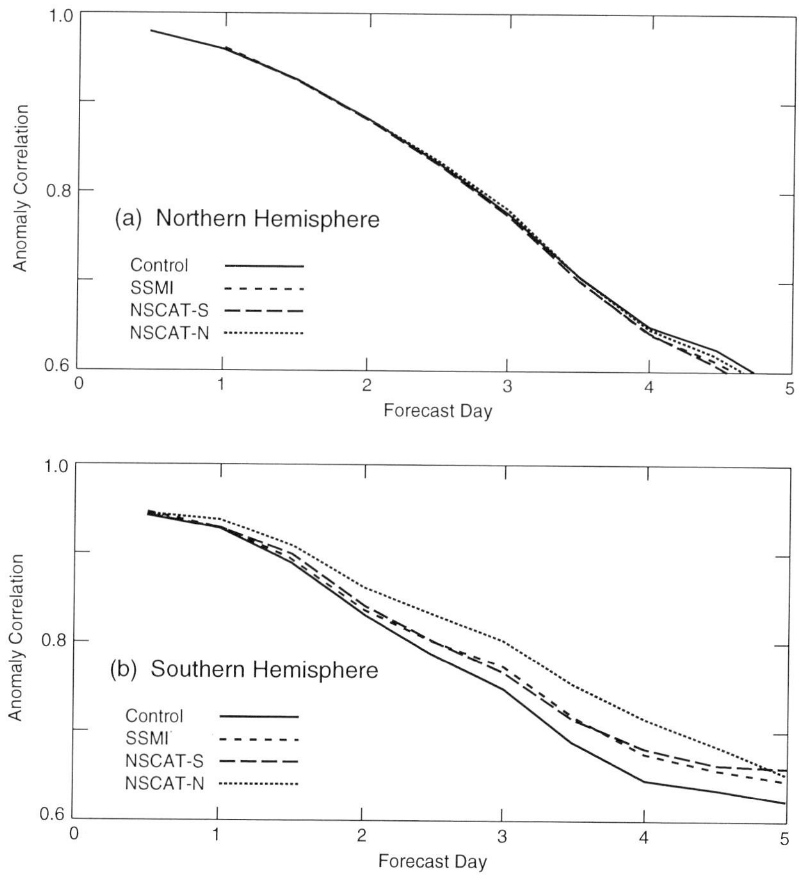

Figure 7. Anomaly correlation of the 2° × 2.5° GEOS-1 sea level pressure averaged over eight forecasts for the (a) Northern Hemisphere extratropics (30°–86°N, 360° longitudes) and (b) Southern Hemisphere extratropics (86°–30°S, 360° longitudes).

analyses is overwhelmingly positive (not shown). Assimilation of SSMI or NSCAT-S winds also results in significant improvement in forecast accuracy in the Southern Hemisphere extratropics (Figure 7b); however, the SSMI or NSCAT-S impact is about half that of the NSCAT-N impact through most of the forecast period. NSCAT-S and SSMI wind speeds are of comparable utility.

In Experiment 2, the GEOS-2 model was used and the adjusted surface wind field is computed at the time of the NSCAT-N measurement. Spin-up time, assimilation time period, and forecast times were identical to those in Experiment 1. Comparing Figures 7 and 8, the accuracy of the GEOS-2 Control is improved significantly relative to GEOS-1.

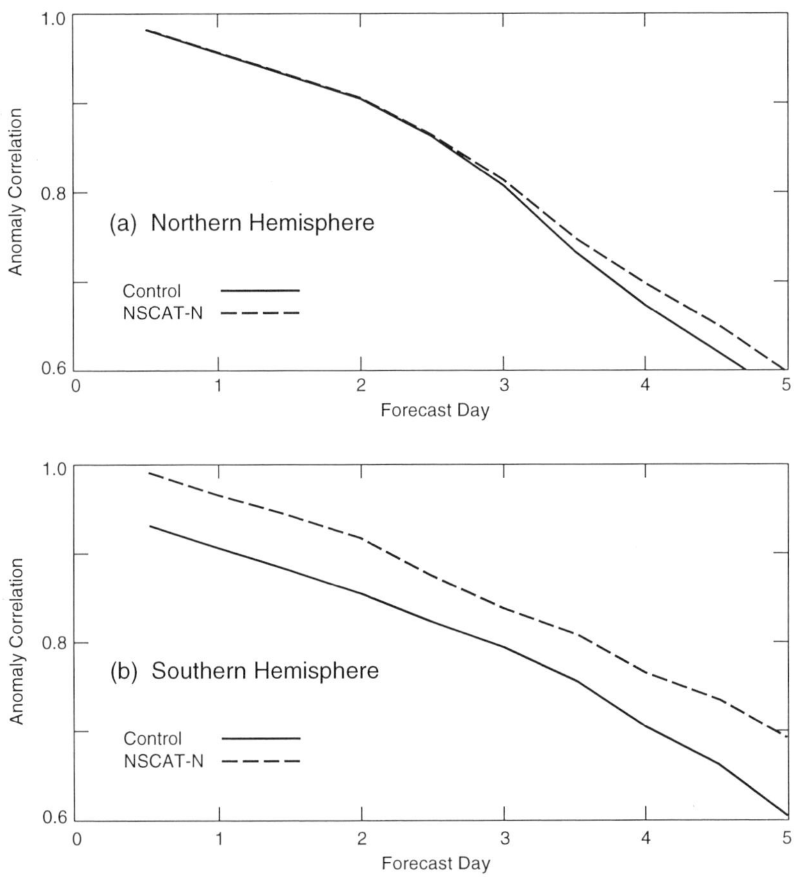

Figure 8. Anomaly correlation of the 2° × 2.5° GEOS-2 sea level pressure averaged over eight forecasts for the (a) Northern Hemisphere extratropics (30°–86°N, 360° longitudes) and (b) Southern Hemisphere extratropics (86°–30°S, 360° longitudes).

For example, the three-day GEOS-2 Control forecast accuracy is equal to the three-day NSCAT-N GEOS-1 accuracy. Despite this substantial improvement in the Control, the impact in the Southern Hemisphere extratropics of NSCAT-N data with GEOS-2 is comparable to the impact obtained with GEOS-1. The poor performance of the GEOS-2 Control in the Southern Hemisphere extratropics at 0.5–1.5 days is due to using NSCAT-N for verification. However, except for this initial time period, the choice of analysis for verification has negligible effect. In both GEOS-1 and GEOS-2, a 24-hour extension of useful forecast skill results from the assimilation of NSCAT-N winds. In the Northern Hemisphere extratropics, a positive impact of NSCAT-N data is obtained with GEOS-2. The asynoptic treatment of NSCAT-N data in GEOS-2 is responsible for this improved impact.

3.4 Impact on synoptic events

Two synoptic cases are described to illustrate the impact of NSCAT-N data on NWP. The assimilation of either ERS or SSMI data would yield similar impacts, although the NSCAT-N impact is generally larger and more frequent (not shown).

Using 2° × 2.5° GEOS-1, the 96-hour Control and NSCAT-N sea level pressure forecasts in the Southern Hemisphere extratropics were made from 00 UT 28 October 1996. Comparison of the Control and NSCAT-N forecasts with the 00 UT 1 November 1996 ECMWF verifying analysis (Figure 9) shows that NSCAT-N data produced a very significant improvement in the cyclonic circulation. In particular, the central pressure and structure of the intense cyclone south of Africa improved very substantially, and the position error is reduced by over 500 km. Other impacts include formation of a weaker cyclone near 47°S, 5°W, and the sea level pressure ridge to the southwest of this cyclone. Impacts such as this are typical in the Southern Hemisphere extratropics.

The impact of NSCAT-N data, assimilated both synoptically and asynoptically, over the North Pacific is examined with 2° × 2.5° GEOS-2. Sixty-hour forecasts were made from initial conditions on 00 UT 22 September 1996. The ECMWF sea level pressure analysis on 12 UT 24 September 1996 is used for verification. The forecast with NSCAT-N data treated asynoptically has the best agreement with the ECMWF analysis (Figure 10). When NSCAT-N is treated synoptically, the time displacement of NSCAT-N data relative to the analysis time produces an incorrect prediction of the cyclone. Accurate data used at the wrong time leads to a negative impact. In contrast, the asynoptic assimilation makes very effective use of the available scatterometer observations.

4. Conclusions

Scatterometer observations improve the definition of the surface wind field over the ocean and, therefore, can aid operational forecasting of meteorological and oceanographic parameters. In addition, these data make possible very short-range forecasting

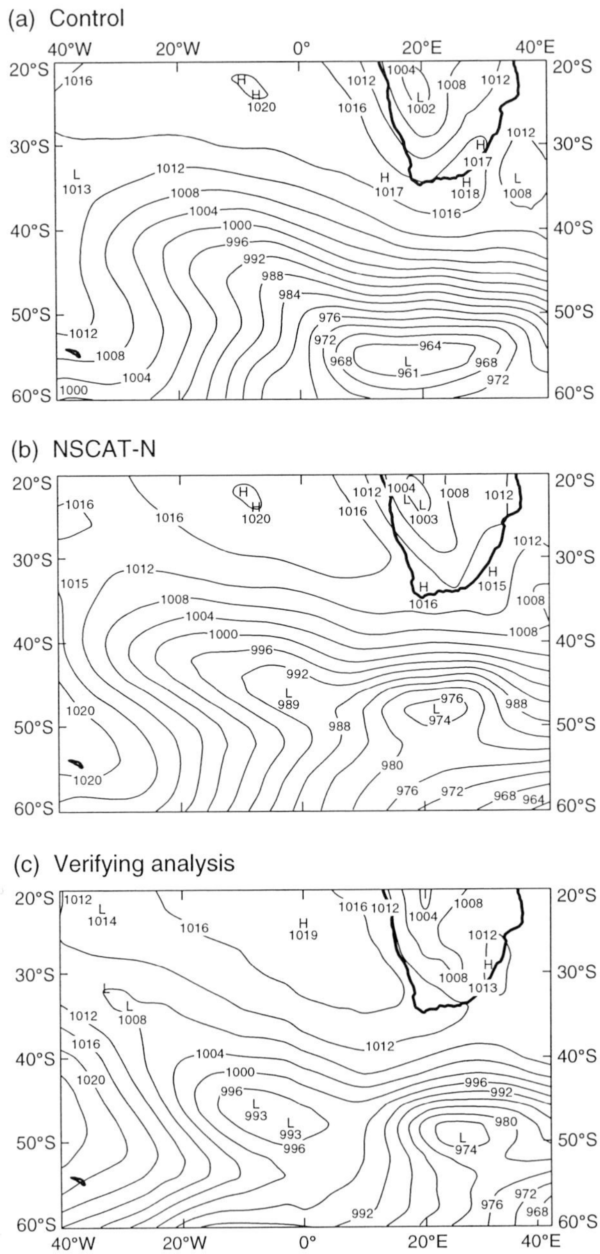

Figure 9. (a) Control and (b) NSCAT-N 2° × 2.5° GEOS-1 sea level pressure forecasts. (c) ECMWF verifying analysis.

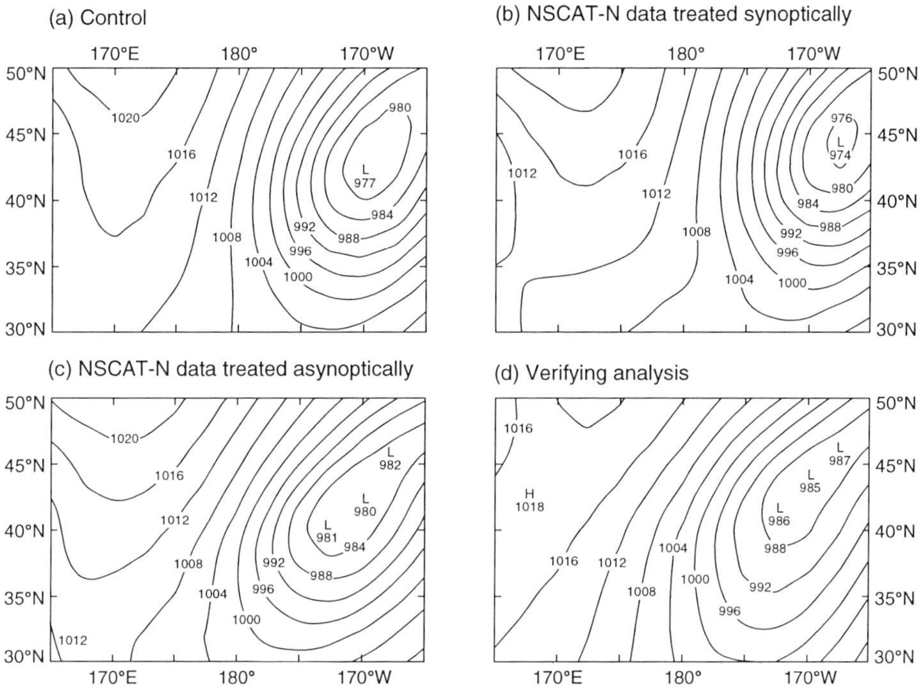

Figure 10. Sixty-hour 2° × 2.5° GEOS-2 sea level pressure forecasts for (a) Control, (b) synoptic NSCAT-N data, (c) asynoptic NSCAT-N data, and (d) ECMWF verifying analysis.

and nowcasting of intense, small-scale or rapidly evolving storms and associated surface waves, which might otherwise be unobserved. A particular advantage of the scatterometer is that it is unaffected by clouds or light precipitation. Heavy precipitation affects the Ku-band SeaSat scatterometer and NSCAT data (Guymer et al. 1981), but does not affect the C-band ERS data. These advantages enabled National Oceanic and Atmospheric Administration (NOAA) National Weather Service Marine Prediction Center forecasters to improve operational analyses, forecasts, and warnings over the oceans for six months during the NSCAT mission (Sienkiewicz, personal communication 1997).

Satellite surface wind data are also vital for understanding the coupled climate system, and for providing accurate air-sea boundary conditions for studies of both the atmosphere and ocean on a variety of spatial scales. Exchanges of energy, latent and sensible, and momentum between the atmosphere and ocean are nonlinearly modulated by surface wind speed. In addition, satellite surface wind data are useful in the validation of numerical analyses and forecasts.

Our studies have shown that the use of scatterometer data in modern data assimilation systems produces realistic global surface wind analyses. However, intermittent updating introduces substantial errors. The adjusted surface wind field must be calculated with the first-guess field at the time of the observation, as is done with GEOS-2, or interpolated to the time of the observation, as is done at ECMWF. More frequent data insertion, using time-continuous assimilation or VAR, is advantageous. In addition, the data assimilation system must take account of the errors of scatterometer surface winds, and accurately extend the vertical influence of the surface wind data.

Assimilation of satellite surface winds in NWP systems yields substantial modifications to surface wind analyses over the ocean and generally improved weather analyses and forecasts. The atmospheric general circulation model and data assimilation system used in our experiments are not as advanced as current operational systems, or GEOS-3. Therefore, the impact of scatterometer data on more advanced systems might be smaller than that described by our experiments, even if care is taken to accommodate the characteristics of satellite surface wind data. Alternatively, we note that the information in the small spatial scales of satellite surface wind data is not useful in our experiments, but might be useful in more advanced systems.

The impact of scatterometer wind data is proportional to the areal coverage. For example, the impact of NSCAT, with a 1200-km swath, is approximately twice that of ERS, with a 500-km swath. Directional information is of equal importance to speed information. For example, the impact of SSMI wind speed is approximately equal to the impact of NSCAT wind speed, and both are approximately equal to one half the impact of NSCAT vector wind. By implication, SeaWinds, with an 1800-km swath, has the potential to have even a greater impact, if the quality of its measurements is similar to NSCAT.

Acknowledgments. Data used in the research reported here were provided by the Jet Propulsion Laboratory (JPL) Physical Oceanography Distributed Active Archive Center (PO.DAAC), Remote Sensing Systems (RSS), the National Center for Environmental Prediction (NCEP), and the European Centre for Medium-Range Weather Forecasts (ECMWF). The authors would like to thank our many colleagues at GSFC, NCEP, JPL, AER, ECMWF, and KNMI who have contributed to research on the use of scatterometer data for NWP. We would particularly like to acknowledge D. Offiler for providing subroutines for ERS-1 processing, and P. Woiceshyn for modifying and implementing these subroutines at the Goddard Space Flight Center. In addition, we would like to acknowledge S. Bloom, E. Brin, J. Ardizzone, J. Terry, J. C. Jusem, D. Bungato, and W. Gemmil for their respective contributions to these experiments. This research was supported by the NSCAT Project and by NASA Headquarters Office of Earth Science.

References

Anderson, D., A. Hollingsworth, S. Uppala, and P. Woiceshyn, A study of the use of scatterometer data in the European Centre for Medium-range Weather Forecasts operational analysis-forecast model: Quality assurance and validation, *J. Geophys. Res.*, *96*, 2619–2634, 1991.

Anderson, D., A. Hollingsworth, S. Uppala, and P. Woiceshyn, A study of the use of scatterometer data in the European Centre for Medium-range Weather Forecasts operational analysis-forecast model: Data impact, *J. Geophys. Res.*, *96*, 2635–2647, 1991.

Andrews, P. L., and R. S. Bell, Optimizing the United Kingdom Meteorological Office data assimilation for ERS-1 scatterometer winds, *Mon. Wea. Rev.*, *126*, 736–746, 1998.

Atlas, R., Recent SeaSat scatterometer data impact studies at GLA, In Research Activities in Atmospheric and Oceanic Modeling, edited by G. Boer, CAS/JSC Report No. 18, TD 263, World Meteorological Organization, Geneva, 1.28–1.30, 1988.

Atlas, R., Atmospheric observations and experiments to assess their usefulness in data assimilation, *J. Meteorol. Soc. Japan*, *75*, 111–130, 1997.

Atlas, R., S. C. Bloom, R. N. Hoffman, J. V. Ardizzone, and G. Brin, Space-based surface wind vectors to aid understanding of air-sea interactions, *Eos, Trans. Amer. Geophys. Un.*, *72*, 201–208, 1991.

Atlas, R., S. C. Bloom, R. N. Hoffman, E. Brin, J. Ardizzone, J. Terry, D. Bungato, and J. C. Jusem, Geophysical validation of NSCAT winds using atmospheric data and analyses, *J. Geophys. Res.*, *104*, 11405–11424, 1999.

Atlas, R., R. N. Hoffman, and S. C. Bloom, Surface wind velocity over the oceans, In *Atlas of Satellite Observations Related to Global Change*, edited by R.J. Gurney, J. L. Foster, and C. L. Parkinson, Cambridge University Press, Cambridge, 129–139, 1993.

Atlas, R., R. N. Hoffman, S. C. Bloom, J. C. Jusem, and J. Ardizzone, A multiyear global surface wind velocity data set using SSM/I wind observations, *Bull. Amer. Meteorol. Soc.*, *77*, 869–882, 1996.

Atlas, R., E. Kalnay, and M. Halem, Impact of satellite temperature soundings and wind data on numerical weather prediction, *Opt. Eng., 24,* 341–346, 1985.

Atlas, R., and A. Pursch, Model sensitivity to low-level wind specification, In NASA Tech. Memo. 84983, Global Modeling and Simulation Branch Research Review, Goddard Space Flight Center, Greenbelt, Maryland, 22–24, 1983.

Atlas, R., J. Terry, and J. Woollen, Simulation studies of the effect of low level wind data on Southern Hemisphere analyses and numerical forecasts. In Research Review, Modeling and Simulation Branch, Laboratory for Atmospheres, Goddard Space Flight Center, Greenbelt, Maryland, 53–56, 1985.

Attema, E. P. W., The active microwave insturment on board the ERS-1 satellite, *Proc. IEEE*, *79*, 791–799, 1991.

Badran, F., S. Thiria, and M. Crepon, Wind ambiguity removal by the use of neural network techniques, *J. Geophys. Res., 96,* 20521–20530, 1991.

Baker, W. E., R. Atlas, M. Halem, and J. Susskind, A case study of forecast sensitivity to data and data analysis techniques, *Mon. Wea. Rev., 112,* 1544–1561, 1984.

Baker, W. E., R. Atlas, E. Kalnay, M. Halem, P. M. Woiceshyn, S. Peteherych, and D. Edelmann, Large-scale analysis and forecast experiments with wind data from the SeaSat-A scatterometer, *J. Geophys. Res., 89,* 4927–4936, 1984.

Bloom, S. C., and R. Atlas, Analysis of space-based ocean surface wind speed data at GLA. In Proc. ECMWF/EUMETSAT Workshop: The Use of Satellite Data in Operational Numerical Weather Prediction: 1989–1993, European Centre for Medium-Range Weather Forecasts, Reading, England, 205–220, 1990.

Bloom, S. C., and R. Atlas, Assimilation of satellite surface wind speed data and its impact. In Preprints, Ninth Conference on Numerical Weather Prediction, American Meteorological Society, Boston, 412–415, 1991.

Brown, R. A., On satellite scatterometer capabilities in air-sea interaction, *J. Geophys. Res.*, *91*, 2221–2232, 1986.

Cane, M. A., V. J. Cardone, M. Halem, and I. Halberstam, On the sensitivity of numerical weather prediction to remotely sensed marine surface wind data: A simulation study, *J. Geophys. Res.*, *86*, 8093–8106, 1981.

Duffy, D., R. Atlas, T. Rosmond, E. Barker, and R. Rosenberg, The impact of SeaSat scatterometer winds on the Navy's operational model, *J. Geophys. Res.*, *89*, 7238–7244, 1984.

Duffy, D. G., and R. Atlas, The impact of SeaSat-A scatterometer data on the numerical prediction of the *Queen Elizabeth II* storm, *J. Geophys. Res.*, *91,* 2241–2248, 1986.

Endlich, R. M., D. E. Wolf, C. T. Carlson, and J. W. Maresca, Oceanic wind and balanced pressure-height fields derived from satellite measurements, *Mon. Wea. Rev.*, *109*, 2009–2016, 1981.

Francis, R., G. Graf, P. G. Edwards, M. McCaig, C. McCarthy, P. Dubock, A. Lefebvre, B. Pieper, P.-Y. Pouvreau, R. Wall, F. Wechsler, J. Louet, and R. Zobl, The ERS-1 spacecraft and its payload, *ESA Bull.*, *65*, 27–48, 1991.

Freilich, M. H., and R. S. Dunbar, A preliminary C-band scatterometer model function for the ERS-1 AMI instrument, In Proc. First ERS-1 Sym., European Space Agency, Paris, 79–83, 1993.

Gloersen, P., and F. T. Barath, A scanning multichannel microwave radiometer for Nimbus-G and SeaSat-A, *IEEE J. Ocean Eng.*, *2*, 172–178, 1977.

Goerss, J. S., and P. A. Phoebus, The Navy's operational atmospheric analysis, *Wea. Forecast.*, *7*, 232–249, 1992.

Grantham, W. L., E. M. Bracalente, W. L. Jones, and J. W. Johnson, The SeaSat-A satellite scatterometer, *IEEE J. Ocean Eng.*, *2*, 200–206, 1977.

Guymer, T. H., J. A. Businger, W. L. Jones, and R. H. Stewart, Anomalous wind estimates from the SEASAT scatterometer, *Nature*, *294*, 735–737, 1981.

Hoffman, R., SASS wind ambiguity removal by direct minimization, *Mon. Wea. Rev.*, *110*, 434–445, 1982.

Hoffman, R. N., SASS wind ambiguity removal by direct minimization: Use of smoothness and dynamical constraints, *Mon. Wea. Rev.*, *112*, 1829–1852, 1984.

Hoffman, R. N., A preliminary study of the impact of the ERS-1 C-band scatterometer wind data on the ECMWF global data assimilation system, *J. Geophys. Res.*, *98*, 10233–10244, 1993.

Hollinger, J. P., J. L. Pierce, and G. A. Poe, SSMI instrument evaluation, *IEEE Trans. Geosci. Remote Sensing*, *28*, 781–790, 1990.

Ingleby, N. B., and R. A. Bromley, A diagnostic study of the impact of SEASAT scatterometer winds on numerical weather prediction, *Mon. Wea. Rev.*, *119*, 84–103, 1991.

Jones, W. L., L. C. Schroeder, and J. L. Mitchell, Aircraft measurements of the microwave scattering signature of the ocean, *IEEE J. Ocean Eng.*, *2*, 52–61, 1977.

Katsaros, K. B., and R. A. Brown, Legacy of the SeaSat mission for studies of the atmosphere and air-sea-ice interactions, *Bull. Amer. Meteorol. Soc.*, *72*, 967–981, 1991.

Lenzen, A. J., D. R. Johnson, and R. Atlas, Analysis of the impact of SeaSat scatterometer data and horizontal resolution on GLA model simulations of the *QE II* storm, *Mon. Wea. Rev.*, *121*, 499–521, 1993.

Moore, R. K., and W. J. Pierson, Worldwide oceanic wind and wave prediction using a satellite radar-radiometer, *J. Hydronaut.*, *5*, 52–60, 1971.

Naderi, F. M., M. H. Freilich, and D. G. Long, Spaceborne radar measurement of wind velocity over the ocean: An overview of the NSCAT scatterometer system, *Proc. IEEE*, *79*, 850–866, 1991.

O'Brien, J., R. Kirk, L. McGoldrick, J. Witte, R. Atlas, E. Bracalente, O. Brown, R. Haney, D. E. Harrison, D. Honhart, H. Hurlburt, R. Johnson, L. Jones, K. Katsaros, R. Lambertson, S. Peteherych, W. Pierson, J. Price, D. Ross, R. Stewart, and P. Woiceshyn, Scientific opportunities using satellite surface wind stress measurements over the ocean: Report of the Satellite Surface Stress Working Group, N.Y.I.T. Press, Nova University, Ft. Lauderdale, 153 pp, 1982.

Parrish, D. F., J. C. Derber, R. J. Purser, W.-S. Wu, and Z.-X. Pu, The NCEP global analysis system: Recent improvements and future plans, *J. Meteorol. Soc. Japan*, *75*, 359–365, 1997.

Pierson, W. J., The measurement of the synoptic scale wind over the ocean, *J. Geophys. Res.*, *88*, 1683–1708, 1983.

Price, J. C., The nature of multiple solutions for surface wind speed over the oceans from scatterometer measurements, *Remote Sensing Environ.*, *5*, 47–54, 1976.

Schroeder, L. C., W. L. Grantham, E. M. Bracalente, C. L. Britt, K. S. Shanmugam, F. J. Wentz, D. P. Wylie, and B. B. Hinton, Removal of ambiguous wind directions for a Ku-band wind scatterometer using three different azimuth angles, *IEEE Trans. Geosci. Remote Sensing*, *23*, 91–100, 1985.

Schubert, S. D., R. B. Rood, and J. Pfaendtner, An assimilated dataset for earth science applications, *Bull. Amer. Meteorol. Soc.*, *74*, 2331–2342, 1993.

Schultz, H., A circular median filter approach for resolving directional ambiguities in wind fields retrieved from spaceborne scatterometer data, *J. Geophys. Res.*, *95*, 5291–5304, 1990. (Errata on p. 9783)

Shaffer, S. J., R. S. Dunbar, S. V. Hsiao, and D. G. Long, A median-filter-based ambiguity removal algorithm for NSCAT, *IEEE Trans. Geosci. Remote Sensing*, *29*, 167–174, 1991.

Stewart, R. H., SeaSat: Results of the mission, *Bull. Amer. Meteorol. Soc.*, *69*, 1441–1447, 1988.

Stoffelen, A., and D. Anderson, Ambiguity removal and assimilation of scatterometer data, *Quart. J. Roy. Meteorol. Soc.*, *123*, 491–518, 1997.

Stoffelen, A., and D. Anderson, Scatterometer data interpretation: Estimation and validation of the transfer function CMOD4, *J. Geophys. Res.*, *102*, 5767–5780, 1997.

Stoffelen, A., and D. Anderson, Scatterometer data interpretation: Measurement space and inversion, *J. Atmos. Oceanic Tech.*, *14*, 1298–1313, 1997.

Stoffelen, A. C. M., and G. J. Cats, The impact of SeaSat-A scatterometer data on high-resolution analysis and forecasts: The development of the *QE II* storm, *Mon. Wea. Rev.*, *119*, 2794–2802, 1991.

Tweles, S., and H. B. Wobus, Verification of prognostic charts, *Bull. Amer. Meteorol. Soc.*, *35*, 455–463, 1954.

Wentz, F. J., and D. K. Smith, A model function for the ocean-normalized radar cross section at 14 GHz derived from NSCAT observations, *J. Geophys. Res.*, *104*, 11499–11514, 1999.

Wilheit, T. T., The effect of wind on the microwave emission from the ocean's surface at 37 GHz, *J. Geophys. Res.*, *84*, 4921–4926, 1979.

Wurtele, M. G., P. M. Woiceshyn, S. Peteherych, M. Borowski, and W. S. Appleby, Wind direction alias removal studies of SEASAT scatterometer-derived wind fields, *J. Geophys. Res.*, *87*, 3365–3377, 1982.

Yu, T. W., and R. D. McPherson, Global data assimilation experiments with scatterometer winds from SEASAT-A, *Mon. Wea. Rev.*, *112*, 368–376, 1984.

Robert Atlas, Mail Code 910.3, Goddard Space Flight Center, National Aeronautics and Space Administration, Greenbelt, MD 20771, U.S.A. (email, atlas@dao.gsfc.nasa.gov; fax, +1-301-614-6297)

Satellites, Oceanography and Society
edited by David Halpern

Chapter 5

Combining altimeter observations and oceanographic data for ocean circulation and climate studies

Silvia L. Garzoli and Gustavo J. Goni
Atlantic Oceanographic and Meteorological Laboratories, National Oceanic and Atmospheric Administration, Miami, Florida

Abstract. Calibrating Topography Experiment (TOPEX)/Poseidon (T/P) altimeter data to inverted echo sounder measurements is one of the methodologies developed to advance the study of the South Atlantic Ocean. The mass transport of the main boundary currents can be monitored using dynamic height time series obtained from altimeter-derived sea surface height anomalies. A 2-layer model yields upper-layer thickness and surface dynamic height of the ocean in areas of strong vertical stratification to identify and track anticyclonic rings, such as those shed from the Agulhas retroflection, which play an important role in the interocean exchange of heat and mass. Interannual variability of boundary currents, derived from analysis of the first five years of T/P data, is discussed. A combination of T/P data and in-situ observations provides significant contributions to the study of ocean dynamics for climate applications.

1. Introduction

Climate is a decisive factor in determining the distribution of populations and food production. Extreme climatic variability, such as the El Niño phenomenon and monsoon rains, can have tremendous social and economic impacts. The need to understand the wind-driven and thermohaline ocean circulation of the South Atlantic Ocean and its interaction with the adjacent oceans for improved climate prediction is, by now, well accepted.

Wind-driven circulation is defined by the effect of wind stress on the sea surface, while thermohaline circulation is driven by the buoyancy flux between the ocean and the atmosphere. According to Gordon (1986) and Schmitz (1995), thermohaline circulation may be of equal or greater importance to the global circulation system than wind-driven circulation because it relates the full volume of the global ocean with the atmosphere to form a global circulation network of mass and heat transports. The classic model of the conveyor belt (Broecker 1991) indicates that the North Atlantic trades cold, deep water for warm, upper water from the South Atlantic. This is a peculiarity of the Atlantic because it is the only ocean that transfers heat northward across the equator. However, it is not known exactly how much heat is transported into the North Atlantic, the source of that heat, and what the main routes of the transport are. Most heat exchange takes place by way of boundary currents, which are the major distributors of heat, and by the rings shed at current retroflections. Understanding the variability of boundary currents is important to improve forecast and climate models.

Another fundamental question is, How much heat and salt are transferred from the Indian Ocean to the Atlantic Ocean? One of the major mechanisms for this transfer is the shedding of rings at the Agulhas retroflection. This phenomenon is related to the intensity of the Agulhas Current, which is typically very stable, but can be disrupted dramatically by the passage of the Natal pulse (Lutjerharms and Roberts 1998). The altering effect of the Natal pulse on the path of the Agulhas Current has a direct effect on the interbasin exchange of water south of South Africa, with important environmental consequences. Lutjerharms and de Ruijiter (1996) hypothesized that a change in wind stress over the southern Indian Ocean leads to a higher frequency of the Natal pulse, which would slow down the conveyor belt by reducing the amount of Agulhas water entering the Atlantic Ocean. A diminished water flow would disrupt pelagic fish availability along the southeastern coast of South Africa and cause a significant reduction of rainfall along this shoreline.

Satellite observations, taken in conjunction with in-situ oceanographic measurements, can be used to study ocean variability and its impact on global climate. In particular, altimeter observations opened a new venue for large-scale and mesoscale studies of ocean dynamics (Zlotnicki 1993). Methodologies have been developed to use ocean observations as ground truth for the altimeter (e.g., Teague et al. 1995; Kelly and Watts 1994). In studies of the ocean dynamics of the South Atlantic, especially in the western and eastern South Atlantic subtropical gyre, in-situ observations were used in conjunction with altimeter data.

A schematic of the circulation in the South Atlantic is presented in Figure 1. The circulation in the western South Atlantic is characterized by the encounter of the warm, southward-flowing Brazil Current with the northward flow of the cold Malvinas Current, a branch of the Circumpolar Current. The encounter of these currents takes place at approximately 38°S, creating a strong thermohaline front called the Confluence Front (Gordon and Greengrove 1986). On the eastern boundary of the South Atlantic subtropical gyre is the Benguela Current, a broad, northward flow along the African coast (Stramma and Peterson 1990). In the vicinity of 30°S, the Benguela Current is confined between the African continental shelf and the Walvis Ridge, located from 2° to 4°E (Reid 1989). The Benguela Current is the main conduit of Indian Ocean water into the Atlantic (Garzoli and Gordon 1996). The global-scale thermohaline circulation brings warmer, saltier water into the South Atlantic from the Indian Ocean at the southern tip of Africa as compensation for the colder, southward-flowing North Atlantic deep water. The shedding

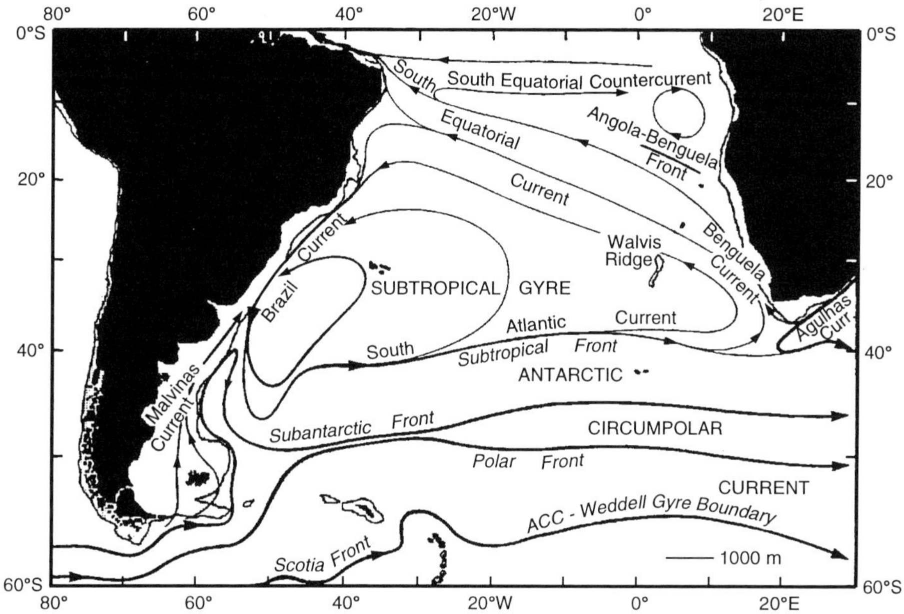

Figure 1. Schematic representation of the upper-layer circulation of the South Atlantic (Peterson and Stramma 1991).

of large rings at the retroflection of the Agulhas Current is the main mechanism of inter-basin exchange. The Benguela Current extension, or South Equatorial Current, which constitutes the equatorward segment of the subtropical gyre, bifurcates as it approaches the easternmost tip of South America, forming the Brazil Current to the south and the North Brazil Current to the north. The South Atlantic Current closes off the circulation of the subtropical gyre in the south.

An inverted echo sounder (IES) is a bottom-deployed instrument that records the time an acoustic signal takes to reach the surface and return (Watts and Rossby 1977). An IES is sometimes equipped with a pressure sensor. The travel time is proportional to the integrated temperature over the water column. In most oceanic regions it is also proportional to the variability of the dynamic height at a particular depth and to the depth of an isotherm representative of the main thermocline (e.g., Garzoli 1993; Garzoli and Bianchi 1987). IES data present an opportunity to determine possible relationships between the sea surface height (SSH) anomaly, measured by altimeter, and dynamic height and thermocline depth variations obtained with the IES.

The first attempt to correlate satellite altimeter and IES data was made with geodetic satellite (Geosat) data of the Gulf Stream, where the correlation between variations of SSH and dynamic height anomalies was 0.92 (Kelly 1991; Kelly and Watts 1994). In the

Kuroshio region, Teague et al. (1995) reported correlations of 0.70–0.96 between SSH anomalies computed from Topography Experiment (TOPEX)/Poseidon, named T/P, data and IES-pressure data. Correlation coefficients were higher closer to the Kuroshio Current, where SSH variations were larger.

To quantify the observed variability with absolute values of oceanic parameters, such as volume transport or depth of the upper layer, Goni et al. (1996) formulated a 2-layer model that related IES-derived upper-layer thickness and altimeter-derived SSH observations. The model was applied for the first time in the southwestern Atlantic to determine the transports of the Brazil and Malvinas Currents at their confluence. The model upper-layer depth was used to detect mesoscale oceanic rings and track their trajectories. Previous to this, altimeter data were used to follow the trajectories of rings across the ocean (Gordon and Haxby 1990; Byrne et al. 1995), but this procedure can be misleading in regions of very complex dynamics, such as the Agulhas retroflection region.

2. Ocean Transports

A series of studies was started in November 1988 to determine the possibility of monitoring the entire South Atlantic boundary current system with altimetry. The first attempt took place in the southwestern Atlantic, with the objective of studying the variability of the Confluence Front and the transports associated with the two merging currents. An array of 10 IES instruments was deployed near the Brazil-Malvinas Confluence during November 1988 and recovered during February 1990 (Garzoli 1993). IES-derived surface dynamic heights were obtained for 15 months and analyzed in terms of geostrophic velocities and transports. Goni et al. (1996) extrapolated the IES results to 1986 with Geosat SSH anomaly data.

A 2-layer model ocean (Figure 2) had thermocline deviations, which were compensated by variations of sea height. The resultant pressure gradients were associated with the baroclinic structure of the circulation. The upper and lower layers were assumed to be in motion, with the upper layer extending from the sea surface to the depth of the main thermocline, assumed to be the 10°C isotherm in the eastern South Atlantic. The lower layer is assumed to be tied to the barotropic component of the pressure gradient. In the 2-layer model, the upper-layer thickness, h_1, is (Goni et al. 1996)

$$h_1(x,y,t) = \bar{h}_1(x,y) + h'_1(x,y,t) = \bar{h}_1(x,y) + \frac{\eta'(x,y,t)}{\varepsilon(x,y)} + B'(x,y) \qquad (1)$$

where h_1 is the mean upper-layer thickness, h'_1 is the upper layer thickness anomaly, η' is the altimeter-derived sea height anomaly, B' is the barotropic contribution to the sea surface height anomaly, and ε is proportional to the reduced gravity, g', where $g' = \varepsilon\, g$, and g is the acceleration of gravity. The parameters ε and B' are computed from the linear

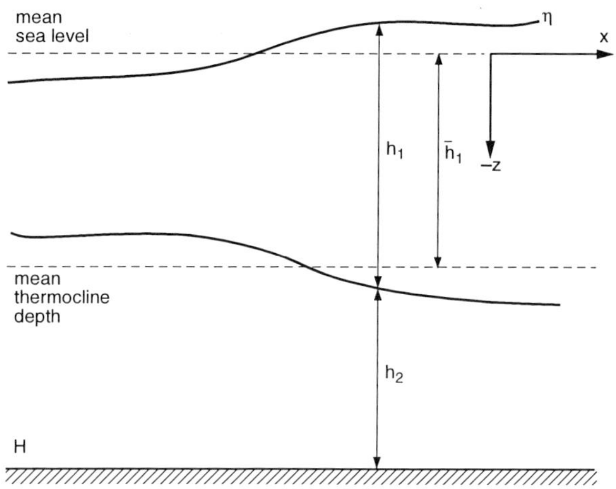

Figure 2. Schematic of an ocean model with finite depth, H, in a region of vertical density stratification, with a sharp vertical gradient between different water masses. The interface is given by the thermocline, which is approximately 8°C in the western South Atlantic and 10°C in the eastern South Atlantic. The water column is divided into upper and lower layers with thickness h_1 and h_2, respectively, with $\bar{h}_1$ being the mean upper-layer thickness; η is sea height anomaly (Goni et al. 1996).

regression between altimeter-derived SSH and IES-derived upper-layer thickness anomalies. When there are no simultaneous observations of these two parameters, ε can be computed from climatological data. The parameter ε is defined as

$$\varepsilon = \frac{g'}{g} = \frac{\rho_2 - \rho_1}{g * \rho_2}$$

where ρ_1 and ρ_2 are the mean densities of the upper and lower layers, respectively. The value of B' remains unknown and is set to zero. In this formulation, the geostrophic velocity in the upper layer, $\bar{v}_1$, is (Goni et al. 1996)

$$\bar{v}_1(x, y, t) = \frac{g'}{f} \nabla h_1(x, y, t) \tag{2}$$

where f is the Coriolis parameter. The baroclinic (S_{cl}) and barotropic (S_{tr}) transports are (Goni et al. 1996)

$$S_{cl}(x, y, t) = \frac{g'(x, y)}{2f} \Delta h_1^2(x, y, t) \tag{3}$$

$$S_{tr}(x, y, t) = \frac{Hg}{f} \Delta B(x, y, t) \tag{4}$$

where H is the mean ocean depth and B is the barotropic component of SSH.

Time series of baroclinic and barotropic transports were computed in the western South Atlantic Ocean using equations (3) and (4) with IES and satellite data (Goni et al. 1996). The barotropic sea surface height component, B, was obtained using mass conservation (Goni et al. 1996). The combination of altimeter and IES data provided the values of the parameters ε and B'. The agreement between the transports obtained with altimeter and IES data was very good. The parameters derived from the altimeter and IES data were then used to extend the results beyond the 15 months of IES observations to the 3 years of the Geosat mission, 1986–1989. Figure 3 indicates that the Brazil Current transport in the upper layer decreased by 5 Sv (1 Sv = 10^6 m^3 s^{-1}) from 35° to 37°S, and that there is a significant barotropic contribution to the upper-layer transport in the Confluence Front. A very good agreement was also found between results from this method and those derived from the conservation of potential vorticity.

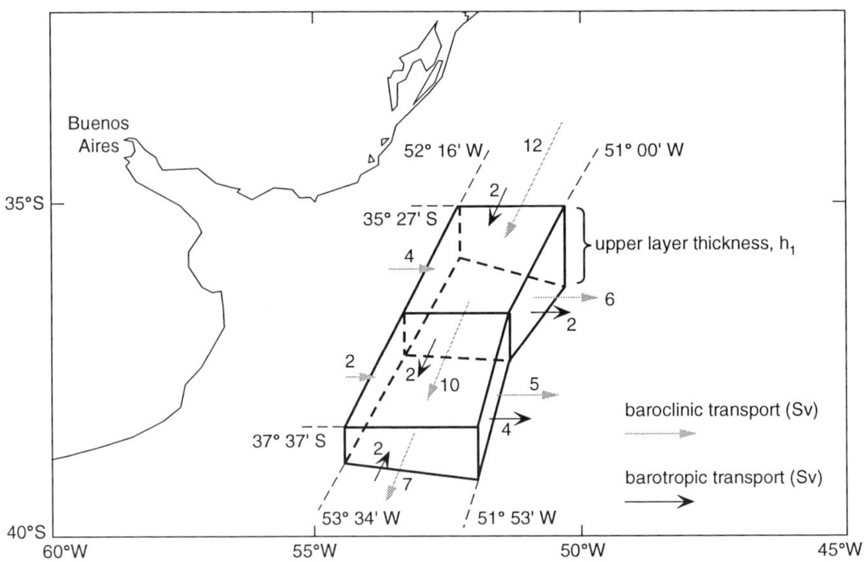

Figure 3. Schematic of mean upper-layer transports in the western South Atlantic. Grey arrows show mean baroclinic transports derived from 1986–1989 Geosat altimeter data using the 2-layer model approximation in the Brazil-Malvinas Confluence region. Mean transport of the southward-flowing Brazil Current decreases from 12 to 7 Sv in 200 km. Barotropic transports in the upper layer are estimated using conservation of mass and are shown by the black arrows (Goni et al. 1996).

A second investigation combining altimeter data with IES was made during the Benguela Sources and Transport Experiment (BEST). The study focused on the origin and transport variability of the Benguela Current (Garzoli and Gordon 1996). Fieldwork began in June 1992 and consisted of IES and current meter moorings and hydrographic surveys. To further explore the possibility of combining IES and altimeter data, the instruments were deployed as close as possible to T/P ground tracks. T/P, which became operational in October 1992, repeats its ground tracks every 9.91 days. The T/P SSH anomalies are computed every 9 km along the ground track with a 4-cm accuracy.

Prior to BEST, studies of the Benguela Current were based on single realizations of hydrographic sections (Sverdrup et al. 1942; Fu 1981; Stramma and Peterson 1990), and the sources of the Benguela Current were unknown. An array of IES instruments equipped with pressure sensors were deployed for 18 months (Figure 4) to measure the Benguela Current transport (Garzoli and Gordon 1996). One of the key results was that

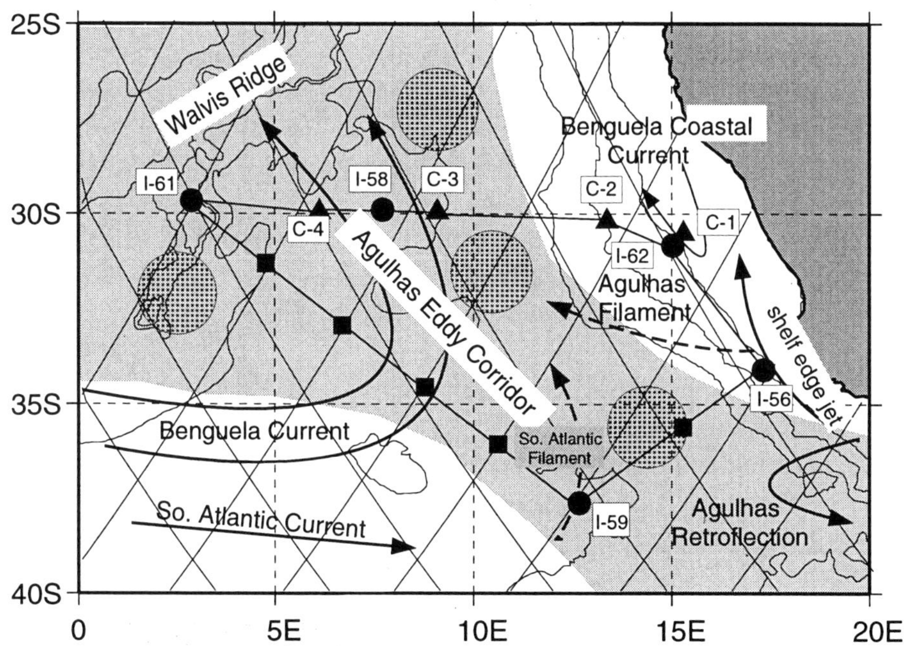

Figure 4. Schematic of the circulation of the southwestern Atlantic. Large circles with dotted pattern indicate the presence of warm rings. Shaded areas indicate the assumed ring corridor. Symbols indicate the location of the IES (●) and current meter moorings (▲) deployed during the BEST experiment. Squares are interpolation points from altimeter data (adapted from Garzoli et al. 1999).

mean transport of the Benguela Current was 13 Sv, of which 50% was derived from the central South Atlantic, 25% from the Indian Ocean, and 25% from a blend of Agulhas and tropical Atlantic waters.

To extend the BEST results in time and space, Garzoli et al. (1997) showed that correlation coefficients computed between T/P SSH and IES surface dynamic height data were 0.77–0.95. Transports were computed using two different methods. The first method mirrored the one used to calculate transports from IES data: a linear relation between travel time and surface dynamic height is obtained using conductivity-temperature-depth (CTD) data collected previously in the region and during the three BEST cruises (Garzoli 1993). From surface dynamic height series, geostrophic transports are calculated,

$$T_r\ (x, y, t) = \alpha V_g\ (x, y, t) \Delta x\ \Delta z \tag{5}$$

where $\alpha = 0.4$, an empirically determined constant (Garzoli and Gordon 1996), V_g is the geostrophic velocity derived from differences in surface dynamic height, Δx is the distance between stations, and Δz is the depth. The second method used the Goni et al. (1996) 2-layer model. Agreement was found between the two methods at 30°S (differences are ±0.82 Sv in the mean), but at 37°S the 2-layer model overestimated the transports by 30%. The mass transport calculated for the BEST period was extended in time to the first three years of T/P and in space to improve the spatial resolution of the moorings. A surface dynamic height field was created from T/P data, and 3-year time series were obtained at the sites indicated in Figure 4. One of the key findings of BEST was that while the main transport of the Benguela Current remained approximately constant, the percentages of the different water types in the current changed.

3. Results

3.1 Generation and propagation of rings

The large and energetic anticyclonic rings shed at the Agulhas retroflection have a well-marked signature in the thermocline field (Duncombe 1991). After being shed, the rings are characterized by a depression of isotherms at 500- to 700-m depths, with a homogeneous upper layer at about 16°C. Thermocline depressions can be detected with IES data scaled to the depth of the thermocline (Duncombe et al. 1996). During BEST, three anticyclonic rings were observed. The method used to locate the rings was to survey the region with expendable bathythermographs. Once a thermostad related to the presence of a ring was detected, a hydrographic survey was performed.

In earlier cruises, the search for rings was done blindly, because satellite images of sea surface temperature (SST) in this region are rare due to cloud coverage. Altimeter-derived SSH anomaly fields can be very misleading, because warm rings do not always exhibit their distinctive, positive SSH anomaly values as they translate through this region of very high SSH variability. On the other hand, the upper-layer thickness, derived from the mean climatological upper-layer thickness field and the SSH anomaly, provides an absolute field that can be used to survey and monitor the rings. A comparison of T/P SSH anomaly and upper-layer thickness is shown in Figure 5. The Brazil-Malvinas Confluence region in the west, and the Agulhas retroflection and Benguela Current in the east, exhibit higher SSH anomalies associated with frontal motions and with the generation and translation of warm rings. The location of the Brazil-Malvinas Confluence Front and the warm rings associated with the Brazil Current and Agulhas Current are more easily identified in the upper-layer thickness than in the SSH anomaly. The Brazil-Malvinas Front is a strong discontinuity in the upper-layer thickness. It is associated with an alternation of positive and negative SSH anomalies, which makes it impossible to use SSH to detect the front. Warm Brazil Current rings are located south of the Brazil-Malvinas Confluence and Agulhas rings translate along the Agulhas ring corridor. A study of the generation and propagation of Agulhas rings in October 1992–October 1995 (Goni et al. 1997) detected 17 rings with T/P data and revealed new boundaries of the ring corridor. In the mean, the rings had a radius of 116 km, a volume anomaly of 26×10^{12} m^3, a mean speed of 9 cm s^{-1}, and an available potential energy of 24×10^{15} joules. On average, each ring contributed 1 Sv to the transport of water from the Indian Ocean to the Atlantic Ocean.

In the Benguela Current Experiment to study the path of the Benguela Current, the ship was guided toward ring locations (Figure 6) using upper-layer thickness computed with real-time T/P data. This method considerably reduced the amount of ship time required to survey the rings. Lagrangian floats and drifters were deployed along 30°S and 7°W, and at the core of three Agulhas rings (Garzoli et al. 1999). The extensive surveys conducted inside each of the rings allowed a detailed comparison between the ring characteristics measured by direct observations and by altimetry. The altimeter could determine the depth of the 10°C isotherm, representative of the ring's core size, with an error of 25–85 m. Differences between the radii of maximum tangential velocities from direct measurements and altimeter data were 0, 20, and 5 km, respectively, for each of the three rings. The T/P-derived mean translation velocities for the three rings were 9, 8, and 8 cm s^{-1}, which are comparable to those measured from drifters deployed inside each ring (5.5, 6.5, and 6.3 cm s^{-1}). These parameters allow the calculation of the heat and volume anomalies transported by the rings. The volume anomaly is the volume of water contained above the 10°C isotherm and within the radius of maximum velocity. T/P data underestimated the volume transport of each ring (0.4, 0.4, and 0.3 Sv) compared to direct observations (1, 1, and 0.5 Sv); however, T/P data provides a good estimate of the 3-ring mean heat content (0.004 PW) compared to direct observations (0.003 PW). T/P data were used to show that the three rings originated in the retroflection area approximately 7, 14, and 17 months before the rings were surveyed.

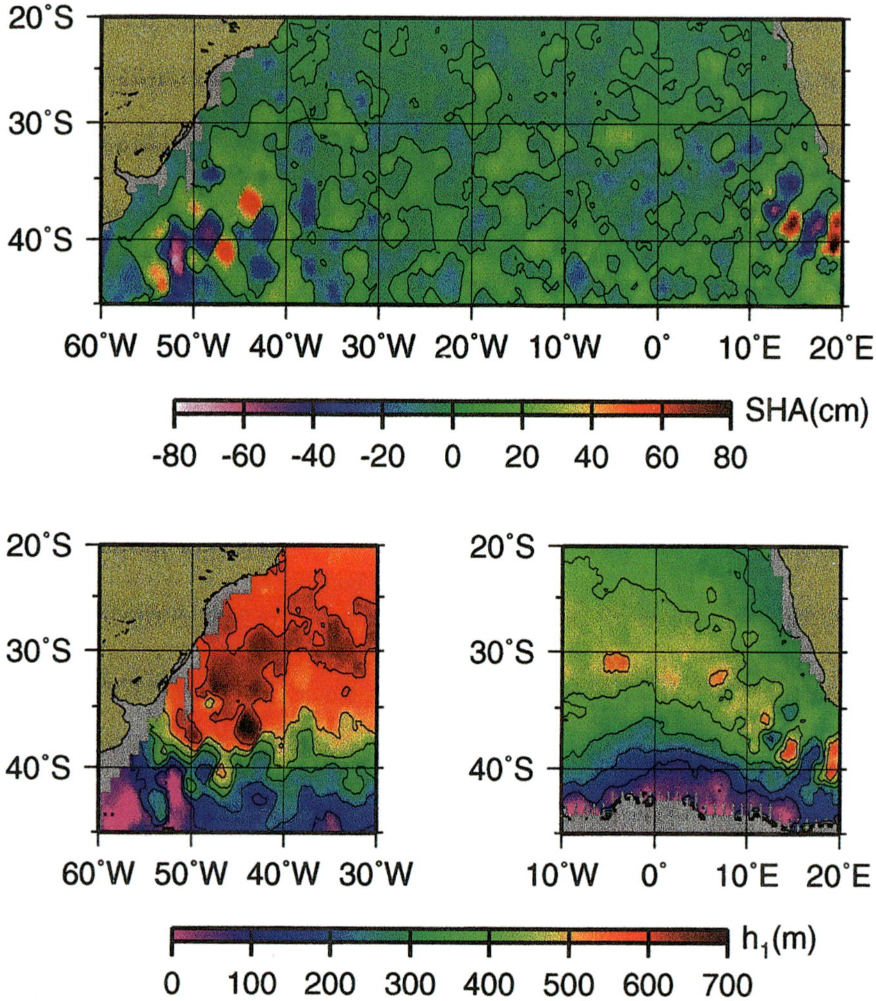

Figure 5. Upper panel shows the sea surface height anomaly field during 5–15 March 1993 in the South Atlantic. The Brazil-Malvinas Confluence region to the west and the Agulhas retroflection and Benguela Current to the east exhibit higher values of anomalies associated with frontal motions and with the generation and traslation of warm rings. Lower panels show T/P-derived upper-layer thickness (see text) for the same regions. In the western basin, it is represented by the 10°C isotherm and in the eastern basin by the 14°C isotherm. The Brazil-Malvinas Front is observed as a strong discontinuity in the upper-layer thickness, which is associated with an alternation of positive and negative sea surface height anomaly values.

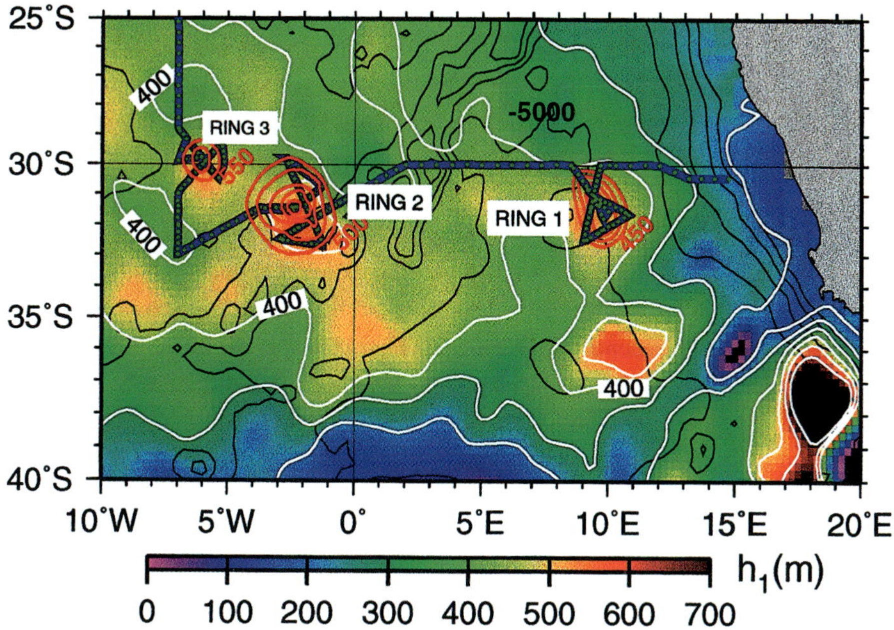

Figure 6. Locations of the three rings as depicted by the altimeter upper-layer thickness map corresponding to 16 September 1997. The cruise track is shown as a blue line. Hydrography-derived depths of the 10°C isotherm for each ring are shown as red contours (interval=50 m). White contours correspond to the T/P-derived depth of the upper layer (interval=100 m). Bathymetry is indicated by black contours. The altimeter data was used to guide the ship during the ring survey (Garzoli et al. 1999).

3.2 Benguela Current

Continuation of the initial three-year T/P mission provided data to study interannual variability of the Benguela Current and its source waters. T/P SSH anomalies are referred to the 1993–1997 mean, and T/P data to May 1998 are considered.

Dynamic height series at the surface, referenced to 1000 m, are used to compute transports (Garzoli et al. 1997). Examples of the transport time series obtained for the upper kilometer are shown in Figure 7 (for complete details, see Garzoli et al. 1997). For the period of overlapping IES and T/P data, a regression was obtained between IES dynamic height and T/P SSH anomaly data. The relationship was used to scale the T/P SSH anomalies to dynamic height anomalies. The dynamic heights are then used to compute geostrophic velocities and mass transports using equation (5). The agreement between the two transports (Figure 7) is very good. Differences in mean values range from −1.5 Sv to 0.9 Sv. The correlation coefficients between the two series during the overlapping period range from 0.70 to 0.82, with a mean of 0.80. The mean root-mean-square (rms) difference is 0.6 Sv. The total transport of the Benguela Current is measured across 30°S between the African continental shelf and the Walvis Ridge (Figure 4). Garzoli and Gordon (1996) determined different components of the Benguela Current. Water from the Indian Ocean, chiefly Agulhas water, enters the observational "box" mostly between the IES56 and IES59 (Figure 4). Between IES62 and IES56 (Figure 4), the water entering the box is a blend of Agulhas water in the form of filaments and tropical water coming from the north in the form of a coastal current. From the southwest, mostly South Atlantic water enters the box between IES59 and IES61. Table 1 and Figure 8 summarize the different components of the Benguela Current for 1993–1997. Differences in the annual transports during 1993–1995 (Table 1) and the Garzoli et al. (1997) values (12, 15, and 14 Sv for 1993, 1994, and 1995, respectively) are due to the fact that calendar years are used in this paper, while Garzoli et al. (1997) defined a year as October to September. Errors in the estimates are ±1 Sv. The total Benguela transport (Table 1) is

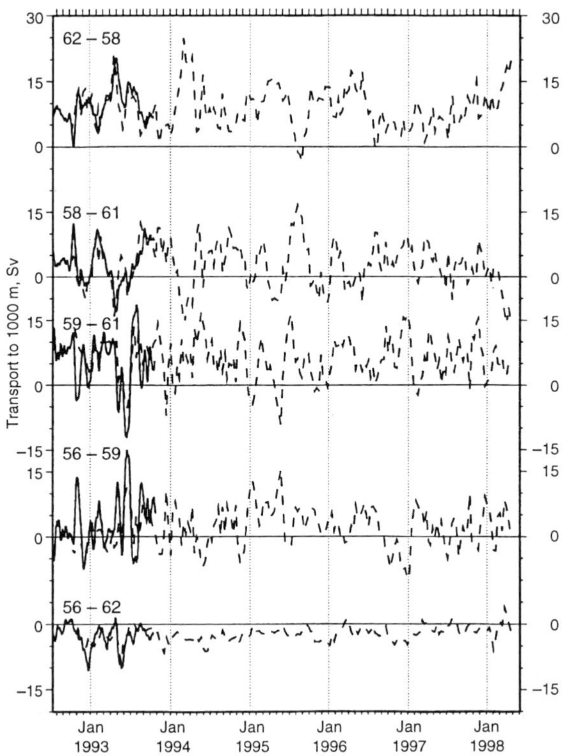

Figure 7. Time series of the upper-kilometer geostrophic transports between the IES locations shown in Figure 4. Dashed line from altimeter data; solid line from IES data.

Table 1. Annual mean geostrophic transports of the upper 1000 m, in Sv (1 Sv = 10^6 m^3 s^{-1}).

	1993	*1994*	*1995*	*1996*	*1997*	*5-year mean*
Total Benguela transport across 30°W	13	13	12	11	10	12
South Atlantic transport	7	7	3	8	7	6
Percent of total	56	57	29	68	72	58
Indian Ocean transport	2	1	6	1	2	2
Percent of total	18	12	51	10	17	20
Tropical Atlantic transport	3	4	2	3	1	3
Percent of total	26	31	20	22	11	22

defined as the transport across 30°S between IES62 and IES61. Differences between the total Benguela transport and the sum of the components are due to the ±1 Sv uncertainties in the calculation. The 1993–1997 mean Benguela transport (12 ± 1 Sv) is 2 Sv lower than the 1993–1995 estimate (14 ± 1 Sv). There was an apparent decrease in transport in 1996 and 1997. During 1993–1997, only once, in 1995, was the major contributor of water to the Benguela Current from the Indian Ocean (51%); for all other years it was from the South Atlantic. Larger contributions from the South Atlantic are observed during 1997 (72%) and 1996 (68%), when the lowest Benguela transport occurred across 30°S (10 and 11 Sv, respectively). In 1993–1997, the South Atlantic contributes 58% to the transport of the Benguela Current, the Indian Ocean contributes 20%, and the mix from the Indian Ocean and tropical Atlantic water adds the remaining 22%. The five-year mean contribution from the Indian Ocean is the lowest due to the rotational character of the flow entering the "box." When a ring crosses the IES56–IES59 line, positive velocities cancel negative velocities and the remainder is the translation velocity. In 1995, the mean flow incoming from the Indian Ocean was larger (Figure 8) because the rotational flow was centered at IES59.

3.3 Agulhas Current

The westward Agulhas transport south of Africa along 19°E can be calculated using equation (3) for 1993–1997. The 2-layer model is used in this region because no in-situ data are available to scale the dynamic height. Values of ε (or g') were obtained from climatology (Levitus et al. 1982). A comparative analysis of both methods to compute transports (Garzoli et al. 1997) concluded that south of 37°S the 2-layer model overesti-

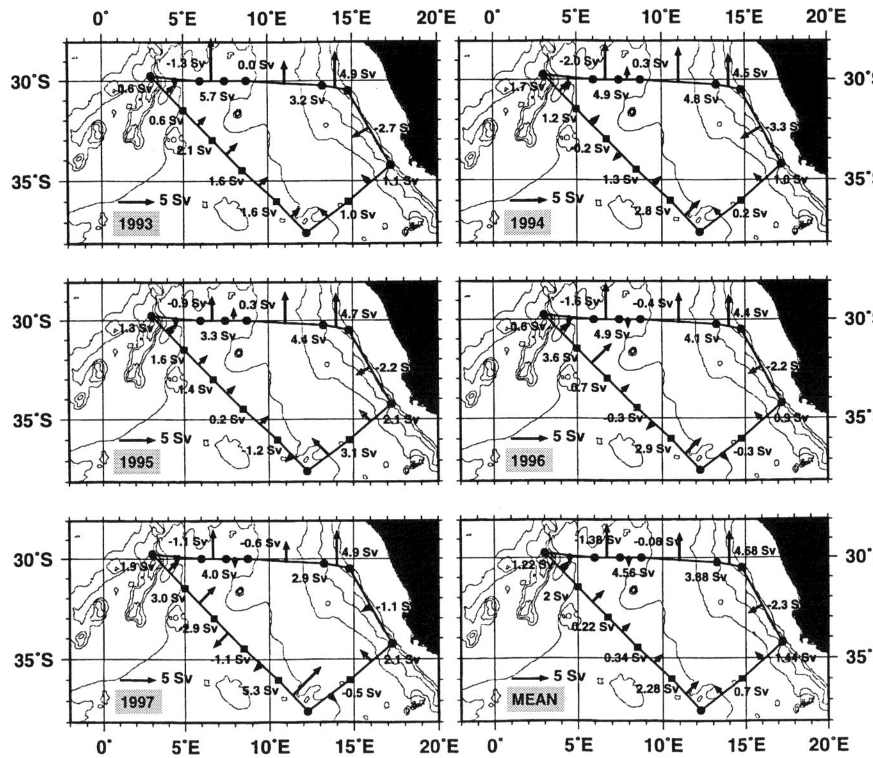

Figure 8. Schematic of the transports (in Sv) from the surface to 1000 m estimated from the altimeter-derived dynamic height series at the location of the BEST moorings (black circles) and at the interpolated locations (black squares). Errors are ±1 Sv.

mated the transports by 20%. Therefore, comparisons between transport estimates north of 37°S and along 19°E must be done only in terms of relative and not absolute values. The upper-layer westward Agulhas baroclinic transport into the Atlantic (Figure 9), calculated using the 2-layer model (Goni et al. 1997), was obtained by summing all T/P-derived westward velocities, starting from the South African coast, until the flow reverses toward the east. Calendar-year upper-layer mean transports are given in Table 2. The

Table 2. Annual intensity of Agulhas transport across 19°E.

	1993	*1994*	*1995*	*1996*	*1997*
Transport, Sv	13	16	17	17	23
Number of rings	4	5	7	5	4
Mean transport per ring, Sv	0.8	1.1	1.0	1.3	2.4

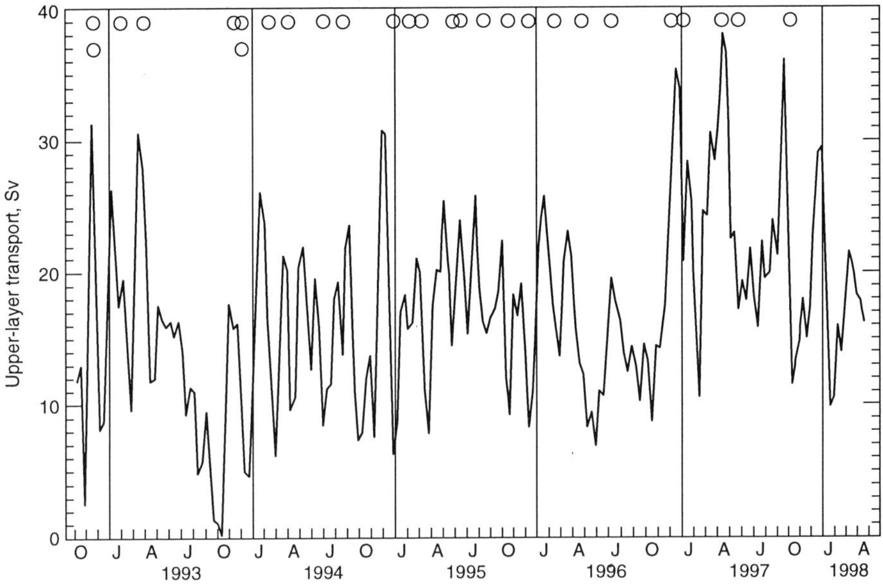

Figure 9. Time series of the upper-layer westward Agulhas transport into the South Atlantic along 19°E. Circles show the times when each of the rings analyzed was first detected.

depth of the upper layer is 450–500 m along 19°E between the African coast and 40°S. The Agulhas transport shows no apparent seasonal signal (Figure 9), as had been suggested by Matano et al. (1998). The transport time series (Figure 9) suggests intermittent shedding of rings by the Agulhas Current. The mean transport of the Agulhas Current (Table 2) was significantly higher during 1997 (23 Sv) than the mean (15.7 Sv) of the previous four years. The number of rings shed per year at the retroflection is given in Table 2. During the first three years of T/P (Goni et al. 1997), a peak in the transport was followed by the shedding of a ring. These results also indicated that in the annual mean, the larger the transport the larger the number of rings shed. However, in 1996 the mean transport was the same as 1995, but the number of rings shed in 1996 was smaller than in 1995. During 1997, a year of maximum transport, the number of rings shed was only four, although each ring transported twice the volume as in the previous years (Table 2). It should be noted that, in this paper, the volume of a ring is calculated the first time the ring is detected after shedding. This is why the rings are larger in volume than the ones calculated by Garzoli et al. (1999), which are surveyed seven to seventeen months after detachment. The rings shown in Figure 9 were tracked from the time of detachment until they reached the Walvis Ridge (Figure 10). Overall, the 1993–1997 ring trajectories fall within the 1993–1995 ring corridor (Figure 10).

4. Discussion and Conclusions

There was an apparent decrease of the Benguela Current annual mean transport, from 13 Sv in 1993 to 10 Sv in 1997. The 1993–1997 mean is 12 Sv, a value comparable to but lower than the previous estimate of 14 Sv. The Benguela Current volume transport is

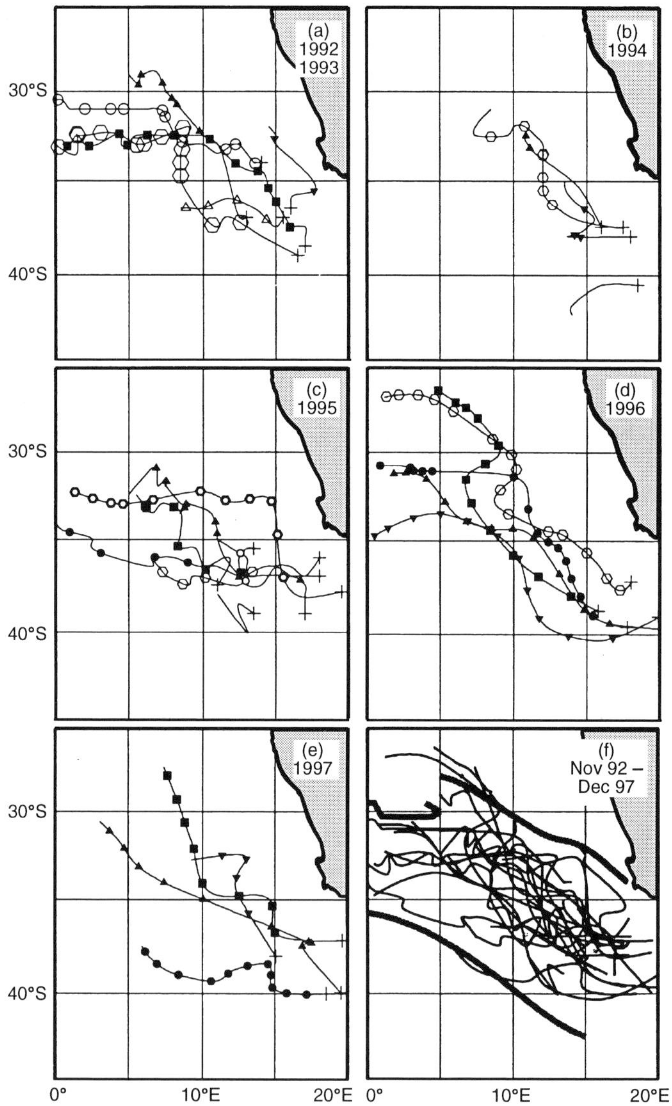

Figure 10. Trajectories of all rings observed with T/P altimeter data during each of the first five years of the T/P mission (panels a–e), and a composite of all trajectories (panel f). Lines indicate the envelope of the ring corridor provided by Goni et al. (1997).

composed of 58% South Atlantic water; 20% Indian Ocean water; and the remaining 22% is a mix of Indian Ocean and tropical Atlantic waters. During 1995, an anomalous year, South Atlantic water was not the main contributor to the Benguela Current transport and a larger than average number of rings was shed, seven total, with an average transport anomaly of 1.0 Sv per ring. During 1997, the Benguela Current transport was lower than average and the South Atlantic was its largest contributor (78%).

An unusually high Agulhas Current transport in 1997 was 6 Sv more than the 1996 value. Although the number of rings shed was low during 1997, each was larger than average and the volume transport was 1.4 Sv greater than the average.

Analysis of T/P data has yet to explain why, when the Agulhas transport is larger, the main contribution to the Benguela Current comes from the South Atlantic and fewer rings are shed at the retroflection. Matano et al. (1998) concluded that there was a seasonal variation of the penetration of the Agulhas Current into the Atlantic Ocean. During austral winter, the current follows the canonical case and extends to approximately 14°E before retroflecting back into the Indian Ocean. During austral summer, an increase in Agulhas transport forces a larger portion of the current to flow over the shallower depths of the Agulhas Plateau (25°–28°E). The topography-trapped flow generates a recirculation cell that displaces the main branch of the retroflection towards the east. Further modeling studies by Matano (personal communication) indicated that a strong Agulhas Current gets trapped by the topography and retroflects further east. Also, bottom-trapping affects the characteristics of the eddies shed at the retroflection, causing the shedding of fewer larger rings. Our observations provide evidence for Matano's results.

To understand the role of the oceans in climate, it is necessary to observe the global ocean. Although moored instruments provide accurate measurements of oceanic parameters, only a finite number of moorings can be deployed. Satellite altimeter data, used in combination with in-situ observations, represents a valuable tool to study global ocean circulation and makes it possible to monitor the transport of boundary currents, including interannual variability of the transports as well as the distributions of mass and heat across different basins. This information is crucial, among other applications, for the tuning and implementation of forecast models.

Acknowledgments. The authors are indebted to Dr. Ricardo Matano, who kindly allowed the use of his unpublished modeling results to interpret the observations. Altimeter data were made available to this work by NASA/JPL. Andreas Roubicek's contribution in the calculations of the ring parameters and ring tracking is kindly acknowledged. Roberta Lusic prepared the manuscript for publication. Support to attend the ICSOS meeting and to prepare this paper was provided by NOAA/AOML/PhOD.

References

Broecker, W. S., The great ocean conveyor, *J. Oceanogr.*, *4*, 79–89, 1991.

Byrne, D. A., A. L. Gordon, and W. F. Haxby, Agulhas eddies: A synoptic view using GEOSAT ERM data, *J. Phys. Oceanogr.*, *25*, 902–917, 1995.

Duncombe Rae, C. M., Agulhas retroflection rings in the South Atlantic Ocean: An overview, *S. Afr. J. Mar. Sci.*, *11*, 327–344, 1991.

Duncombe Rae, C. M., S. L. Garzoli, and A. L. Gordon, The eddy field of the southeast Atlantic Ocean: A statistical census from the Benguela Sources and Transports Project, *J. Geophys. Res.*, *101*, 11949–11964, 1996.

Fu, L.-L., The general circulation and meridional heat transport of the subtropical South Atlantic determined by inverse methods, *J. Phys. Oceanogr.*, *11*, 1171–1193, 1981.

Garzoli, S. L., Geostrophic velocity and transport variability in the Brazil-Malvinas Confluence, *Deep-Sea Res.*, *40*, No. 7, 1379–1403, 1993.

Garzoli, S. L., and A. Bianchi, Time-space variability of the local dynamics of the Malvinas/Brazil Confluence as revealed by inverted echo sounders, *J. Geophys. Res.*, *92*, 1914–1922, 1987.

Garzoli, S. L., and A. L. Gordon, Origins and variability of the Benguela Current, *J. Phys. Oceanogr.*, *101*, No. C1, 879–906, 1996.

Garzoli, S. L., G. J. Goni, A. Mariano, and D. Olson, Monitoring the upper southeastern Atlantic transport using altimeter data, *J. Mar. Res.*, *55*, 453–481, 1997.

Garzoli, S. L., C. M. Richardson, C. M. Duncombe Rae, D. M. Fratantoni, G. J. Goni, and A. J. Rubicek, Three Agulhas rings observed during the Benguela Current Experiment, *J. Geophys. Res.*, in press, 1999.

Goni, G. J., S. Kamholz, S. L. Garzoli, and D. Olson, Dynamics of the Brazil-Malvinas Confluence based on inverted echo sounders and altimetry, *J. Geophys. Res.*, *101*, 16273–16289, 1996.

Goni, G. J., S. L. Garzoli, D. Olson, and O. Brown, Agulhas ring dynamics from TOPEX/Poseidon satellite altimeter data, *J. Mar. Res.*, *55*, 861–883, 1997.

Gordon, A. L., Interocean exchange of thermocline water, *J. Geophys. Res.*, *91*, 5037–5046, 1986.

Gordon, A. L., and C. Greengrove, Geostrophic circulation of the Brazil-Falkland Confluence, *Deep-Sea Res.*, *33*, 573–585, 1986.

Gordon, A. L., and W. F. Haxby, Agulhas eddies invade the South Atlantic: Evidence from GEOSAT altimeter and shipboard conductivity-temperature-depth surveys, *J. Geophys. Res.*, *95*, 3117–3125, 1990.

Kelly, K. A., The meandering Gulf Stream as seen by the GEOSAT altimeter: Surface transport, position and velocity variance from 73° to 46°W, *J. Geophys. Res.*, *96*, 16721–16738, 1991.

Kelly, K. A., and D. R. Watts, Monitoring Gulf Stream transport by radar altimetry and inverted echo sounders, *J. Phys. Oceanogr.*, *24*, 1080–1084, 1994.

Levitus, S., Climatological Atlas of the World Ocean, NOAA Tech. Paper 13, 173 pp, Washington, D.C., 1982.

Lutjeharms, J. R. E., and H. R. Roberts, Eddies at the subtropical convergence south of Africa, *J. Phys. Oceanogr.*, *18*, 761–774, 1998.

Lutjeharms, J. R. E., and W. P. de Ruijiter, The influence of the Agulhas Current on the adjacent coastal ocean: Possible impacts of climate change, *J. Mar. Sys.*, *7*, 321–336, 1996.

Matano, R. P., C. G. Simionato, W. P. de Ruijiter, P. J. van Leeuween, P. T. Strub, D. B. Chelton, and M. G. Schlax, Seasonal variability in the Agulhas retroflection region, *Geophys. Res. Lett.*, *25*, 4361–4364, 1998.

Reid, J., On the total geostrophic circulation of the South Atlantic Ocean: Flow patterns, tracers and transport, *Prog. Oceanogr.*, *23*, 149–244, 1989.

Schmitz, W. J., On the interbasin-scale thermohaline circulation, *Rev. Geophys.*, *33*, 151–173, 1995.

Sverdrup, H. U., M. W. Johnson, and R. H. Fleming, *The Oceans*, Prentice Hall, 1087 pp, 1942.

Stramma, L., and R. G. Peterson, The South Atlantic Current, *J. Phys. Oceanogr.*, *20*, 846–859, 1990.

Watts, D. R., and H. T. Rossby, Measuring dynamic heights with inverted echo sounders: Results from MODE, *J. Phys. Oceanogr.*, *7*, 345–358, 1977.

Teague, W. J., Z. R. Hallock, G. A. Jacobs, and J. L. Mitchell, Kuroshiro sea surface height fluctuations observed simultaneously with the inverted echo sounder, *J. Geophys. Res.*, *100*, 24987–24994, 1995.

Zlotnicki, V., Quantifying time-varying oceanographic signals with altimetry, In *Satellite Altimetry in Geodesy and Oceanography*, edited by R. Rummel and F. Sanso, Springer-Verlag, Berlin, 144–188, 1993.

Silvia L. Garzoli, Atlantic Oceanographic and Meteorological Laboratories, National Oceanic and Atmospheric Administration, 4301 Rickenbacker Causeway, Miami, FL 33149, U.S.A. (email, garzoli@ aoml.noaa.gov; fax +1-305-361-4392)

Satellites, Oceanography and Society
edited by David Halpern

Chapter 6

Remote sensing of oceanic extra-tropical Rossby waves

Paolo Cipollini, David Cromwell, Graham D. Quartly, and
Peter G. Challenor

James Rennell Division for Ocean Circulation and Climate, Southampton Oceanography Centre, Southampton, United Kingdom

Abstract. Long baroclinic Rossby waves play a fundamental role in ocean dynamics and can dramatically affect weather patterns and climate. In recent years, satellite remote sensing has provided a global picture of these propagating signals, which are characterized by very large wavelengths (hundreds or thousands of km) and small surface signatures (a few cm). After presenting the basics of Rossby wave theory, we review the results obtained so far from satellite observations and briefly summarize recent suggested revisions to the standard linear theory of Rossby wave propagation. We then describe the processing and the corrections applied to global datasets of altimeter and sea surface temperature data to make them suitable for Rossby wave analysis. We discuss the importance of Rossby wave studies for the validation of ocean circulation models, and present a brief comparison between satellite-observed and model-derived waves. Finally, we consider possible future developments of Rossby wave research, which will improve our knowledge about how the oceans respond to atmospheric forcing and climatic events.

1. Introduction

Rossby waves play a fundamental role in atmospheric and ocean dynamics. This special type of wave, also known as a planetary wave, owes its existence to forces associated with the shape and rotation of the Earth. The name derives from the pioneering work of Rossby (1939), who built on the theory of planetary waves first formulated by Hough (1897). Oceanic Rossby waves are important for a number of reasons: they maintain the strong western boundary currents, such as the Kuroshio Current and the Gulf Stream; they are the main oceanic response to changes in atmospheric forcing; and they are the only means (other than boundary waves) by which information is transferred westward from the eastern ocean boundary (Gill 1982). Rossby waves can even be responsible for disturbing the position of the western boundary currents. By using an ocean general circulation model and data from the geodetic satellite (Geosat) altimeter, Jacobs et al. (1994) present evidence for the existence of an extra-tropical Rossby wave in the North Pacific, generated by the El Niño event of 1982–83. This wave has also been observed by Jacobson and Spiesberger (1998) in expendable bathythermograph (XBT) data. Jacobs et al. (1994) suggest that after a decade's delay this wave induced a shifting of the Kuroshio Current in the northwest Pacific, which may have affected the climate of North America, and may even have been responsible for the dramatic meteorological events in North America in 1993, such as the flooding of the Mississippi River (McPhaden 1994). In brief, as well as being one of the ways the ocean itself *responds* to climate events, Rossby waves may also be a *generator* of climate change and weather variability.

This paper restricts its attention to Rossby waves outside the tropical band of 10°S to 10°N, referred to as *extra-tropical* Rossby waves. (Tropical Rossby waves are covered elsewhere, e.g., Philander (1990); Delcroix et al. (1991); Lawrence et al. (1998).) At the sea surface, the height amplitude is 1–10 cm, wavelength is 100–1000 km, and speed is 1–10 cm s^{-1} (apart from the barotropic mode, see below). The vertical disturbance in the density field can be described as the combination of different *modes*, which can be depth-independent (barotropic) or depth-dependent (baroclinic), and which propagate at different speeds: the lower the mode number, the greater the speed. The schematic of a Rossby wave in the simplest baroclinic case (first baroclinic mode) is shown in Figure 1. The large-scale undulations at the surface are inversely mirrored in the thermocline, where the amplitude is three orders of magnitude greater (tens of meters).

The focus of this paper is on baroclinic Rossby waves, as these are observable with satellite altimetry. Baroclinic Rossby waves have more important effects on ocean circulation and climate than barotropic Rossby waves, which do not affect the thermocline. After presenting the basics of Rossby wave theory (Section 2), Section 3 reviews the results obtained from satellite observations and then briefly summarizes recently suggested revisions to the standard linear theory of Rossby wave propagation. Section 4 describes the processing and corrections applied to global datasets of altimeter and infrared data to make them suitable for our Rossby wave analysis, whose results are presented in Section 5. Section 6 discusses the importance of Rossby wave studies for the validation of ocean circulation models, and presents a brief comparison between satellite-observed and model-derived Rossby waves. Future developments of Rossby wave research are discussed in Section 7. Many examples in this paper are taken from the Atlantic Ocean at 34°N, where strong Rossby waves are associated with the Azores Current. However, Rossby waves occur worldwide.

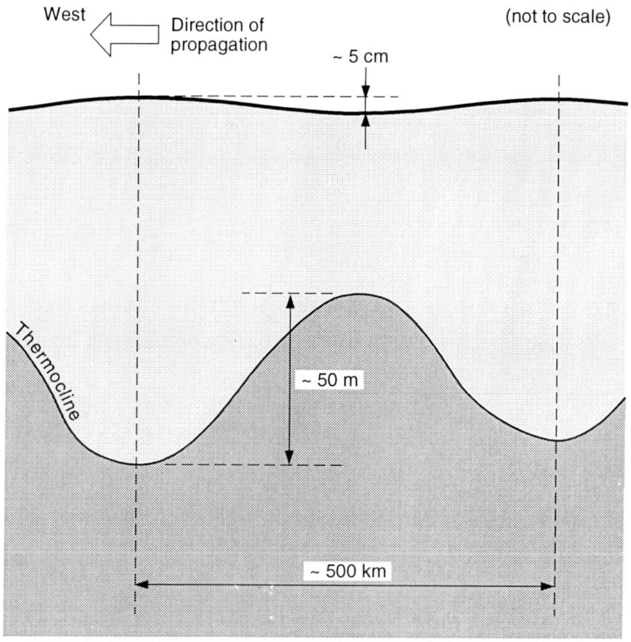

Figure 1. Schematic of a first-mode baroclinic Rossby wave.

2. What Are Rossby Waves?

The Earth rotates on its axis with a constant angular velocity, Ω. At any site on the Earth's surface, the vertical component of angular velocity, $f/2$, is:

$$f/2 = \Omega \sin(\phi) \qquad [1]$$

where f is the Coriolis parameter and ϕ is latitude. Fluid parcels on the rotating Earth possess angular momentum with respect to a fixed frame of reference and, therefore, fluid parcels possess *potential vorticity,* which is conserved. The component of potential vorticity due to the Earth's rotation is f. In the case of a shallow, homogeneous layer of thickness, H, conservation of potential vorticity can be expressed as (e.g., Gill 1982, Equation [11.2.14]):

$$\frac{\partial\left(\zeta - \frac{f\eta}{H}\right)}{\partial t} + \beta v = 0 \qquad [2]$$

where ζ is the relative vorticity of the fluid parcel (i.e., the vertical component of the vorticity relative to the rotating frame), η is the free height of the shallow layer, β is the latitudinal variation in f (i.e., $\beta = df/d\phi = 2\Omega \cos(\phi)$), and v is the meridional (north-south) component of fluid velocity.

According to [2], a parcel of water displaced latitudinally is subject to a restoring force in order for potential vorticity to be conserved. The resultant effect of (1) the restoring force, (2) inertia of the fluid parcel, and (3) the initial disturbance, which may be caused by atmospheric forcing or a change in ocean currents, is a propagating Rossby wave.

Solutions of [2], representing the different *modes* of propagating Rossby waves, take the form:

$$\eta \sim \exp[i(kx + ly - \omega t)] \qquad [3]$$

where k and l are zonal (east-west) and meridional wavenumbers, respectively, and ω is wave frequency. The dispersion relation for zero background mean flow is:

$$\omega = -\frac{\beta k}{(k^2 + l^2 + a^{-2})} \qquad [4]$$

where a is the local deformation radius, which varies with latitude, local density stratification, and mode number. Each baroclinic mode has a deformation radius, i.e., the nth mode baroclinic Rossby radius is:

$$a_n = \frac{c_n}{|f|}, n = 1,2,..., \qquad [5]$$

and c_n is the wave speed of the nth mode baroclinic Rossby wave. The baroclinic Rossby radius is a length scale often associated with boundary currents, fronts, and eddies. Chelton et al. (1998) discuss the geographical variability of the first baroclinic deformation radius.

The minus sign on the right-hand side of [4] indicates that Rossby waves have *westward* phase velocity. In the long wavelength limit, $k^2 + l^2 << a^{-2}$, Rossby waves are non-dispersive. It is worth noting from [4] that Rossby waves have a maximum or cut-off frequency, $\omega_c = -(\beta a)/2$, which occurs when $l = 0$ and $k = a^{-1}$. Alternatively, Rossby waves have a *turning latitude*, beyond which Rossby wave solutions become evanescent, i.e., no propagation exists (Gill 1982, pp. 440–443).

Several mechanisms have been proposed for the generation of propagating baroclinic Rossby waves. These typically involve a disturbance of the ocean, such as by an anomaly in wind stress curl, buoyancy forcing, coastally trapped waves, and reversals in coastal currents. In the eastern North Atlantic Ocean, the most likely forcing mechanism is wind stress fluctuation at the eastern boundary (Polito and Cornillon 1997). The DYNAMO Group (1997) detected Rossby wave propagation in an eddy-permitting model of the North Atlantic that had no wind forcing, suggesting that buoyancy forcing generated the waves.

3. Observations and New Theories

Since the 1970s, oceanographers have analyzed in-situ data in an attempt to verify the existence of oceanic planetary waves, e.g., Bernstein and White (1974, 1977), Emery and

Magaard (1976), Kang and Magaard (1980), Price and Magaard (1980, 1983, 1986), and Kessler (1990). White et al. (1998) present a comprehensive review of the observations. Such studies were difficult owing to the nonsynoptic nature and irregular spatial sampling of the data, and their results are limited to a few areas. However, the advent of high-resolution satellite altimetry heralded a breakthrough in Rossby wave studies.

The satellite radar altimeter, with its repeat sampling and global quasi-synoptic capabilities, is an excellent instrument for investigating a wide range of large-scale oceanographic phenomena having sea surface height signatures. As common to all instrument systems, there are weaknesses. No direct information is gained about the ocean subsurface. Also, altimeter data are one-dimensional along-track. Moreover, given the long period and small surface amplitude of baroclinic Rossby waves, time series of several years, with an accuracy and precision better than 10 cm, are required.

Satellite altimeter observations of Rossby waves began with Geosat, which had a 17-day repeat cycle during November 1986–January 1990. The launch of the first European Remote-sensing Satellite (ERS-1) in July 1991 continued altimeter observations of sea surface topography. However, ERS-1 data suitable for investigations of Rossby waves were confined to April 1992–December 1993 and March 1995–June 1996, when the orbit repeat interval was 35 days. The August 1992 launch of the Topography Experiment (TOPEX)/Poseidon, named T/P, satellite greatly expanded opportunities for Rossby wave studies because of its 10-day repeat cycle and the unprecedented approximate 3-cm root-mean-square accuracy. A second ERS (ERS-2) was launched in April 1995 to yield near-simultaneous data with ERS-1 until June 1996, when data acquisition from ERS-1 was terminated. The Geosat follow-on (GFO) satellite launched in February 1998 on a 17-day repeat orbit has been encumbered with problems. At the time of preparation of the paper, both T/P and ERS-2 were operational, and GFO data were in the final phase of calibration and validation before their distribution. Future altimeter measurements are expected from the Jason-1 satellite and Environmental Satellite (ENVISAT), planned for launch in June 2001.

Barotropic Rossby waves were detected by Gaspar and Wunsch (1989) using Geosat data. The first detection of baroclinic Rossby waves with altimetry was also achieved with Geosat data (e.g., Jacobs et al. 1993; Le Traon and Minster 1993; Tokmakian and Challenor 1993). However, tidal aliasing caused by the Geosat orbit was a problem. Any error in the M_2 principal lunar tidal component became a fictitious signal with propagation properties similar to a first baroclinic-mode Rossby wave. Isolating Rossby waves in Geosat data required removal of the aliased tidal component (e.g., tidal models).

The ERS-1/2 and T/P data are largely free of the tidal aliasing problem (Schlax and Chelton 1994). Using T/P data, Chelton and Schlax (1996) (CS) demonstrated Rossby wave propagation in all major ocean basins and showed that the observed Rossby wave speed was significantly higher than that predicted by linear theory applied to climatological datasets. Polewards of 10° latitude, the ratio of CS-derived Rossby wave speeds to theoretical speeds ranged from 0.7 to 4.4 with most values between 1.1 and 2.5. This means that the ocean can respond twice as quickly to climate events as was previously supposed!

Before the study of CS, a few reports had suggested the necessity to include baroclinic mean zonal flows, which can be thought of as modifying β, to calculate linear Rossby wave speeds, e.g., Kang and Magaard (1980) and Price and Magaard (1983). In a modeling study of the eastern North Atlantic Ocean, Herrmann and Krauss (1989) showed that an *effective* β, which included effects of zonal mean currents, lateral shear, and bottom topography, was equal to 0.2–1.8 of planetary β, with topography having the dominant effect.

The global analysis of CS renewed studies of Rossby wave theory. Killworth et al. (1997) found that the interaction of Rossby waves with the baroclinic mean zonal flow yielded a faster phase speed, which matched the observations much better than the standard theory. Dewar (1998) analyzed the effect of flow on wave propagation to confirm the results of Killworth et al. (1997).

Alternative explanations to the results of CS exist. Qiu et al. (1997) suggested that wind forcing and eddy dissipation of Rossby waves caused the higher Rossby wave speed. White et al. (1998) argued that the observations could be explained by resonance arising from atmosphere-ocean coupling. Killworth and Blundell (1999) indicated that slopes in ocean bottom topography could increase or decrease the speed of Rossby waves with respect to the standard theory over moderate distances.

Satellite altimeter observations have also shown, consistent with the standard theory, that Rossby waves travel faster in western oceanic basins where the thermocline is deeper, compared to eastern basins in the same ocean at the same latitude (e.g., Tokmakian and Challenor 1993; Cipollini et al. 1996; CS).

Rossby waves have also been observed with sea surface temperature (SST) data. Assuming that SST can be regarded as a proxy for surface density, and that surface density is correlated with subsurface density, then perturbations in the sea surface height (SSH) and, likewise, in the density field, should also occur as perturbations in SST. Halliwell et al. (1991) detected a significant correlation between the signatures of westward propagating fluctuations in both Geosat SSH and National Oceanic and Atmospheric Administration (NOAA) Advanced Very-High Resolution Radiometer (AVHRR) SST data in the Sargasso Sea. Another area of active Rossby wave propagation is near the Azores Front in the northeast Atlantic: energetic, propagating baroclinic Rossby waves were observed with T/P and ERS-1 altimeter data and ERS-1 Along-Track Scanning Radiometer (ATSR) SST data (Cipollini et al. 1997a, 1997b). Hughes et al. (1998) used ERS-1 SST and sea surface slope data to reveal eastward and westward propagating baroclinic Rossby waves in the Antarctic Circumpolar Current (ACC). The eastward propagation was probably caused by the significant eastward barotropic mean flow of the ACC, whose speed exceeds the westward speed of propagating Rossby waves. The observations confirmed theoretical predictions that the ACC is 'supercritical' with respect to the propagation of baroclinic Rossby waves. Just as CS demonstrated the global nature of Rossby waves in SSH data, recent work by Hill et al. (2000), using almost five years of ATSR data, revealed their existence in the SST anomaly field. In some mid-latitude locations the SST-derived speeds were in agreement with the observations of CS, but elsewhere, especially in lower latitudes, propagation speeds were slower, which might correspond to higher-order baroclinic modes. This is consistent with the results of Cipollini et al. (1997a, 1997b) at 34°N in the North Atlantic, where, with concurrent use of SSH and SST, the first three baroclinic modes were detected; the slowest mode had the strongest SST amplitude.

4. Processing Satellite Data to Observe Rossby Waves

This section examines data from two different types of instruments—altimeters, which measure SSH over a 6- to 8-km footprint, and the ERS-1 ATSR infrared radiometer, which measures SST—and provides a brief description of the instruments, corrections, and interpolation onto a regular grid. Further details are in Cipollini et al. (1997a, 1997b).

4.1 Sea surface height

The T/P satellite carries two altimeters, one named "TOPEX" and the other "Poseidon" (Fu et al. 1994). Careful intercalibration has produced a single, combined T/P altimetric dataset. At each latitude the satellite ground track is 2.8° apart in longitude (see Figure 2a), which is repeated every 9.92 days (typically called a '10-day repeat cycle'). A set of corrections is applied for orbit errors, atmospheric delays, tides, and sea state effects (Cipollini et al. 1997a). The rms accuracy of the SSH retrieval is about 2–3 cm (Cheney et al. 1994). A three-year mean height profile is calculated for 1993–1995, and SSH anomalies are computed relative to that profile. Each cycle of data is then interpolated onto a 1° by 1° grid. The interpolation, which uses a weighted mean of all the data within 200 km of a grid point, reduces instrument and correction errors while leaving the larger scale variability relatively unaffected. CS suggests that SSH fluctuations with an amplitude of ~1 cm might be observable.

The ERS platform carries a number of different sensors, including an altimeter and an infrared radiometer (Francis et al. 1991). Due to orbit changes during the ERS-1 mission, this paper only examines ERS-1 data for April 1992–December 1993, when the satellite was in a 35-day repeat orbit. ERS-2 data are used from the beginning of its mission in April 1995. Although the temporal sampling is poorer than that of T/P, the spatial coverage (Figure 2b) is correspondingly better. ERS altimeter data were treated in a similar way to those from T/P, except that referencing was done to the relevant mean ERS SSH

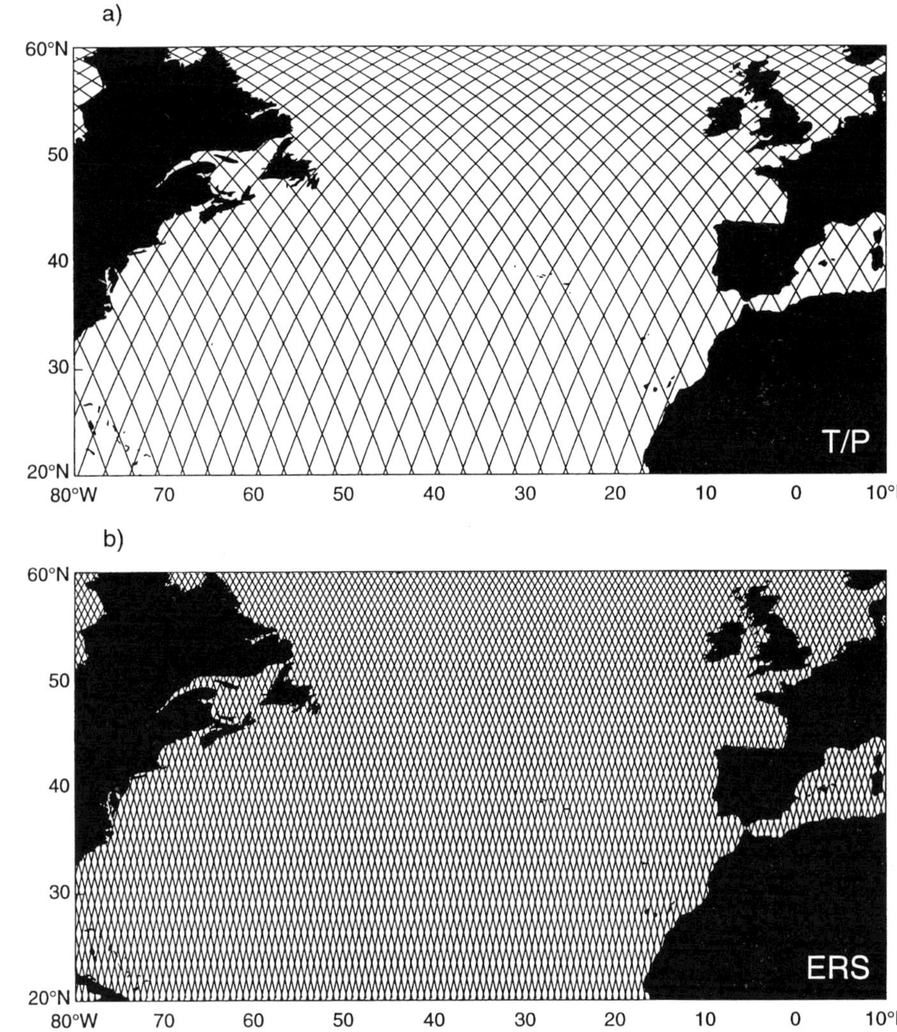

Figure 2. Altimeter ground tracks over the North Atlantic: (a) T/P sampling grid; (b) ERS-1 (35-day repeat) and ERS-2 sampling grid.

profile. The DGM-E04 gravity model was used to correct orbit errors (Scharroo and Vissier 1998), which makes the accuracy of ERS-derived SSH comparable to that of T/P and perfectly suitable for the observation of Rossby waves.

4.2 Sea surface temperature

The ATSR (Mutlow et al. 1994) was an experimental multi-channel radiometer on ERS-1, and is also on-board ERS-2. The accuracy of 1-km × 1-km ERS-1 ATSR data is

0.3 K (Forrester and Challenor 1995). We used the 0.5° × 0.5° average SST product distributed by Rutherford Appleton Laboratory.

5. Results

Whether dealing with SSH or SST, the domain of gridded data is a three-dimensional cuboid in longitude, latitude, and time (Figure 3), and the general approach to analyze it would be to use three-dimensional signal processing techniques (see below). Keeping in mind that long Rossby waves propagate mainly zonally, we can as a first approximation examine the signature of waves on a longitude-time section at a given latitude. This approach reduces the analysis to two dimensions. An alternative way of visualizing the longitude-time plot (also known as a Hovmöller diagram) is to take a zonal section from every repeat cycle and pile all the sections up, as represented in Figure 4. Evidence of baroclinic Rossby wave propagation appears as diagonal alignments of crests and troughs moving westward with time (Matthews et al. 1992; Périgaud and Delecluse 1992; Le Traon and Minster 1993; Tokmakian and Challenor 1993). CS found baroclinic Rossby waves to be almost ubiquitous in the main ocean basins. Figure 5 shows the longitude-time plot of gridded T/P SSH anomalies along 25°S across all ocean basins. The diagonal

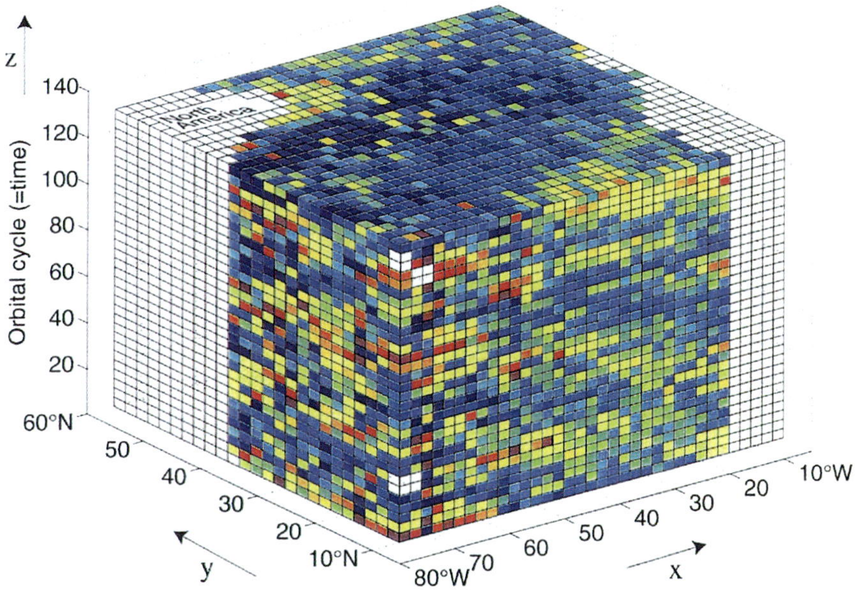

Figure 3. Cuboid of T/P gridded data in the North Atlantic (simplified view, keeping every fourth cycle and every second degree).

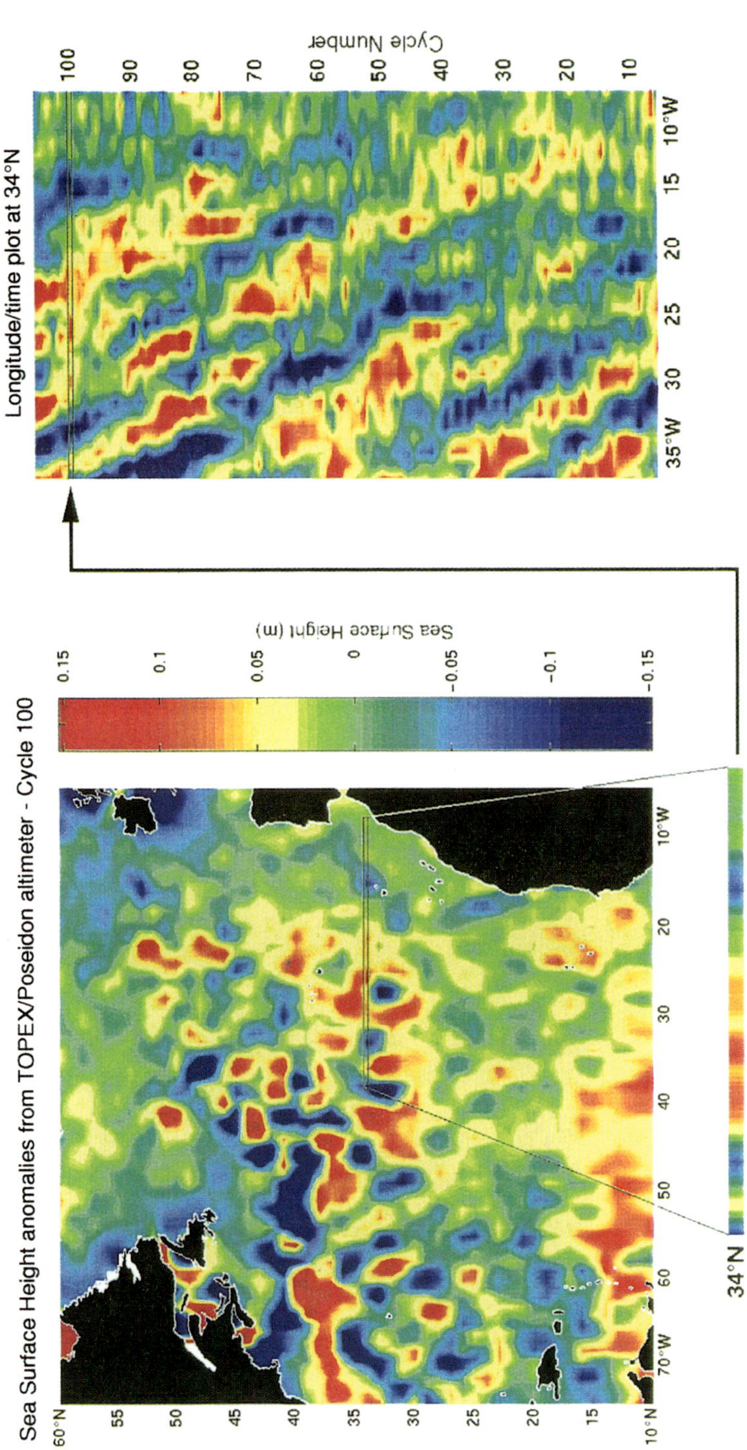

Figure 4. Schematic of the production of a longitude-time plot.

features correspond to Rossby waves, and it is possible to estimate the zonal propagation speeds from the slope of the crests and troughs, which can be done visually or with more objective methods, such as the Radon Transform described below. Note that speed changes with longitude. The most evident trend is a speed increase as the waves travel westward, although in some areas, e.g., around 80°–90°E in the Indian Ocean, it appears that the speed is slightly less than that to the east. In Figure 5, speeds are about 4–5 cm s^{-1}, which is similar to the result predicted by Killworth et al. (1997) for first-mode baroclinic Rossby waves at this latitude.

Hill et al. (2000) demonstrated that baroclinic Rossby waves are detectable in the ATSR SST anomaly field computed with respect to the Global Ocean Surface Temperature Atlas (GOSTA) monthly climatology (Bottomley et al. 1990). In Figure 6, which shows the longitude-time plot of the SST anomaly at 25°S, a number of westward-propagating features can be seen as diagonal lines superimposed on the residual annual signal; e.g., in the Indian Ocean, waves appear to propagate at a speed of about 4 cm s^{-1}.

In several locations, excellent correspondence was found between SSH and SST. In the propagation "waveguide" at 34°N in the Atlantic Ocean (Figure 7), the main propagation speed of individual crests and troughs in the different datasets was about 2.7 cm s^{-1} (Cipollini et al. 1997a, 1997b). Figure 7 displays zonal gradients of SSH and SST instead of SSH and SST anomaly fields in order to filter out the annual steric effect and enhance evidence of the zonal variability.

A number of two-dimensional data processing techniques can be applied to the longitude-time diagrams in order to study the properties of the propagating waves, and we limit our analysis to two common methods: the two-dimensional Fourier Transform (2D-FT) and the two-dimensional Radon Transform (2D-RT). Both map the longitude-time plot onto a transformed space, in this case the wavenumber-frequency space for the 2D-FT and a hybrid velocity-projected coordinate space for the 2D-RT.

In the case of the 2D-FT, mapping is implemented by the Fast Fourier Transform algorithm (2D-FFT) and highlights different spectral components of longitude-time plots that appear as peaks in the wavenumber-frequency spectrum. This method has been used before for the detection of Rossby waves (e.g., Le Traon and Minster 1993; Tokmakian and Challenor 1993). An advantage of this technique is detection of single components of propagating waves, which may correspond to distinct baroclinic modes (Cipollini et al. 1997a; Subrahmanyam et al. 2000). The 2D-FFT of the longitude-time plot in Figure 5 for the South Pacific Ocean from 120° to 90°W is shown in Figure 8. Note the large amplitude fluctuation with 3000-km wavelength and two-year period, named LW1; its second harmonic is named LW2. The spectral peak, with wavelength of ~600 km and period of ~180 days, corresponds to a first-mode baroclinic Rossby wave propagating westward at about 3.9 cm s^{-1}, which is consistent with that predicted by Killworth et al. (1997).

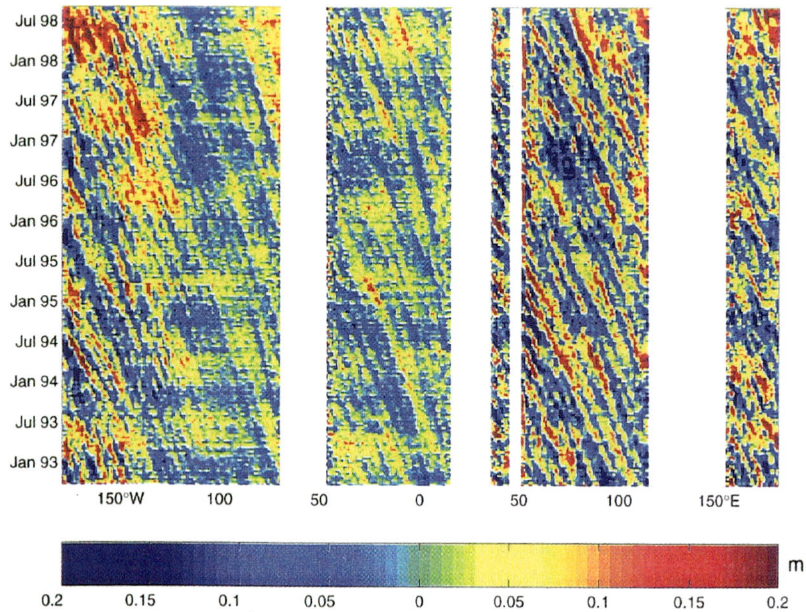

Figure 5. Global longitude-time plot of T/P SSH anomalies at 25°S.

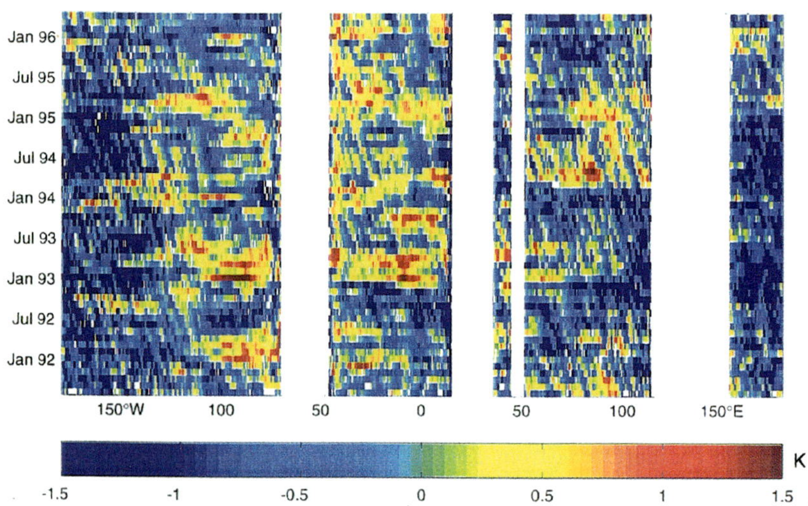

Figure 6. Global longitude-time plot of ATSR SST anomalies at 25°S. Data have been processed as in Hill et al. (2000).

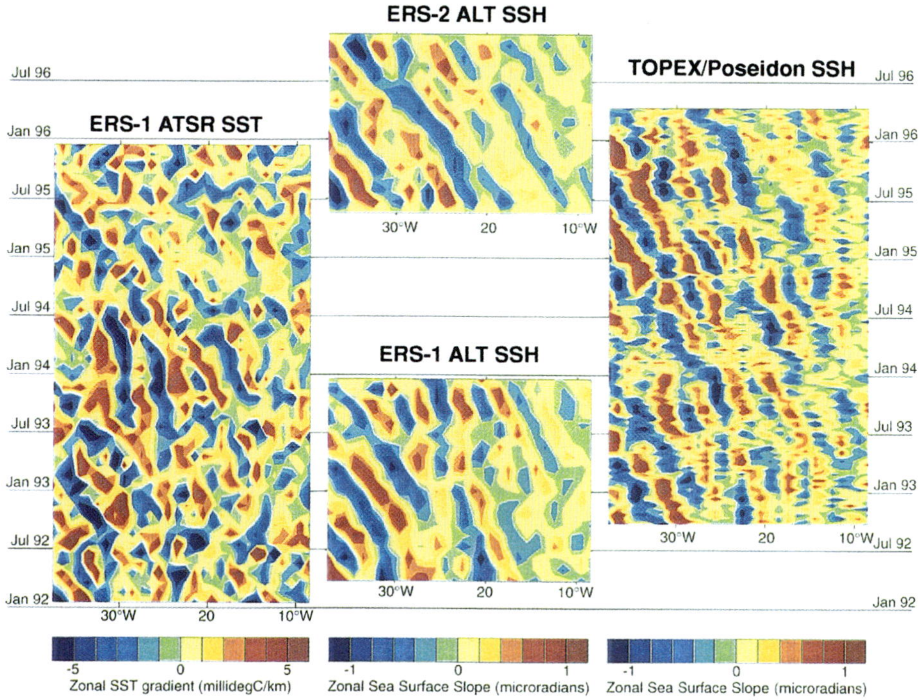

Figure 7. Interpolated longitude-time plots of zonal gradients of SSH anomaly and SST anomaly from T/P, ERS-1/2 altimeters, and ERS-1 ATSR radiometer. ATSR data were regridded to a $1° \times 1°$ grid for consistency with the altimeter processing.

A disadvantage of the 2D-FT (and 2D-RT) method is that uniform propagation characteristics are assumed over the region and time interval. Because the resolution in the wavenumber-frequency spectrum is inversely proportional to the lengths of the longitude and time intervals, the selection of the space-time domain is a trade-off between spectral resolution and space-time homogeneity.

The Radon Transform is a more appropriate method to analyze propagation speed (Radon 1917; Deans 1983; CS) than using the wavelength and period of propagating waves. As illustrated in Figure 9, the 2D-RT at a given angle, θ, is the projection of the longitude-time plot along a direction normal to θ. Lines of equal θ in longitude-time and wavenumber-frequency coordinates are lines of constant speed. Thus, computing the square of the 2D-RT of the longitude-time plot for different values of θ is equivalent to computing the spectral energy along lines of constant speed, and is a straightforward method to find the value of the speed for which the energy is maximum.

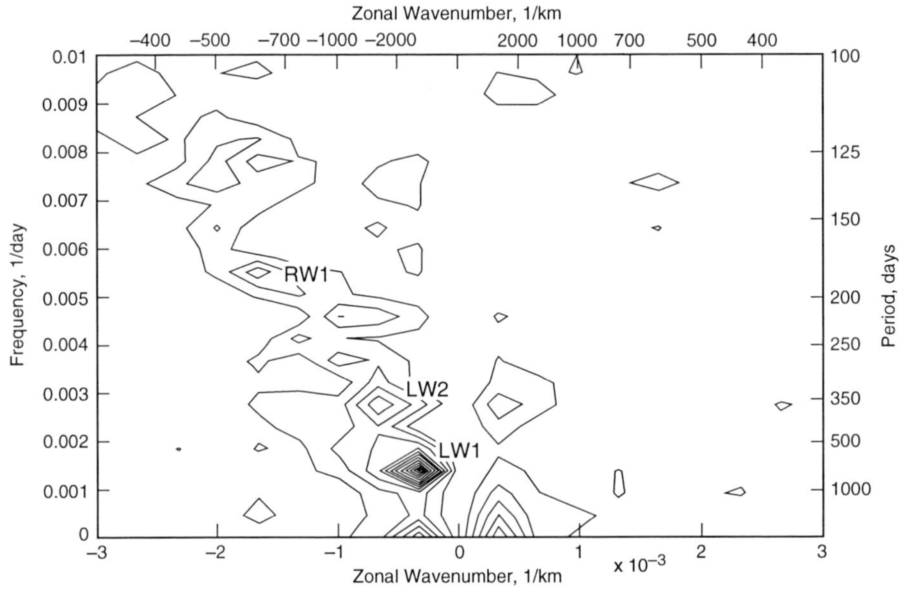

Figure 8. Interpolated 2D-FFT of a subset of the longitude-time plot in Figure 5 (longitude=120° to 90°W, all cycles). The diagram has been flipped horizontally so that westward-propagating signals are on the left. Contour interval is 5 m^2 × degree longitude × 10 days.

The 2D-RT is best applied to data regions where the speed is constant. Because the speed of Rossby waves varies much more with longitude than with time, the longitude band must be limited. Figure 10 shows the zonal speed of the maximum energy computed with the 2D-RT method applied to 11°-longitude × 219-cycle subsets of T/P data. Apart from a few additional energetic areas (e.g., Gulf Stream, Brazil Current), the speeds resemble those computed by Killworth et al. (1997, Figure 10). The agreement is quite good around 20°–25° latitude in both hemispheres, where the measured speeds were greater than those estimated with standard linear theory (CS; Killworth et al. 1997). In the tropical ocean, the satellite-derived speeds were lower than those predicted by Killworth et al. (1997). Perhaps, in the tropics, the 2D-RT preferentially selects higher-order modes of propagation because the fixed-longitude span at low latitudes is just a fraction of the Rossby wavelength. The reduced accuracy under these conditions, which also was reported by Polito and Cornillon (1997), is mitigated in our case by the availability of almost six years of data. Alternately, the theory may overestimate phase speeds at low latitudes. However, at latitudes poleward of 30°, the 2D-RT-analyzed speeds were gener-

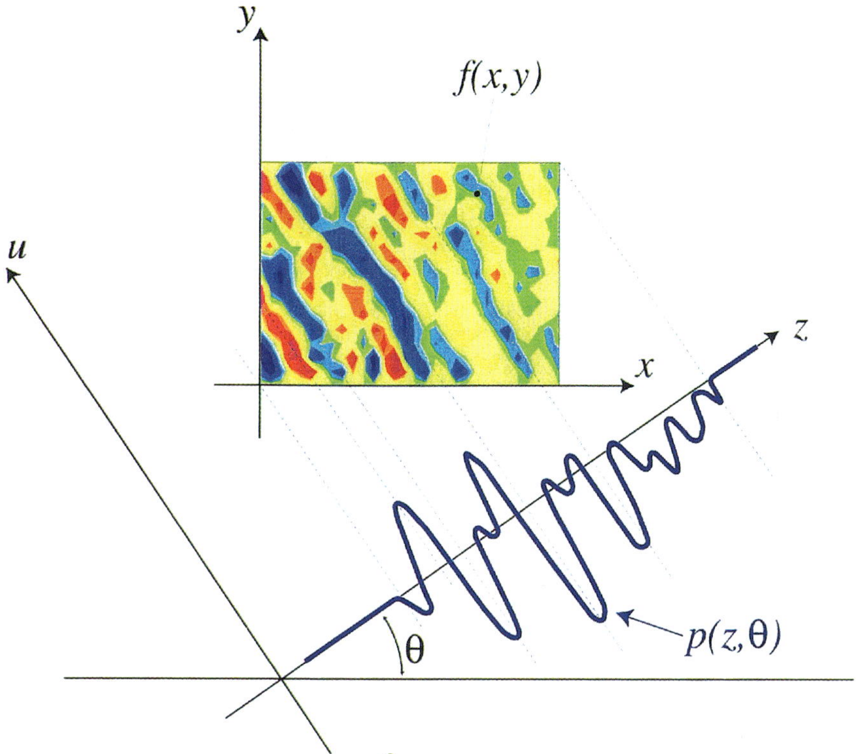

$$p(z,\theta) = \int_u f(x,y)\, du \qquad \begin{array}{l} x = z\cos\theta - u\sin\theta \\ y = z\sin\theta + u\cos\theta \end{array}$$

Figure 9. Schematic of the 2D-RT.

ally greater than those expected from linear theory and those of Killworth et al. (1997). The latitude-dependent differences are being investigated, including the influence of bottom topography (Killworth and Blundell 1999). Also, at high latitudes the Rossby wavelength, e.g., ~300 km for the semiannual period at 40°N, would be approximately the same or smaller than twice the spacing between satellite ground tracks, resulting in spatial aliasing.

Longitude-time plots reveal only the *apparent* zonal speed, as shown in Figure 11. Moreover, longitude-time plots do not give any indication of possible deviations in direction from a pure westward propagation, which is the subject of recent theoretical studies (Killworth and Blundell 1999). To study the directional properties of Rossby waves and estimate the true speed, it is important to use cuboids of data, not just slices. Consequently, three-dimensional techniques, such as the 3D-FFT and 3D-RT, are necessary.

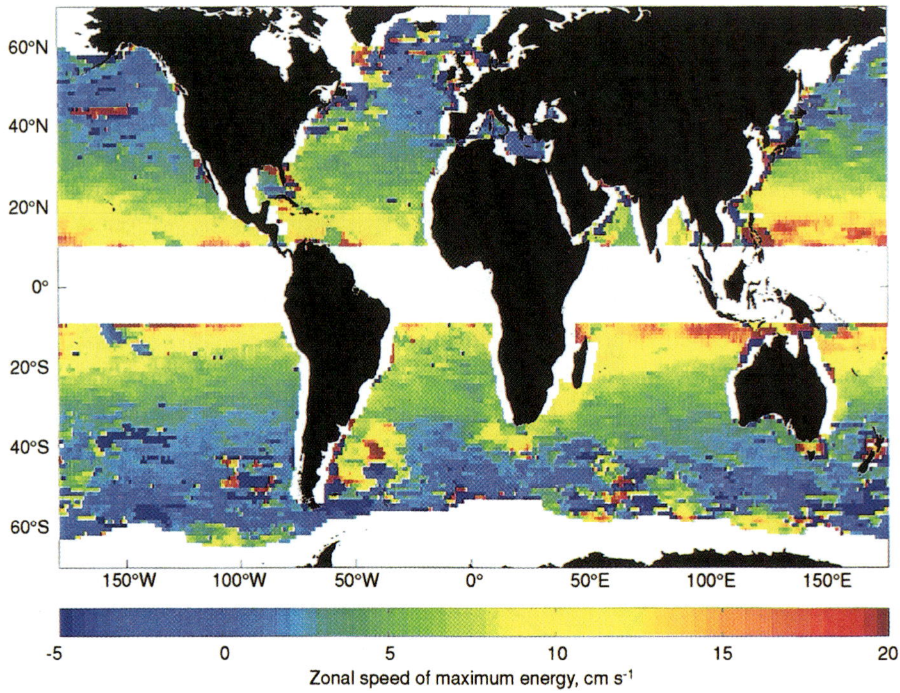

Figure 10. Main zonal propagation speed from the 2D-RT analysis of T/P subsets of 11° longitude × 219 cycles.

The 3D-RT of a cuboid of data is the projected sum of the data on a plane whose orientation is defined by two angles, θ and φ. In our longitude × latitude × time dimensions, the 3D-RT of a 9° × 9° × n (n is the number of available satellite cycles) subcuboid is computed at every location. Values θ_m and φ_m that maximize the 3D-RT energy are determined. These angles identify the plane orthogonal to the direction of propagation of crests and troughs in the subcuboid, and yield speed and direction of the largest-amplitude propagating wave. In the vicinity of 34°N, 40°W (Figure 12), there is a distinct spectral peak at $\theta_m = 13°$ corresponding to a propagation speed of 2.5 cm s^{-1}; $\varphi_m = 180°$ represents westward motion. The same scheme can be applied to every location in an ocean region to estimate the local wave direction. In the North Atlantic, nearly all propagation directions of the maximum energy of T/P fluctuations were within 20° of westward (Figure 13). No simple effects of topography on the direction of propagation can be observed in the figure. In the most energetic waveguide, around 34°N, Rossby wave speeds seem to decrease approaching the Mid-Atlantic Ridge at about 38°W, and increase after passing it. This observation is consistent with the results of Killworth and Blundell (1999).

6. Rossby Waves in Models

The development of realistic ocean models will improve our understanding of ocean circulation and enable accurate predictions of the ocean's response to changes in environmental conditions. In addition to accurately reproducing the known mean circulation and seasonal variation, it is clearly important that models contain an accurate representation of Rossby waves, because they transmit "information" from one side of an ocean basin to the other and introduce a significant time lag into the response of the ocean. The generation, propagation, and speed of Rossby waves in models is an important indication of their ability to respond to remote forcing and the time required to reach a new equilibrium. Indeed, when models are "spun up" from initial conditions, there are usually a myriad of strong fluctuations propagating throughout the basins. However, these are for anomalous conditions; the appropriate stage for examining the properties of Rossby waves is when the model is in quasi-equilibrium with the atmospheric forcing.

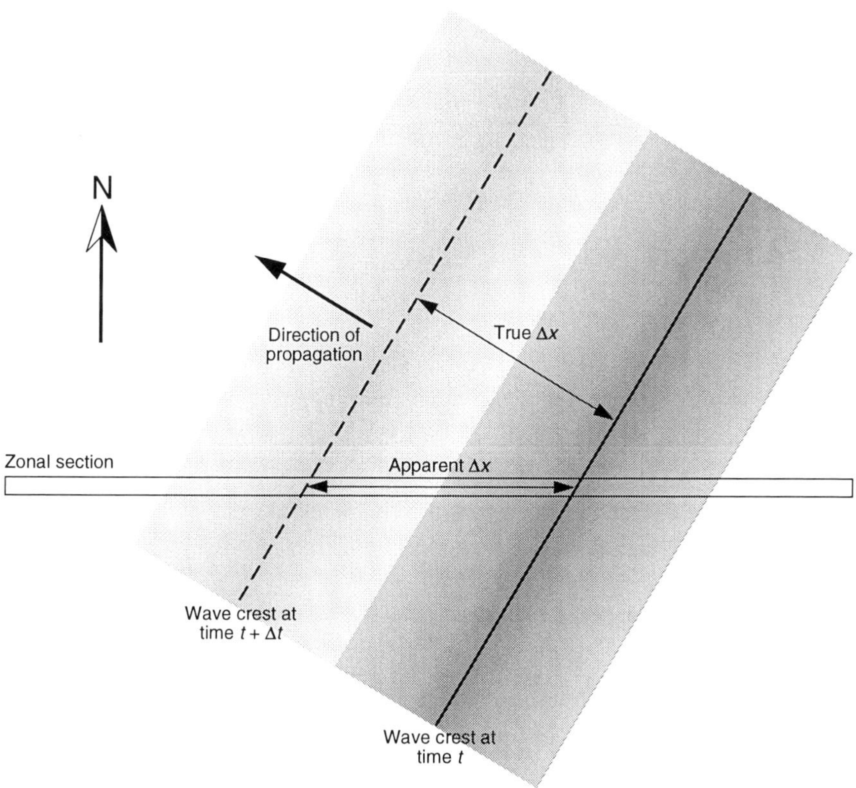

Figure 11. Explanation of apparent speed when taking zonal sections. The apparent speed is $\frac{\Delta x_{apparent}}{\Delta t} \geq \frac{\Delta x_{true}}{\Delta t}$.

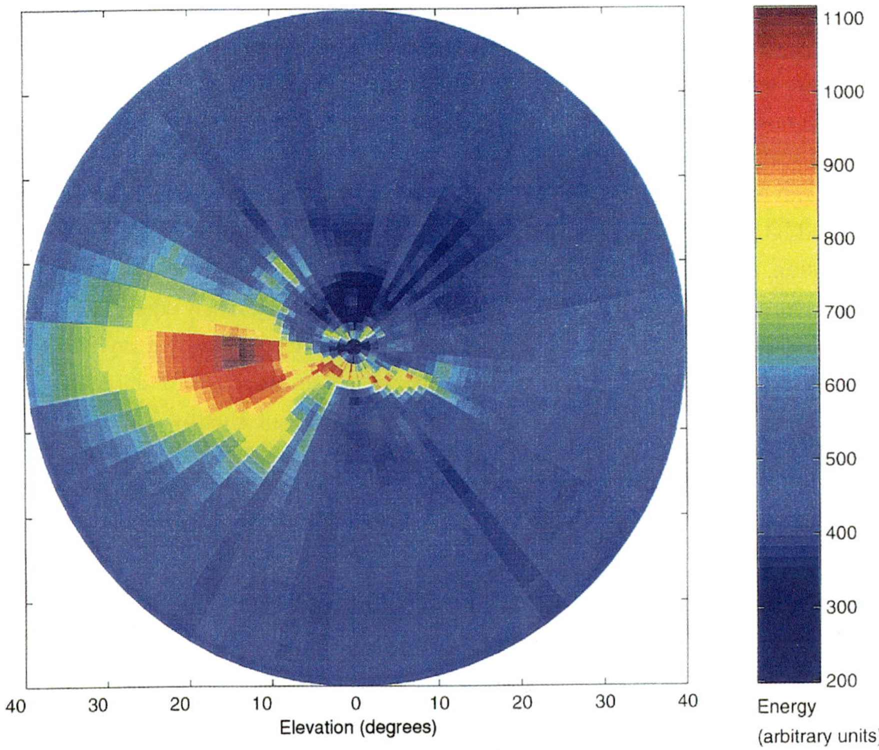

Figure 12. Polar plot of 3D-RT energy at 34°N, 40°W in the Atlantic.

Barnier (1988) investigated the generation and propagation of Rossby waves using a simple, two-layer quasi-geostrophic model of a flat-bottom North Atlantic Ocean. A realistic wind field generated Rossby waves in the center of the basin and at the eastern edge. The addition of a Mid-Atlantic Ridge prevented baroclinic Rossby waves traveling across the basin from east to west, but led to a greater production of Rossby waves near the Mid-Atlantic Ridge.

Oceanographic variables simulated with the 1/3° × 1/3° Atlantic Isopycnic Model (AIM, New et al. 1995) after 20 years of "spin up" exhibited strong gyral circulation, with branching of the North Atlantic Current to produce an eastward-flowing Azores Current. The model Azores Current at 30°W had a mean transport of 10.6 Sv (1 Sv = 1×10^6 m^3 s^{-1}) between 32° and 35°N (New et al. 2000), which is in good agreement with hydrographic observations (Gould 1985).

Dynamic heights from AIM have been interpolated onto a 1° × 1° grid to ensure that the analysis techniques applied here are identical to those used on the altimeter and SST data described in earlier sections. Propagating features can be found at most latitudes in the model simulation; an example at 36°N is shown in Figure 14a. Magnitudes are much weaker than those observed with altimetry (cf. Figure 7) and there is usually an annual cycle, which may be related to wind forcing in the model. The Rossby wave energy in the model is strongest near 38°N, in contrast to being strongest at 34°N for T/P data. At 34°N the model output shows both eastward and westward propagating features, which converge near 25°W (not shown). This convergence is not present in the satellite data, does not correlate with any known bathymetric features, and may simply be a model artifact.

East of the Mid-Atlantic Ridge, the model's Rossby wave speeds in the 30°–38°N region were 40% lower than those derived from T/P data. It is not known why there is a

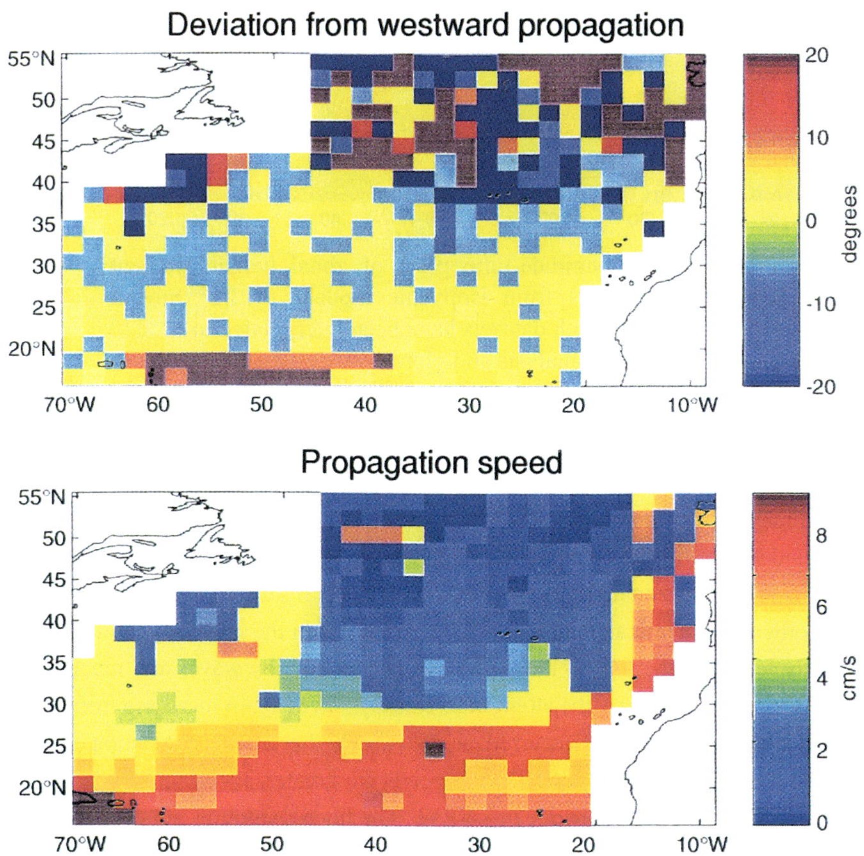

Figure 13. Speed and deviation of main propagating signal in the North Atlantic from 3D-RT analysis of T/P data. The main signal in each location has been selected among only those propagating within ±20° from pure westward.

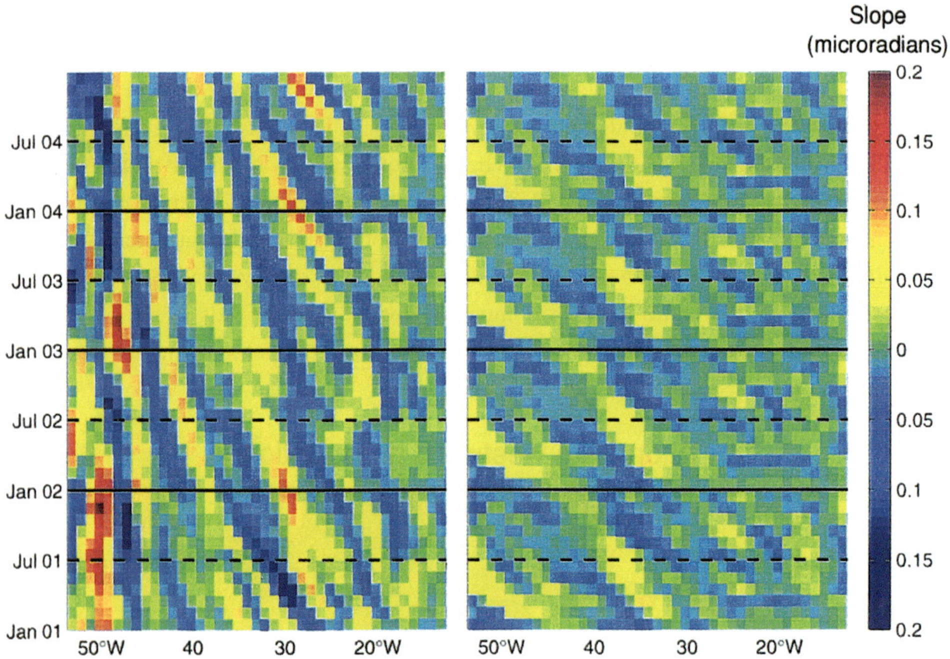

Figure 14. Longitude-time plots of zonal height gradient anomalies at 36°N from isopycnic models: (a) 1/3° model, (b) 4/3° model. Both datasets have been interpolated to 1° × 1° resolution prior to analysis and plotting.

difference. Perhaps the use of surface-referenced potential density as the model vertical coordinate may lead to unrealistically low values of model thermal wind shear. This is currently under investigation.

Figure 14b shows zonal height gradient anomalies simulated with the 4/3° × 4/3° AIM, but otherwise identical conditions. The variability in the model dynamic heights is significantly less than the 1/3° × 1/3° AIM run, and its Rossby waves are much weaker. In the coarser-resolution case, the strength of Rossby waves again increases from 32°N to 36°N. Eastward propagation occurred at 38°N. In the coarser-resolution AIM, the Rossby waves are weaker and propagation speeds are greater, showing a slight acceleration crossing the Mid-Atlantic Ridge at ~40°W. The finer-resolution run of AIM shows a marked reduction in propagation speed on crossing the Mid-Atlantic Ridge—this is not consistent with observations (Figure 13b) and may be due to a strong eastward flow advecting the Rossby waves. The monthly climatological forcing generates a very regular seasonal cycle in the coarser-resolution AIM, whereas the finer-resolution run contains interannual variations.

Rossby waves are not explicitly coded into these models; they are simply a series of solutions to the equations of motion on a rotating surface with varying surface forcing. Therefore, the comparison of Rossby waves simulated from models with observations makes a good test of how well the model dynamics are simulating reality. When models and observations disagree, the model code does not require a specific adjustment of "Rossby wave parameters", but requires an effort to address the underlying problems. For example, weak vertical mixing or too few model layers will lead to a poor representation of the stratification and, hence, the vertical shear, while errors in the parameterization of the bottom drag will affect the bottom few layers and, thus, the depth structure of baroclinic modes.

7. Future Research

Many challenges remain to understand the dynamics of Rossby waves. The suggestion (Section 5) that different baroclinic modes of propagation can be observed and their properties estimated with Fourier techniques leads to questions about the relationship between the modes observed in SST and SSH fields. More importantly, it must be shown whether a combination of the two fields can be used to infer additional information about the characteristics of the modes themselves. It should not be forgotten that information obtained about the strength of the different modes of propagation becomes information on the internal structure of the ocean, overcoming the apparent limitation of satellites to observe only the ocean surface.

Another important field of research is the relationship between ocean Rossby waves and atmospheric variables. This is crucial, not only to understand the possible mechanisms for the generation and amplification of Rossby waves by atmospheric forcing, but also to evaluate the effects of ocean-atmosphere coupling on the waves, which can in turn explain part of the discrepancy between theory and observations, as shown by White et al. (1998).

An improvement in analysis of Rossby wave characteristics could be in the use of non-gridded data (Glazman et al. 1996). The 2D-RT and 3D-RT methods can be easily generalized to non-gridded data. One important related issue is the merging of multiple altimeter datasets to improve the SSH field.

Analysis of longitude-time plots of chlorophyll-*a* data from the Ocean Color and Temperature Scanner (OCTS) and Sea-viewing Wide Field-of-view Sensor (SeaWiFS) shows, at some latitudes only, westward-propagating features whose characteristics are consistent with those expected for baroclinic Rossby waves (Cipollini et al. 1999). The propagation speed increases equatorward as expected for Rossby waves. An initial comparison with altimeter data shows that, in a few cases, the waves propagate at similar speed in the altimeter and ocean-color datasets. In other regions, the relationship is more complex and needs further investigation. One possible explanation is that biology could be more sensitive to higher order baroclinic modes because of the different vertical motion (and nutrient upwelling) induced by those modes. An effect of Rossby waves on phytoplankton growth would imply enhanced (or reduced) removal of CO_2 from the atmosphere and thus an effect on the global carbon cycle.

The results obtained so far are stimulating many intriguing questions and opening up a number of possible paths for further research on Rossby waves. All recent breakthroughs in the study of this important phenomenon have been made possible (or at least prompted) by the availability of large and accurate datasets from satellite instruments, especially altimeters. The effects of these discoveries on the knowledge of the complex mechanisms that regulate the climate of our planet are still not completely understood, but are certainly significant, and are the subject of intense investigation by many research groups. This could revolutionize the way we predict weather and climate change. In conclusion, research into Rossby waves is another field where satellites demonstrate their potential for science and usefulness to society.

Acknowledgments. We thank NASA and AVISO for the provision of TOPEX/Poseidon altimeter data, ESA for ERS altimeter data, and Rutherford Appleton Laboratory for ATSR data. We are grateful to Peter Killworth, Jeff Blundell, Trevor Guymer, and Adrian New for their kind help and suggestions, to Ian Robinson and Katy Hill for provision of the ATSR SST anomalies, Helen Snaith for support of the altimeter processing software, Stefano Raffaglio for his help in T/P data analysis, and Yanli Jia for provision of AIM data. Finally, we thank the two anonymous reviewers whose comments helped to improve the quality of the paper.

References

Barnier B., A numerical study on the influence of the Mid-Atlantic Ridge on nonlinear first-mode baroclinic Rossby waves generated by seasonal winds, *J. Phys. Oceanogr.*, *18*, 417–433, 1988.

Bernstein, B., and W. White, Time and length scales of baroclinic eddies in the central North Pacific Ocean, *J. Phys. Oceanogr.*, *4*, 123–126, 1974.

Bernstein, B., and W. White, Zonal variability in the distribution of eddy energy in the mid-latitude North Pacific Ocean, *J. Phys. Oceanogr.*, *7*, 123–126, 1977.

Bottomley, M., C. K. Folland, J. Hsiung, R. E. Newell, and D. E. Parker, *Global Ocean Surface Temperature Atlas (GOSTA)*, joint project, United Kingdom Meteorological Office and the Massachusetts Institute of Technology, 20 pp, Her Majesty's Stationery Office, Norwich, England, 1990.

Chelton, D. B., and M. G. Schlax, Global observations of oceanic Rossby waves, *Science, 272*, 234–238, 1996.

Chelton, D. B., R. A. de Szoeke, M. G. Schlax, K. E. Naggar, and N. Siwertz, Geographical variability of the first baroclinic Rossby radius of deformation, *J. Phys. Oceanogr.*, *28*, 433–460, 1998.

Cheney, R., L. Miller, R. Agreen, N. Doyle, and J. Lillibridge, TOPEX/Poseidon: the 2-cm solution, *J. Geophys. Res.*, *99*, 24555–24563, 1994.

Cipollini, P., D. Cromwell, and G. D. Quartly, Variability of Rossby wave propagation in the North Atlantic from TOPEX/Poseidon altimetry, *Proceedings of IEEE International Geoscience and Remote-Sensing Symposium (IGARSS '96)*, *1*, Lincoln, Nebraska, 27–31 May 1996, 91–93, 1996.

Cipollini, P., D. Cromwell, M. S. Jones, G. D. Quartly, and P. G. Challenor, Concurrent altimeter and infrared observations of Rossby wave propagation near 34°N in the northeast Atlantic, *Geophys. Res. Lett.*, *24*, 889–892, 1997a.

Cipollini, P., D. Cromwell, M. S. Jones, G. D. Quartly, and P. G. Challenor, The potential of ERS for the detection of Rossby waves in the Northeast Atlantic, *Third ERS Symposium: Space at the Service of Our Environment*, 1473–1478, ESA, Florence, Italy, 1997b.

Cipollini, P., P. G. Challenor, D. Cromwell, T. H. Guymer, S. Raffaglio, The detectability of Rossby waves in ocean colour data, *Southampton Oceanography Centre Internal Document*, No. 44, Southampton, United Kingdom, 29 pp, 1999.

Deans, S. R., *The Radon Transform and Some of Its Applications*, John Wiley, New York, 1983.

Delcroix, T., J. Picaut, and G. Eldin, Equatorial Kelvin and Rossby waves evidenced in the Pacific Ocean through Geosat sea level and surface current anomalies, *J. Geophys. Res.*, *96* (supplement), 3249–3262, 1991.

Dewar, W., On too-fast baroclinic planetary waves in the general circulation, *J. Phys. Oceanogr.*, *28*, 1739–1758, 1998.

DYNAMO, Dynamics of North Atlantic Models: Simulation and assimilation with high resolution models. *Berichte aus dem Institut für Meereskunde an der Christian-Albrechts-Universität Kiel*, No. 294, Kiel, Germany, 334 pp, 1997.

Emery, W., and L. Magaard, Baroclinic Rossby waves as inferred from temperature fluctuation in the eastern Pacific, *J. Mar. Res.*, *34*, 365–385, 1976.

Forrester, T. N., and P. G. Challenor, Validation of ATSR sea surface temperatures in the Faeroes region, *Int. J. Remote Sensing*, *16*, 2741–2753, 1995.

Francis, R., G. Graf, P. G. Edwards, M. McCraig, C. McCarthy, P. Dubock, A. Lefebvre, B. Pieper, P.-Y. Pouvreau, R. Wall, F. Wechsler, J. Louet, and R. Zobl, The ERS-1 spacecraft and its payload, *ESA Bull.*, *65*, 26–48, 1991.

Fu, L.-L., E. J. Christensen, C. A. Yamarone Jr., M. Lefebvre, Y. Menard, M. Dorrer, and P. Escudier, TOPEX/Poseidon mission overview, *J. Geophys. Res.*, *99*, 24369–24381, 1994.

Gaspar, P., and C. Wunsch, Estimates from altimeter data of barotropic Rossby waves in the northwestern Atlantic Ocean, *J. Phys. Oceanogr.*, *19*, 1821–1844, 1989.

Gill, A. E., *Atmosphere-Ocean Dynamics*, Academic Press, San Diego, 1982.

Glazman, R., A. Fabrikant, and A. Greysukh, Statistics of spatial-temporal variations of sea surface height based on TOPEX altimeter measurements, *Int. J. Remote Sensing*, *17*, 2647–2666, 1996.

Gould, W. J., Physical oceanography of the Azores Front, *Prog. Oceanogr.*, *14*, 167–190, 1985.

Herrmann, P., and W. Krauss, Generation and propagation of annual Rossby waves in the North Atlantic, *J. Phys. Oceanogr.*, *19*, 727–744, 1989.

Halliwell, G. R., Y. J. Ro, and P. Cornillon, Westward-propagating SST anomalies and baroclinic eddies in the Sargasso Sea, *J. Phys. Oceanogr.*, *21*, 1664–1680, 1991.

Hill, K. L., I. S. Robinson, and P. Cipollini, Propagation characteristics of extratropical planetary waves observed in the ATSR global sea surface temperature record, *J. Geophys. Res.*, in press, 2000.

Hough, S., On the application of harmonic analysis to the dynamical theory of the tides, Part I, on Laplace's 'oscillations of the first species', and on the dynamics of ocean currents, *Phil. Trans. Roy. Soc. Lond.*, *A*, *189*, 201–257, 1897.

Hughes, C. W., M. S. Jones, and S. Carnochan, Use of transient features to identify eastward currents in the Southern Ocean, *J. Geophys. Res.*, *103*, 2929–2943, 1998.

Jacobs, G. A., W. J. Emery, and G. H. Born, Rossby waves in the Pacific Ocean extracted from Geosat altimeter data, *J. Phys. Oceanogr.*, *23*, 1155–1175, 1993.

Jacobs, G. A., H. E. Hurlburt, J. C. Kindle, E. J. Metzger, J. L. Mitchell, W. J. Teague, and A. J. Wallcraft, Decade-scale trans-Pacific propagation and warming effects of an El Niño anomaly, *Nature*, *370*, 360–363, 1994.

Jacobson, A. R., and J.L. Spiesberger, Observations of El Niño-Southern Oscillation-induced Rossby waves in the northeast Pacific using in-situ data, *J. Geophys. Res.*, *103*, 24585–24596, 1998.

Kang, Q., and L. Magaard, Annual baroclinic Rossby waves in the central North Pacific, *J. Phys. Oceanogr.*, *10*, 1156–1167, 1980.

Kessler, W., Observations of long Rossby waves in the northern tropical Pacific, *J. Geophys. Res.*, *95*, 5183–5218, 1990.

Killworth, P. D., D. B. Chelton, and R. de Szoeke, The speed of observed and theoretical long extra-tropical planetary waves, *J. Phys. Oceanogr.*, *27*, 1946–1966, 1997.

Killworth, P. D., and J. R. Blundell, The effect of bottom topography on the speed of long extra-tropical planetary waves, *J. Phys. Oceanogr.*, *29*, 2689–2710, 1999.

Lawrence, S. P., M. R. Allen, D. L. T. Anderson, and D. T. Llewellyn-Jones, Effects of subsurface ocean dynamics on instability waves in the tropical Pacific, *J. Geophys. Res.*, *103*, 18649–18663, 1998.

Le Traon, P.-Y., and J.-F. Minster, Sea level variability and semiannual Rossby waves in the South Atlantic Subtropical Gyre, *J. Geophys. Res.*, *98*, 12315–12326, 1993.

Matthews, P. E., M. A. Johnson, and J. J. O'Brien, Observation of mesoscale ocean features in the northeast Pacific using Geosat radar altimetry data., *J. Geophys. Res.*, *97*, 17829–17840, 1992.

McPhaden, M. J., The eleven-year El Niño?, *Nature*, *370*, 326, 1994.

Mutlow, C. T., A. M. Závody, I. J. Barton, and D. T. Llewellyn-Jones, Sea surface temperature measurements by the along-track scanning radiometer on the ERS-1 satellite: Early results, *J. Geophys. Res.*, *99*, 22575–22588, 1994.

New, A. L., R. Bleck, Y. Jia, R. Marsh, M. Huddleston, and S. Barnard, An isopycnic model study of the North Atlantic: Part 1, Model experiment, *J. Phys. Oceanogr.*, *25*, 2667–2699, 1995.

New, A. L., Y. Jia, M. Coulibaly, and J. Dengg, On the role of the Azores Current in the ventilation of the North Atlantic Ocean, *Prog. Oceanogr.*, submitted, 2000.

Périgaud, C., and P. Delecluse, Annual sea level variations in the southern tropical Indian Ocean from Geosat and shallow-water simulations., *J. Geophys. Res.*, *97*, 20169–20178, 1992.

Philander, S. G., *El Niño, La Niña, and the Southern Oscillation*, 293 pp, Academic Press, 1990.

Polito, P. S., and P. Cornillon, Long baroclinic Rossby waves detected by TOPEX/Poseidon, *J. Geophys. Res.*, *103*, 3215–3235, 1997.

Price, J., and L. Magaard, Rossby wave analysis of the baroclinic potential energy in the upper 500 meters of the North Pacific, *J. Mar. Res.*, *38*, 249–264, 1980.

Price, J., and L. Magaard, Rossby wave analysis of subsurface temperature fluctuations along the Honolulu-San Francisco great circle, *J. Phys. Oceanogr.*, *23*, 258–268, 1983.

Price, J., and L. Magaard, Interannual baroclinic Rossby waves in the midlatitude North Atlantic, *J. Phys. Oceanogr.*, *16*, 2061–2070, 1986.

Qiu, B., W. Miao, and P. Müller, Propagation and delay of forced and free baroclinic Rossby waves in off-equatorial oceans, *J. Phys. Oceanogr.*, *27*, 2405–2417, 1997.

Radon, J., Über die Bestimmung von Funktionen durch ihre Integralwerte längs gewisser Mannigfaltigkeiten, *Berichte Sächsische Akademie der Wissenschaften. Leipzig, Math.–Phys. Kl.*, *69*, 262-267, 1917. English translation in S. R. Deans, *The Radon Transform and Some of Its Applications*, John Wiley, New York, 204–217, 1983.

Rossby, C.-G., Planetary flow patterns in the atmosphere, *Quart. J. Roy. Meteorol. Soc.*, *66* (suppl.), 68–97, 1939.

Scharroo, R., and P. Vissier, Precise determination and gravity field improvement for the ERS satellites, *J. Geophys. Res.*, *103*, 8113–8127, 1998.

Schlax, M. G., and D. B. Chelton, Aliased tidal errors in TOPEX/Poseidon sea surface height data, *J. Geophys. Res.*, *99*, 24761–24775, 1994.

Subrahmanyam, B., I. S. Robinson, J. R. Blundell, and P. G. Challenor, Rossby waves in the Indian Ocean from TOPEX/Poseidon altimeter and model simulations, *Int. J. Remote Sensing*, in press, 2000.

Tokmakian, R. T., and P. G. Challenor, Observations in the Canary Basin and the Azores Frontal region using Geosat data, *J. Geophys. Res.*, *98*, 4761–4773, 1993.

White, W. B., Y. Chao, and C.-K. Tai, Coupling of biennial oceanic Rossby waves with the overlying atmosphere in the Pacific basin, *J. Phys. Oceanogr.*, *28*, 1236–1251, 1998.

Paolo Cipollini, James Rennell Division for Ocean Circulation and Climate; Southampton Oceanography Centre, European Way, Southampton SO14 3ZH, United Kingdom. (email, cipo@soc.soton.ac.uk; fax, +44-23-8059-6400)

Satellites, Oceanography and Society
edited by David Halpern

Chapter 7

A study of meddies using simultaneous in-situ and satellite observations

Paulo B. Oliveira, Nuno Serra, Armando F. G. Fiúza, and Isabel Ambar
Instituto de Oceanografia, Universidade de Lisboa, Lisbon, Portugal

Abstract. Data from the National Oceanic and Atmospheric Administration (NOAA) and Topography Experiment (TOPEX)/Poseidon (T/P) satellites, a surface drifter, and subsurface floats, named RAFOS, at a depth of ~1000 m in the Mediterranean Undercurrent were used to investigate Mediterranean Water eddies, or meddies, off the southwest coast of Portugal. Analysis of RAFOS trajectories and simultaneous distributions of NOAA Advanced Very-High Resolution Radiometer (AVHRR) sea surface temperature (SST) and T/P sea level anomalies revealed that meddies frequently have a signature in the surface thermal field. Characteristics of these signatures include clockwise wrapping of filament structures around the eddy periphery and higher sea level anomalies of ~10 cm. A sequence of six SST images, taken between 14 March and 2 May 1994, was compared with contemporaneous 11-day trajectories of selected RAFOS floats to investigate the eddy-like features of the SST field in relation to eddy motion in the Mediterranean Undercurrent. The structure and evolution of submesoscale cyclonic and anticyclonic structures identified in the SST field were found to reflect mid-depth flow structures with similar features, as revealed by the RAFOS trajectories. This approach yielded relevant clues to the interpretation of the movement of meddies: namely, their irregular trajectories with stops and sudden changes in direction that could be attributed to their interaction with other vortices.

1. Introduction

The history of the study of ocean eddies demonstrates how our understanding of marine physics is determined by the tools available to researchers. Although it is now well known that open ocean circulation is dominated by eddies, it only became possible to study them in the early 1970s. At that time, self-recording instruments and mooring technology had been developed to provide time-series measurements at selected positions and depths in the ocean for periods far longer than a ship could possibly remain in one location (Wunsch 1981). The study of the ocean eddy field was also strongly influenced by the advent of satellites. Remote-sensing techniques made possible the generation of large-scale synoptic maps with surface evidence of eddies and the geographical distribution of eddy statistics (Robinson 1983). Examples of the first descriptions of the Gulf Stream meanders using satellite-borne infrared (IR) sensors are the works by Rao et al. (1971, cited in DeRycke and Rao (1973)) and Stumpf and Rao (1975); Huang et al. (1978) and Cheney and Marsh (1981) described the same subject using radar altimeter data.

Since those early studies, many works have been published using satellite data to investigate eddy activity in the ocean. Recently, there has been an increasing contribution from satellite altimetry due to its all-weather capability and its ability to provide nearly synoptic, basin-wide observations of the dynamical state of the ocean. IR images continue to be very useful to describe eddy fields on a regional scale.

The use of remotely sensed data to study eddies near the eastern boundary of the northeast Atlantic started with Dickson and Hughes (1981), who provided the first synoptic description of mesoscale eddy activity over the Biscay Abyssal Plain; the basis for this description was a single IR image of May 1979 from the Television and Infrared Observation Satellite (TIROS-N) Advanced Very-High Resolution Radiometer (AVHRR). Using several sea surface temperature (SST) images of the Bay of Biscay obtained from the National Oceanic and Atmospheric Administration (NOAA) AVHRR on board the NOAA-10 and NOAA-11 satellites, Pingree and Le Cann (1992) presented a study of the formation and early stages of development of anticyclonic mesoscale eddies with a core of slope water. These structures, which they called "swoddies" for slope water oceanic eddies, can readily be observed in satellite IR imagery.

Apart from the study of upper ocean eddies, there has been a growing interest in the study of isolated, submesoscale, subsurface vortices containing water of Mediterranean origin. Named "meddies" for Mediterranean Water eddies by McDowell and Rossby (1978), they are now known to be frequent features in the northeast Atlantic at intermediate depths.

Mediterranean Water (MW) flows through the Strait of Gibraltar into the eastern North Atlantic, where it forms a warm and salty tongue that extends westward from the Iberian Peninsula (Figure 1). This tongue is a unique feature of the North Atlantic subtropical gyre at intermediate depths (centered at about 1000 m) and has been traditionally interpreted to be a result of advection and/or eddy diffusion processes (Needler and Heath 1975). The discovery of meddies in the eastern North Atlantic (e.g., Armi and Stommel 1983; Armi et al. 1989; Pingree and Le Cann 1993; Prater and Sanford 1994), which are able to transport water of Mediterranean origin over thousands of kilometers with very little mixing, has challenged the classical interpretation of how MW is spread in the Atlantic. This challenge was first taken up by Armi and Stommel (1983), who suggested that the Mediterranean outflow does not inject salty water only at the eastern boundary; point sources distributed in some unknown way, beginning at the boundary and extending thousands of kilometers into the North Atlantic basin, could contribute as well. From calculations of the salt flux divergence in a triangular area to the east of the Mid-Atlantic Ridge,

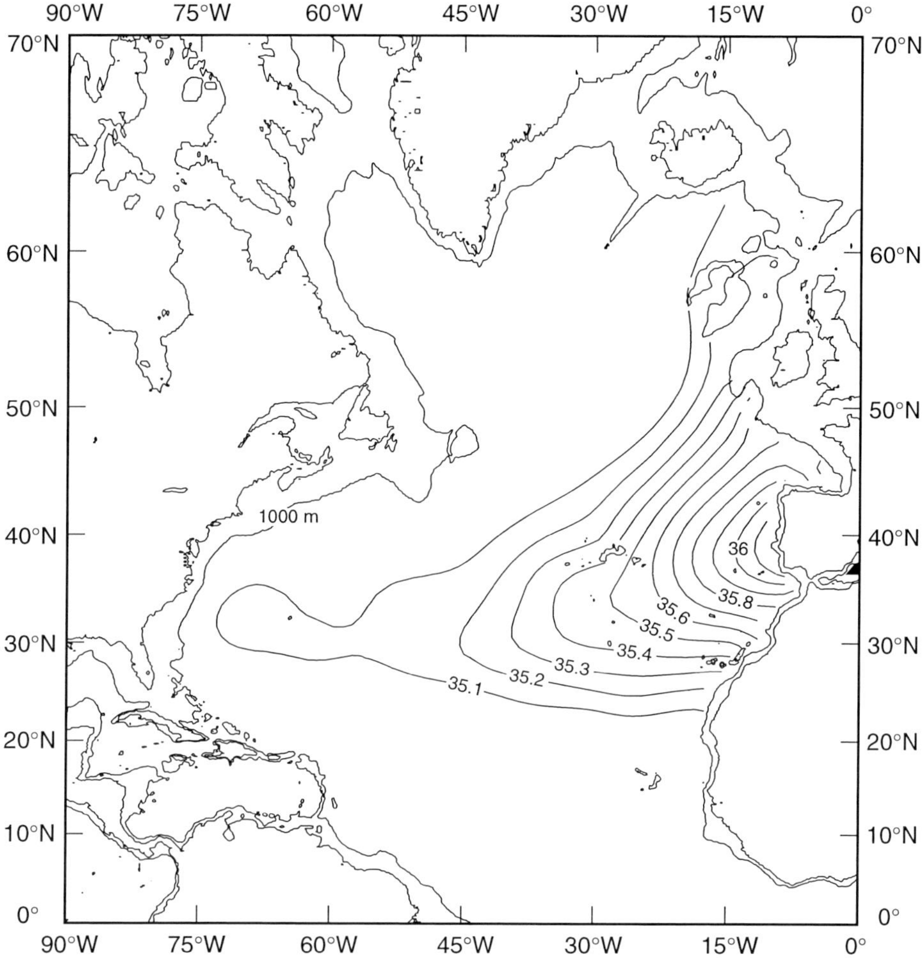

Figure 1. Mean salinity distribution at 1000 m in the North Atlantic from Levitus et al. (1994). Also, 1000-m depth contour is shown. A high-salinity tongue of mixed Mediterranean Water extends from southwestern Portugal across most of the North Atlantic.

Armi and Stommel (1983) concluded that this flux divergence could be balanced by the decay of just three meddies per year.

Meddies are important to maintain the large-scale MW tongue. Richardson et al. (1989) concluded that the annual salt flux due to 8–12 meddies is about 25% of the salt anomaly flux out of the Mediterranean, and suggested that meddies must play an important role in determining the location and shape of the salt tongue. Bower et al. (1997), using data from RAFOS floats, estimated that 15–20 meddies form each year off Portugal and indicated that newly-formed meddies remaining near the continental slope cause fluid particles approaching from upstream to be deflected off-slope, thus representing the indirect role of meddies in the deflection of MW away from the Iberian Peninsula. A subsurface float tracked acoustically in the SOund Fixing And Ranging (SOFAR) channel is called a RAFOS (SOFAR spelled backwards) float (Rossby et al. 1986). Arhan et al. (1994) suggested that the meddy contribution to total Mediterranean salt along a quasi-meridional hydrographic section centered at about 15°W is ~20%. Assuming the salt content is advected by meddies at an average speed of 1 cm s^{-1} and the remaining 80% is transported at a background speed of 0.2 cm s^{-1}, Arhan et al. (1994) concluded that the meddy contribution to salt transport would reach 55%. Mazé et al. (1997) concluded that meddies and other eddies are responsible for virtually all of the westward salt flux across 12.5°W.

The use of satellite data to study meddies has been explored relatively little due to the supposedly weak influence of such structures on the surface flow, raising questions of whether or not the surface signal generated by meddies is strong enough to be detected by remote sensing (Marshall 1988).

Several studies indicate that meddies are associated with surface features. Käse and Zenk (1987) argued that, near the generation site, meddies should have surface vorticity signals. Using satellite-tracked surface drifters, Schultz Tokos et al. (1994) confirmed the surface signal of a meddy found in the vicinity of Josephine Seamount (37°N, 14°W), showing that the trajectories of a surface drifter and a RAFOS float described the translation of the same meddy for three months. Stammer et al. (1991) reported that Geodetic satellite (Geosat) sea surface height (SSH) is significantly correlated with dynamic topography of the layer containing the core of the Mediterranean outflow and the maximum velocities of the baroclinic eddies. Pingree and Le Cann (1993) showed that a "smeddy" (shallow Mediterranean Water eddy) at 700-m depth on the southern Tagus Abyssal Plain was identifiable in satellite IR imagery. Further evidence of the surface signature of meddies in thermal images was presented by Pingree (1995).

To evaluate the role of meddies in maintaining the basin-wide salt distribution and understand the influence of meddies on the intensity of the subtropical gyre, it is important to study their formation rate, geographical distribution, movement, and decay. The following sections present the results of a research program conducted at the Institute of Oceanography, University of Lisbon, in the frame of the Canary Islands Açores Gibraltar Observations (CANIGO) Project to study meddies using both satellite and in-situ data. The research was carried out using NOAA AVHRR SST data, Topography Experiment (TOPEX)/Poseidon, named T/P, SSH distributions, and particle trajectories at the level of the MW from RAFOS floats and at the surface from drifters. The three objectives were to study: (1) the relationship between meddy trajectories and surface currents; (2) the surface thermal signature of meddies; (3) meddy expression in SSH. The results are expected to contribute to the development of a satellite-based system for monitoring meddies off the coast of Portugal.

2. Data Description and Processing Methods

In order to investigate surface signatures of meddies and their detectability with satellite-borne instruments, data obtained from a set of isobaric RAFOS floats deployed during A Mediterranean Undercurrent Seeding Experiment (AMUSE) (Bower et al. 1997) were used to provide in-situ evidence of meddies. AMUSE was aimed at the direct observation of meddy formation and the spreading pathways of MW into the North Atlantic. AMUSE also included an observational component based on the successful seeding, on a weekly basis, of a total of 49 RAFOS floats in the Mediterranean Undercurrent off southern Portugal, at depths of about 1100 m. Deploying the floats upstream of Cape St. Vincent allowed Bower et al. (1997) to confirm the hypothesis that the sharp change in the direction of the bathymetry around the cape induces a separation of the undercurrent and a subsequent shedding of meddies, a mechanism first proposed by D'Asaro (1988). The float data span a period of 21 months, from May 1993 to February 1995, with higher density of floats during the whole of 1994. All floats recorded temperature, pressure, and acoustic information every 8 hours. The duration of each reconstructed trajectory varied from 9 to 333 days. Details of data processing are described in Bower et al. (1997) and Hunt et al. (1998). Another contemporaneous RAFOS float (mdl37) from the Meddyphore experiment (Richardson et al. 1999) was also used to provide further in-situ evidence of a meddy in the study region.

To test the kinematic signature of meddies at the surface, the trajectory of a satellite-tracked surface drifter of the World Ocean Circulation Experiment/Tropical Oceans Global Atmosphere (WOCE/TOGA) type was deployed off the northern coast of Portugal during the Multidisciplinary Oceanographic Research in the Eastern Boundary of the North Atlantic (MORENA) project. This drifter was part of a set deployed in November 1993 to study the structure and variability of the Portugal coastal current system (Fiúza 1996). Several of these drifters were found to describe mesoscale-to-submesoscale anticyclonic trajectories after leaving the coastal zone (Martins 1997); one (no. 694) passed in the vicinity of a meddy identified with an AMUSE float on 1 February 1994.

High-resolution picture transmission (HRPT) data from the NOAA-11 and NOAA-12 satellites were received at the Space Oceanography Facility of the Instituto de Oceanografia. Data from the AVHRR instrument for 1994 were extracted from the HRPT archive and processed to create a database of IR images for the region covered by the trajectories of the surface drifter and RAFOS floats, extending from 36°–40°N, 8°–14°W.

The brightness temperature from AVHRR channel 4 (11.5–12.5 μm) was used as an estimate of the SST. This single-channel approach, used in studies concerned with SST patterns rather than actual temperature values (e.g., Emery et al. 1986; Frouin et al. 1990; Kahru et al. 1995), is based on the assumption that horizontal spatial scales of atmospheric moisture structure are much greater than those of SST features (Emery et al. 1986). This approach has the advantage of not introducing a noise amplification factor inherent to atmospheric correction algorithms that utilize combinations of different IR channels (Barton 1995). The noise in the SST image is thus kept at the 0.12°C level of the radiometric resolution of the single channel.

The images were navigated to within ±1 km (1 pixel) using satellite ephemeris data and adjusted interactively for timing and satellite attitude (roll, pitch, and yaw) errors. Clouds were identified using individually tuned thresholds for absolute values of AVHRR channels and their spatial variability in 3-pixel × 3-pixel boxes.

Color palette mapping, which was chosen to enhance the thermal structures of interest, shows warm areas in red and cold areas in blue. Cloudy areas and land are represented in white. In order to better delineate the relevant SST features of interest, track their evolution from image to image, and compare them with the trajectories of floats, gradient distributions were produced using an isotropic operator (Simpson 1990).

The positions of major fronts and eddy-like features were digitized from the visual superimposition of IR images and gradient distributions. The location and direction of the rotation of cyclonic and anticyclonic eddies found in the SST field were inferred from the wrapping of colder/warmer water around a warmer/colder core. To produce sketches representative of the eddies of major interest, ellipses were fitted to eddy structures using estimates of the location of their centers, and of their zonal and meridional boundaries as delineated by their thermal edges. Therefore, these sketches indicate estimates of the sizes of the eddies at the surface.

The T/P altimeter data used in this study span the year 1994 and cover the region of interest in the northeast Atlantic (34–41°N, 6–20°W); the data were supplied by the Collecte Localisation Satellites (CLS) Space Oceanography Division, Toulouse, France (AVISO 1997; Le Traon et al. 1995; Le Traon and Ogor 1998). T/P data included corrections for instrumental errors, tropospheric and ionospheric effects, surface waves, tidal influences, and the inverse barometer effect. Sea level anomalies (SLA) were generated from the corrected distance of the sea surface above the reference ellipsoid using a repeat-track analysis method. For a given pass of the satellite (one-half revolution), the corrected data were resampled every 7 km on a regular grid using cubic splines; a mean value for that pass (consisting of the 3-year mean from January 1993 to December 1995) was subtracted. Each SLA profile along the satellite pass was then low-pass filtered with a cutoff wavelength of 50 km to remove the short wavelength/high-frequency variability. The small repeat period of the T/P satellite results in ground tracks that are too coarsely spaced (2.8° or ~245 km at the mean latitude of the study area) to resolve the eddy field (Greenslade et al. 1997); an analysis based on SLA profiles was used instead of areal distributions.

With the aim of investigating the relationship between RAFOS trajectories and satellite data, float trajectories were divided into 11-day periods, which were then superimposed on contemporaneous IR images and SLA profiles. In this latter case, the intersections of the T/P groundtracks with the RAFOS trajectories were investigated to verify if the SLA anomaly was positive, i.e., consistent with an anticyclonic feature.

3. Results

The initial behavior of the majority of RAFOS floats after their deployment near 36.5°N, 8.5°W was to travel westward against the continental slope south of Portugal, while recording high temperatures (>12°C) characteristic of the Mediterranean Undercurrent. Reaching Cape St. Vincent, some of the floats began to perform clockwise circular motions, indicating that they were entrained in newly formed meddies. During AMUSE, nine floats registered six meddy formation events in the vicinity of Cape St. Vincent and three at the Estremadura Promontory (Bower et al. 1997). Furthermore, six other floats revealed the presence of meddies as they were caught in the periphery of these structures. Figure 2 presents the trajectories of the most representative meddy floats superimposed on the Levitus et al. (1994) climatological mean salinity distribution at 1000 m.

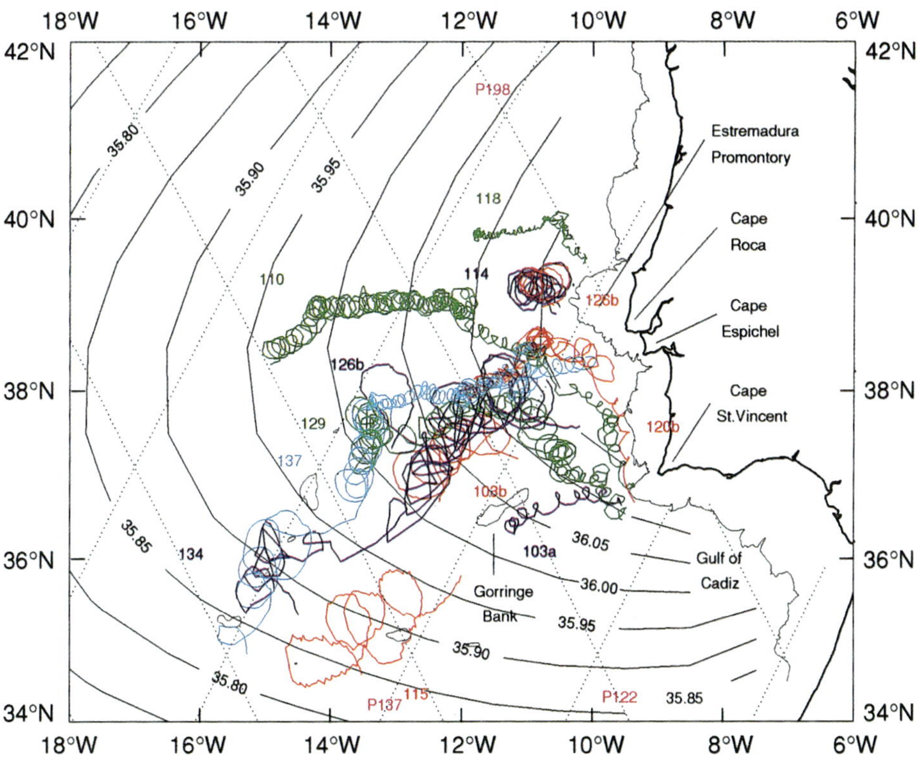

Figure 2. RAFOS float trajectories in the northeast Atlantic at ~1000-m depth, from AMUSE (11 floats) and Meddyphore (float 137) experiments, superimposed on the Levitus et al. (1994) mean climatological salinity distribution at that level. The 1000-m bathymetric contour (thin solid line) and the T/P groundtracks (dotted lines) are also shown.

Three different types of surface signatures of the meddy tracked by RAFOS float am118 are revealed through a satellite-tracked surface drifter trajectory (Figure 3), satellite altimetry (Figure 4), and satellite thermometry (Figure 5). This particular meddy, which was formed at the Estremadura Promontory and traveled northward close to the continental slope, began to move westwards into the open ocean after reaching 40°N (Figure 4b). RAFOS float am118 followed this meddy, which had a mean translation speed of 1.4 cm s^{-1} at approximately 1000 m during slightly more than four months.

In Figure 3, the trajectory of RAFOS float am118 between 3 December 1993 and 21 February 1994 is plotted in 20-day segments. The float performed anticyclonic loops

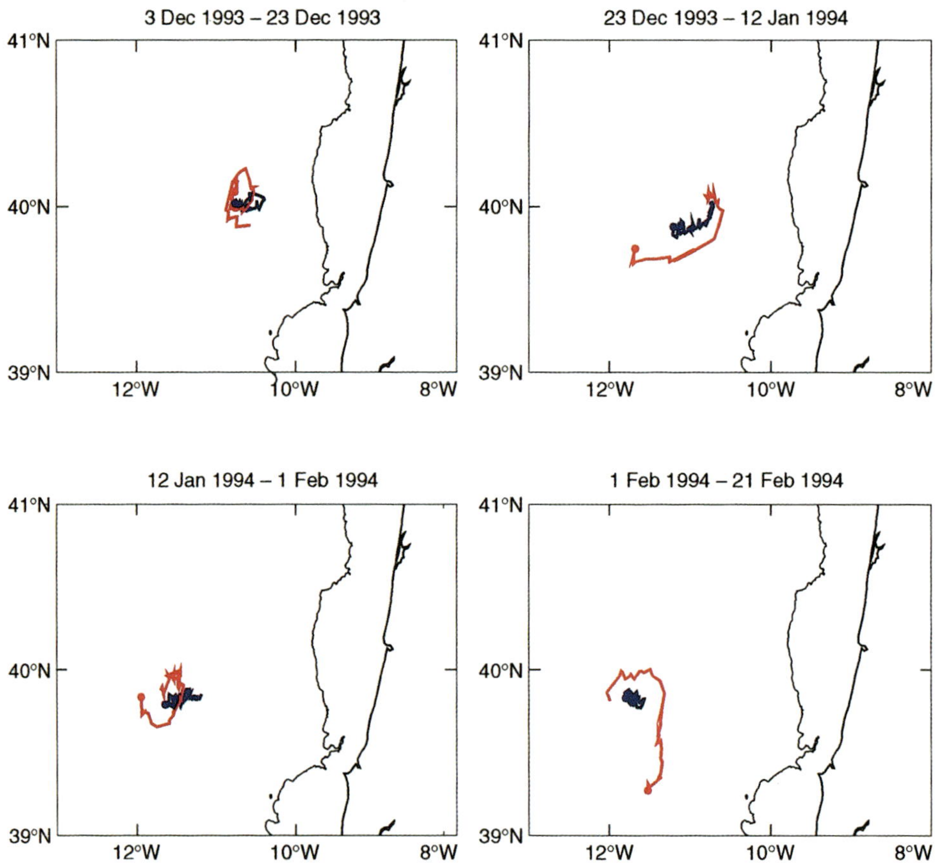

Figure 3. Simultaneous trajectories of RAFOS am118 (blue) and surface drifter 694 (red) during four sequential 20-day intervals, covering the period 3 December 1993–21 February 1994. The small dots indicate the last position in each 20-day segment of trajectory.

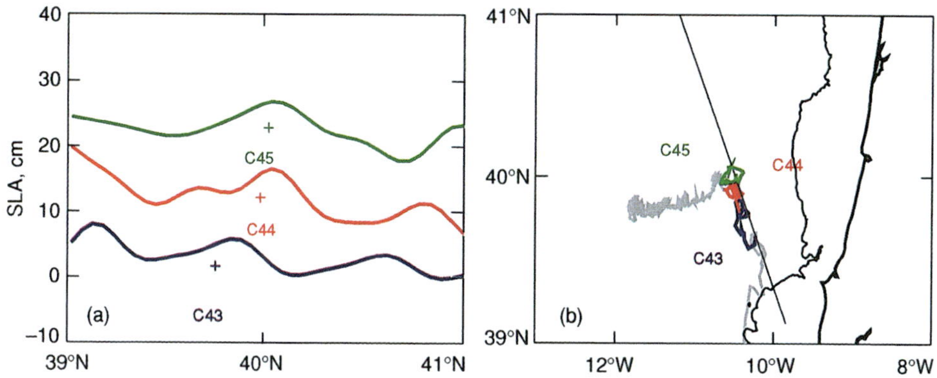

Figure 4. (a) Low-pass filtered (50-km cutoff wavelength) SLA profiles from T/P cycles C43 (blue), C44 (red), and C45 (green) during 13 November–12 December 1993. For clarity, each profile is offset by 10 cm from the previous one; a cross indicates the latitude of the center of the meddy on the day of each T/P pass as estimated from the RAFOS trajectory. (b) T/P ground track and the trajectory of RAFOS am118; the trajectory segments correspond to the three 10-day periods relative to T/P cycles C43, C44, and C45, and are color-coded as in (a).

with a radius of only 3 km and a period of rotation of 3 days while it moved to the west. The mean azimuthal velocity computed for this float (~8.5 cm s^{-1}) was consistent with solid-body rotation, showing that the float was situated near the center of the core of the meddy. While describing a trajectory compatible with a meddy-like vortex, the float measured a consistently high temperature (~11.9°C), another indication that it was located within the meddy's core. The kinematic signature of the meddy on the sea surface was revealed by the anticyclonic trajectory of a surface drifter (in red, Figure 3). Each anticyclonic loop was completed by the surface drifter in about 20 days; therefore, the rotation period of the meddy signature at the surface was considerably longer than the 3 days found for the solid-body rotation at mid-depth. The mean loop radius of 19 km at the surface and the computed mean azimuthal speed of 18 cm s^{-1} were not consistent with a solid-body rotation assumption.

From mid-November to mid-December 1993 the meddy crossed directly beneath a T/P groundtrack (Figure 4b), providing a good opportunity to detect an altimetric signal at the surface caused by the presence of the vortex. A positive anomaly, consistent with anticyclonic rotation, can be seen in Figure 4a. The SLA profiles for cycles 43–45, from 13 November to 12 December 1993, are depicted in colors that correspond to the piece of the float trajectory with the same color in Figure 4b.

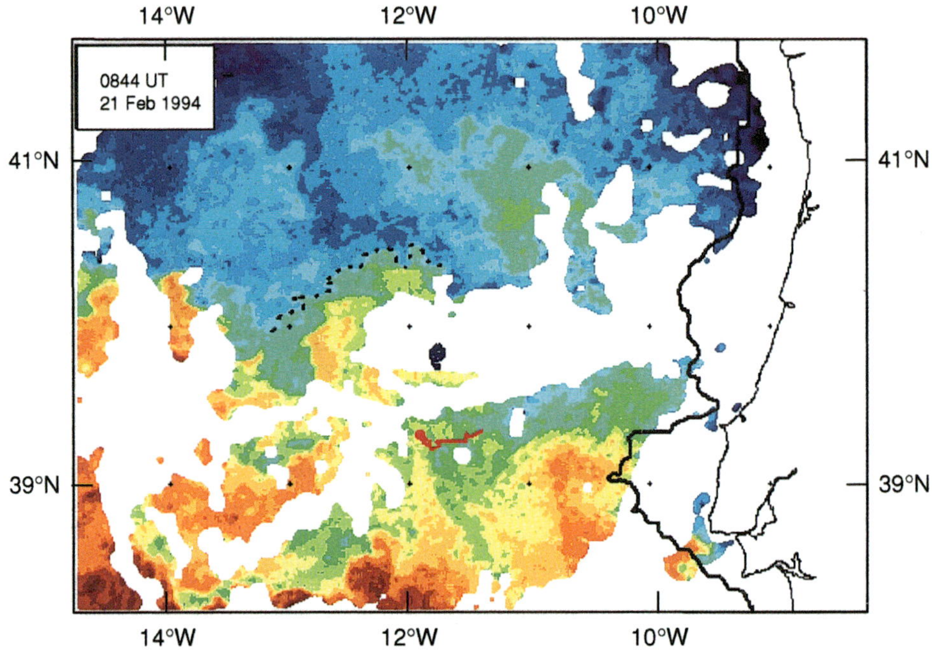

Figure 5. Infrared (AVHRR Channel 4) image from NOAA-12 at 0844 Universal Time (UT) 21 February 1994, with superimposed 11-day RAFOS am118 (blue) and surface drifter 694 (red) trajectories. The presence of a thermal front is made more visible by the black dots.

Only near the end of the period considered in Figure 3 was it possible to obtain a reasonably cloud-free AVHRR IR image that revealed the surface signature of the same meddy in the thermal field (Figure 5). Both the RAFOS (blue) and the drifter (red) trajectories for the 11-day period (16–26 February) are superimposed in the SST field. The IR image shows the presence of a thermal front (dotted line) delineating a circular pattern along the northern edge of the eddy.

3.1 Meddy signature on sea surface temperature

To investigate the signature of meddies and other features associated with the effect of the Mediterranean outflow on the surface thermal field, selected RAFOS trajectories were split into segments of 11-day periods centered on the image day (i.e., image date ±5 days) and then superimposed on selected SST images (illustrated in the left-hand sides of Figures 6 and 7). The right-hand sides of Figures 6 and 7 show the lines representing the relevant thermal fronts (red), the trajectories of the floats (blue), and the sketches of the major cyclonic (cyan) and anticyclonic (green) features identified on the surface thermal field. The 8°–14°W, 36°–40°N region includes the two sites for meddy formation proposed by Bower et al. (1997): Cape St. Vincent (37°N) and the Estremadura Promontory (39°N).

14 March 1994

A tongue of warm water occurred off the Portuguese west coast between 10° and 12°W on 14 March 1994 (Figure 6, top). The western boundary of the tongue extends northeastward from the southern limit of the image (35.5°N) to 39°N, exhibiting offshore protrusions at 37° and 38°N that are compatible with the presence of two anticyclonic eddies (A2 and A3). Anticyclones identified in the SST field are labeled "A" and cyclones "C," followed by an identification digit. Eddies A2 and A3 are centered at 37.3°N, 12°W and 38.0°N, 11.1°W and have diameters of ~90 km and ~35 km, respectively. The positions and apparent rotations of surface structures A2 and A3 are in agreement with the underlying meddies revealed by the trajectories of RAFOS floats am103b and mdl37, respectively. The western boundary of A2 is identifiable as a band of colder water surrounding a warmer core. A3 is the surface signature of the same meddy studied by Pingree (1995), which he named Pinball due to its irregular trajectory.

Southwest of Cape St. Vincent is a remarkable feature—a cool round patch with a diameter of 13 km bounded by a strong circular thermal front, suggesting the presence of a small cyclonic eddy (C1) located at the tip of a thin filament of cold water rooted near Cape St. Vincent. The trajectories of three RAFOS floats (am106b, am115, and am129) located near the northern segment of this eddy were nearest to C1 two days before the image date, when they all traveled in a quasi-zonal, westward direction in agreement with the expected circulation associated with the northern edge of a cyclone.

24 March 1994

The unusually cloud-free image on 24 March 1994 (Figure 6, center) shows further northward progression of warmer waters in most of the offshore region, co-existing with cold waters on the shelf and along the upper slope. The distribution of the offshore colder waters between 37° and 39°N has the shape of an elongated *S*, with its axis oriented in a northeast-southwest direction. This configuration appears to be associated with surface signatures A2 and A3 of the meddies tracked by RAFOS floats am103b and mdl37, respectively. In addition to the migration of the meddies in opposite directions along the *S* axis, other differences between the SST signatures on 14 and 24 March are: (1) replacement of colder by warmer waters around the western edge of anticyclone A2, which apparently was advected from the Gorringe Bank area; (2) a clearer surface signature of meddy Pinball (A3) by a distinct warm core almost completely enclosed by colder waters.

The northern boundary of the filament-like, warm surface waters extends offshore from the 1000-m bathymetric contour near the Estremadura Promontory to at least 12.5°W. It also shows a remarkable correspondence with the trajectories of RAFOS floats am109 and am119, as if these floats were tracking the surface flow. Floats am114 and am126b rotated clockwise, following a circular path centered at 39.2°N, 10.8°W within the meddy whose surface expression (A4) seems to be responsible for advection of cold shelf waters westwards along its southern boundary at 39.1°N.

The thermal field is characterized by a continuous band of cold water along the coast and extending offshore to 10.5°W. The relationship between RAFOS float trajectories in this region and the surface thermal field ranges from reasonably good agreement (e.g., RAFOS am107 travels eastward along a surface thermal front at 37.3°N) to complete disagreement (e.g., RAFOS am135b travels across the cold water branch at 38°N). Between these two extreme cases is the possible signature of a recently formed meddy tracked by RAFOS am129 that is rotating clockwise beneath the offshore edge of the 36.9°N filament.

Southwest of the circular edge of the filament off St. Vincent, the small cyclonic eddy identified on March 14 can still be seen on 24 March. During the ten-day period separating the two images, this cyclone moved 40 km to the west of its original position; it maintained roughly the same diameter and is surrounded by weaker gradients.

8 April 1994

A reasonably cloud-free SST image was obtained on 8 April 1994 (Figure 6, bottom), 15 days after the previous one. The 8 April image suggests a typical upwelling situation, with a southward jet of cold coastal waters in the south, particularly evident after its separation from the coast near Cape St. Vincent (Fiúza 1983, 1996).

North of 38°N, the trajectories of RAFOS floats am114, am126b, and md137 show the continued presence of two meddies. The northernmost meddy, which was tracked by floats am114 and am126b at average depths of 846 m and 1196 m, respectively, appears to be slightly shifted relative to the position of the corresponding anticyclonic surface flow inferred from the SST pattern (A4). The trajectory of float am114 crossed a well-defined surface front perpendicularly, in a southwesterly direction, from colder to warmer waters, one day before the image date. Float am114 left the warmer side two days later,

Figure 6 (facing page). SST images (AVHRR Channel 4) and sketches of major features on (top) 14 March, (middle) 24 March, and (bottom) 8 April 1994, with superimposed 11-day trajectories of RAFOS floats centered at the dates of each image. Small dots indicate the last position in each trajectory. Major cyclones and anticyclones are represented as cyan and green blobs, respectively, in corresponding sketches. Major thermal fronts are marked in red. Anticyclones are labeled "A" and cyclones "C," each followed by an identification digit that increases from south to north. Note that the first two letters of a float's identification were dropped to minimize label space.

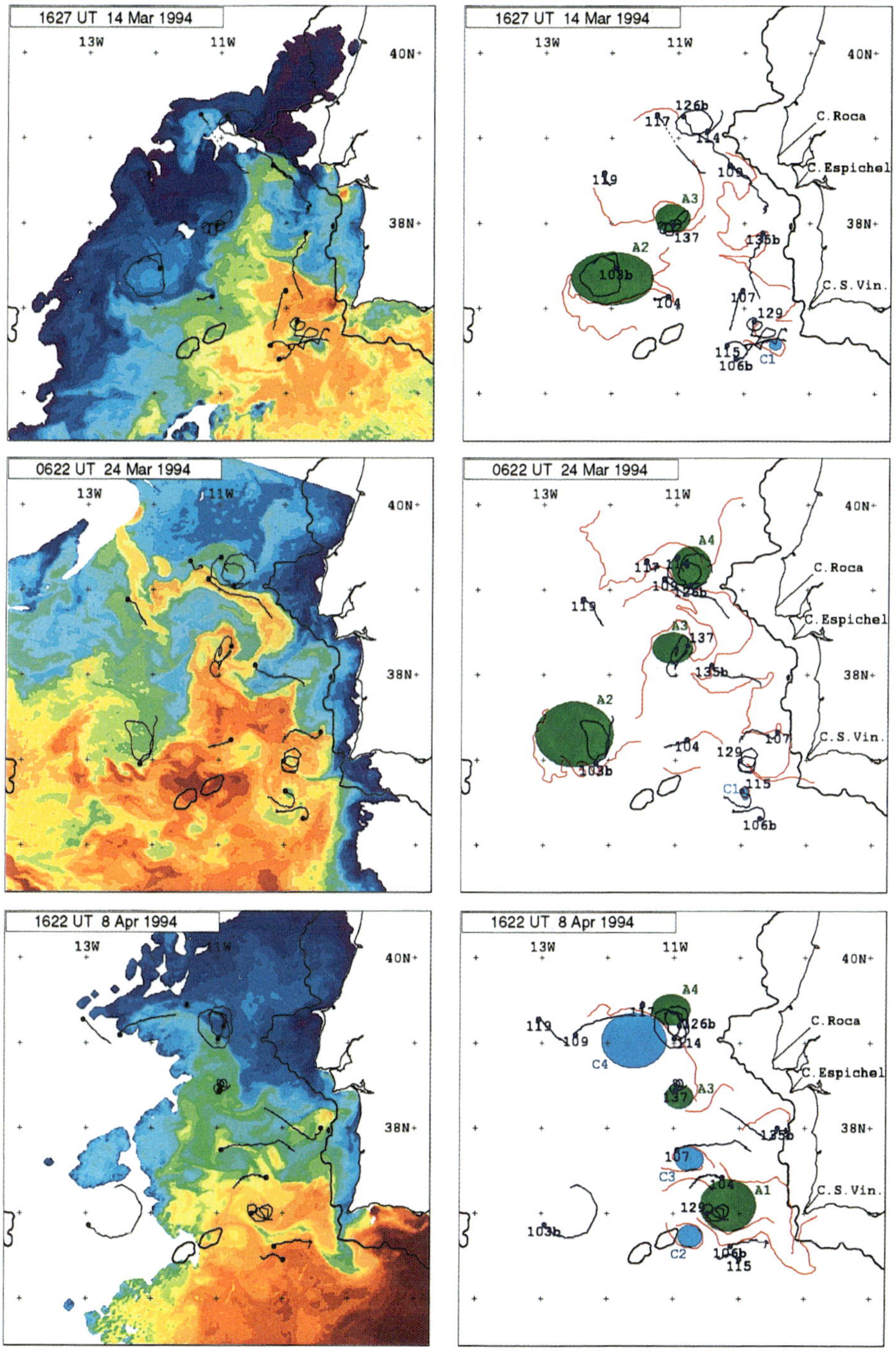
1627 UT 14 Mar 1994
13W
11W
40N
38N
C.Roca
C.Espichel
C.S.Vin.
126b
117
114
109
119
A3
137
A2
135b
103b
104
107
129
115
106b
C1
0622 UT 24 Mar 1994
A4
126b
C3
1622 UT 8 Apr 1994
C4
A1
C2

while float am126b remained for 11 days in the cold side. On 8 April 1994, both floats had velocities parallel to the surface front. The shape of the northwest extension of the SST front indicates the presence of cyclone C4, which is in agreement with the trajectory of float am109. The location of the surface anticyclone A3, in the vicinity of meddy Pinball (float mdl37), was inferred from the southward extension of the SST front at ~30 km to the east of the meddy center, and from the circular-shaped front ~50 km to the southeast of the meddy center.

South of 38°N there are two SST features that can be related to the mid-depth flow described by the RAFOS trajectories: (1) cold-core cyclone C3 bounded by curved fronts, which may correspond to the cyclonic track shown at depth by the trajectory of float am107; (2) warm-core anticyclone A1 revealed by the curvature of the northern edge of the thin, wavy filament at the southern boundary of the meddy tracked by floats am129 and am104. An interesting feature revealed only in the SST field was the cold core cyclone C2 with a radius of approximately 15 km, whose center is located about 45 km to the southwest of the center of the meddy tracked by float am129.

12 April 1994

One of the most striking features observed in the SST image of 12 April 1994 (Figure 7, top) is the excellent agreement between the trajectory of RAFOS am103b and the circular shape of surface front A2. At the same latitude (37°N) but approximately 200 km to the east, the meddy revealed by float aml29 and the curved trajectories of floats am104 and am106b performed several clockwise loops during the 11-day period. The SST signature of this meddy (A1) became more evident due to the presence of the cold water filament rooted at the coast near 37.8°N, which allows a clear identification of the eastern boundary of the meddy's surface signature. As in the 8 April SST image, cyclones C2 and C3 can be identified in the SST field in the vicinity of A1. The northern cyclone C3 has a diameter of about 25 km; its core is located 75 km to the northwest of the meddy core, as estimated from RAFOS am129. The southern cyclone C2 had moved about 15 km northwestward, occupying a position approximately 50 km to the southwest of the meddy, thus slightly increasing its distance from the meddy core.

Between the latitude of Cape Espichel and 38°N, the thermal field reveals the presence of a cold water filament extending southwestward ~150 km from the coast to about 11°W and manifesting the mushroom-like shape characteristic of an eddy pair whose anticyclonic part is A3. Float mdl37, tracking meddy Pinball, evolved underneath this anticyclone.

The 12 April IR image indicates the presence of cyclone C4 centered at 39.1°N, 11.9°W with an estimated diameter of 80 km. The counterclockwise rotation inferred from the SST field is in agreement with the trajectories of RAFOS floats am109 and am117, exhibiting a cyclonic motion. The weak SST gradients in the area prevent an unambiguous tracking of C4. However, comparison of the 8 and 12 April SST images suggests a slight progression

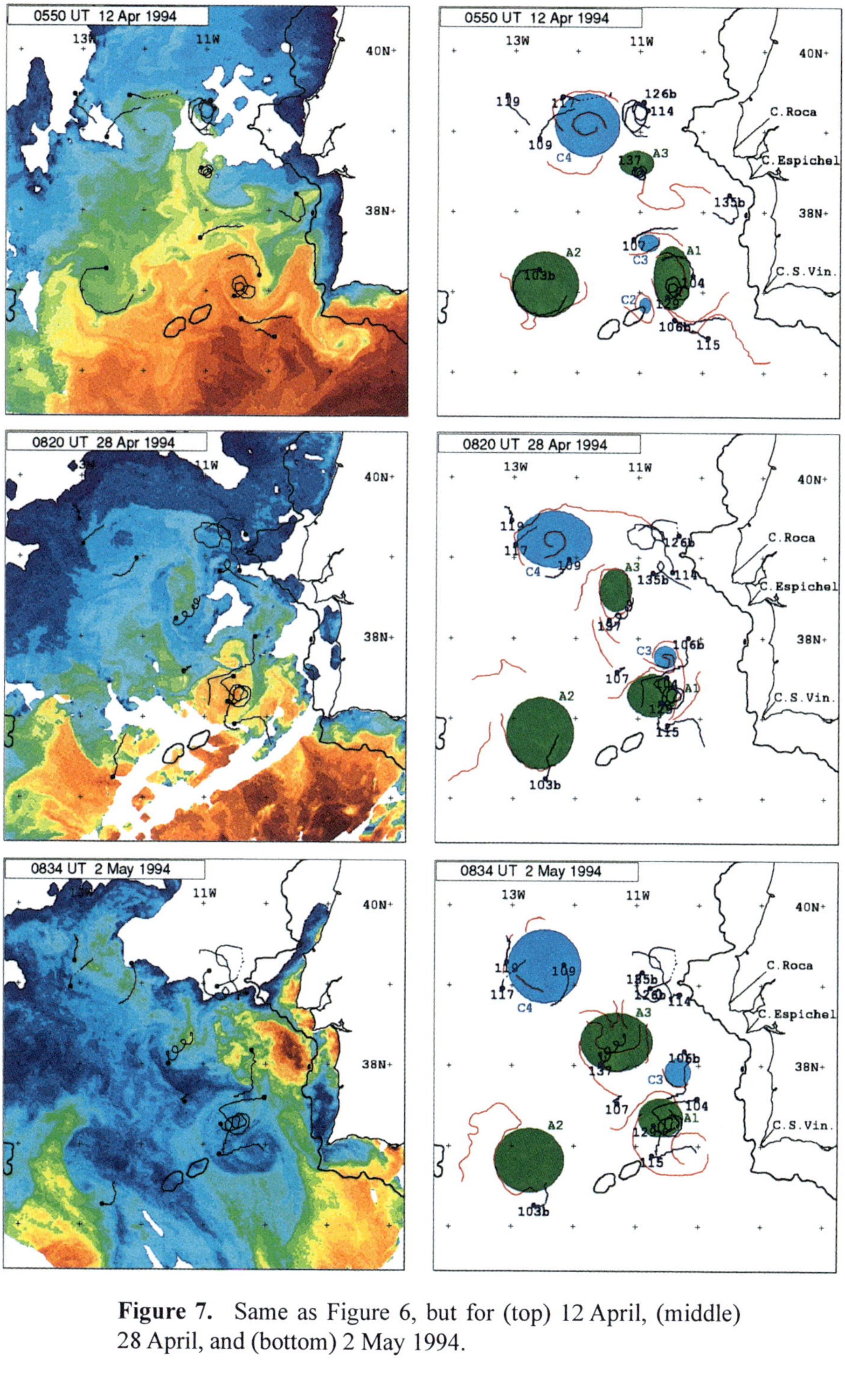

Figure 7. Same as Figure 6, but for (top) 12 April, (middle) 28 April, and (bottom) 2 May 1994.

of the cyclone toward the northeast. This is in good agreement with the trajectory of float am117, which started a westward movement after a period of almost no net displacement.

28 April 1994

The SST distribution on 28 April 1994 (Figure 7, center) suggests a relaxation of upwelling conditions, given the weakening of the signature of the cold water filaments extending offshore from the coast. There are also surface thermal signatures of the five eddies identified on 12 April: three anticyclones (A1, A2, and A3) and two cyclones (C3 and C4). All these eddy-like surface features correspond to similarly shaped and equally directed trajectories of subsurface floats, indicating a remarkable similarity of the circulation at the ocean surface and at the level of MW for both types of vortices.

North of 38°N, the thermal field is dominated by the two structures already identified on 8 and 12 April: the anticyclonic surface signature of meddy Pinball (A3) and of the cyclone (C4) associated with trajectories of floats am109 and am117. Comparison of the 12 and 28 April SST images indicates that this cyclone moved ~40 km to the west in 16 days. The meddy surface expression A3 is characterized by an oval warm water patch surrounded by colder waters advected from the coast; it is shifted ~35 km to the north of the meddy core, estimated from the trajectory of RAFOS float mdl37. The mdl37 trajectory also indicates a relatively fast southwesterly progression of the meddy during 23 April–3 May, contrasting with its relative immobility during the previous period.

South of 38°N on the 28 April SST image is the clearly defined surface cyclone C3, with a radius of about 20 km, located at a distance 45 km north of the core of the meddy estimated from the trajectories of floats am129 and am104. The 8, 12, and 28 April images show that this cyclone traveled eastward with a translation speed of 2.3 cm s^{-1}; this is consistent with the trajectory of float am106b at a mean depth of 1100 m. The meddy SST signature A1 is a warm water pool with ~80-km diameter, centered to the west of the meddy core position estimated from the trajectories of floats aml29 and aml04. The 28 April image also reveals the existence of a strong, rounded front west of Gorringe Bank (see feature A2 in Figure 2), which probably is the surface thermal signature of the meddy previously tracked by float am103b, although this float followed an almost rectilinear path to the south during 23 April–3 May.

2 May 1994

The most clearly identifiable SST front on 2 May 1994 (Figure 7, bottom) is located west of Gorringe Bank. This front demonstrates the persistence of anticyclone A2, which is likely to represent the surface signature of the meddy previously revealed by RAFOS float am103b. East of Gorringe Bank is a mushroom-shaped feature whose axis runs in a northeast-southwest direction. The trajectories of floats aml29 and am104 seem to correspond to anticyclone A1 of this eddy pair, whereas the trajectory of float am115 is somewhat compatible with the cyclone. The A3 surface anticyclone corresponding to meddy Pinball exhibits a more circular shape than before. No shift is found between the centers of the trajectory of float mdl37 and the surface signature of the meddy, which continued its southwestward progression.

North of 39°N and west of 12°W, slightly colder water "bounded" by trajectories of floats am109, am117, and am119 confirmed the presence of cyclone C4 as the one observed on 28 April. The float-determined westward speed of 2.9 cm s^{-1} was similar to that computed from the comparison of the 12 and 28 April SST images.

3.2 Meddy signatures on sea surface topography

To further test the signature of meddies in the SSH measured with the T/P altimeter, individual SLA profiles were selected relative to all periods when the satellite ground tracks crossed the trajectory of a meddy. Figure 8a shows low-pass filtered SLA profiles corresponding to the one-month period from 11 February to 11 March 1994, when the satellite passed over the meddy tracked by RAFOS float am103b (Figure 8b).

In order to quantify the SLA associated with the presence of a meddy, a SLA* is defined as the difference between the SLA at the zero-crossing of the first derivative of the SLA profile at the center of the meddy and the average SLA at the neighboring zero-crossings of the same first derivative, one to the left of the center and another to the right. An estimate of the horizontal extent ("diameter") for the vortex is the distance between

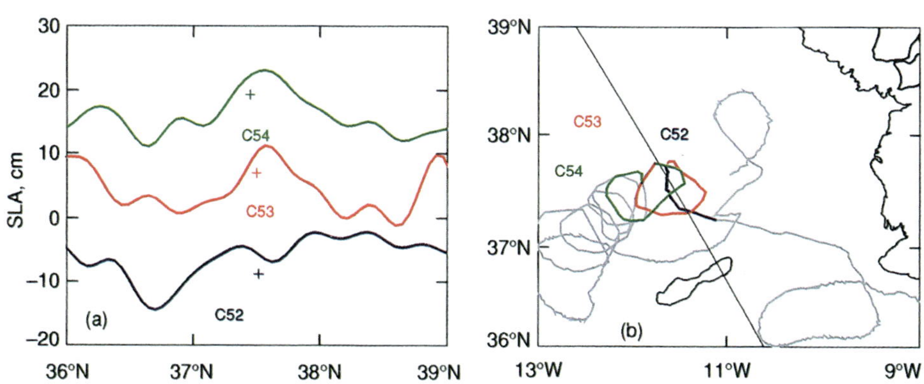

Figure 8. (a) Low-pass filtered SLA profiles from T/P ground track 122 for cycles 52 (blue), 53 (red), and 54 (green) (11 February–11 March 1994). Each profile is offset by 10 cm from the previous on. A cross indicates the latitude of the center of the meddy on the day of each pass, as estimated from the RAFOS float trajectory. (b) T/P ground track and trajectory of RAFOS am103b. The trajectory segments correspond to the three 10-day periods relative to T/P cycles 52–54 and are color-coded as in (a).

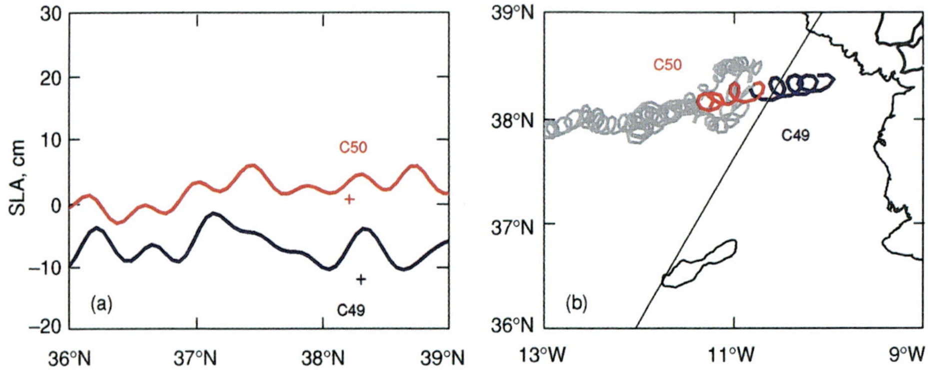

Figure 9. Same as Figure 8, but from T/P ground track 137 for cycles 49 (blue) and 50 (red) (12 January–30 January 1994) and the trajectory of RAFOS md137.

the outer zero-crossings. The most intense SLA* corresponded to T/P cycle 53 for the case of float am103b and was about 11 cm. The diameter of the meddy signature at the surface was about 145 km.

Figure 9 presents similar results for T/P pass 137 during cycles 49 and 50 (12–30 January 1994), corresponding to the case of meddy Pinball, which was followed by float mdl37. During cycle 49, SLA* reached 7 cm; the distance from edge to edge was 67 km.

Figure 10 presents the case of the meddy tracked by float am129. During T/P cycles 61–63 (11 May–8 June 1994), this meddy traveled to the northwest, crossing T/P ground track 137. SLA* reached its highest value (12 cm) in cycle 62 when the meddy was directly below the path of T/P. The horizontal extent of the signature was 111 km.

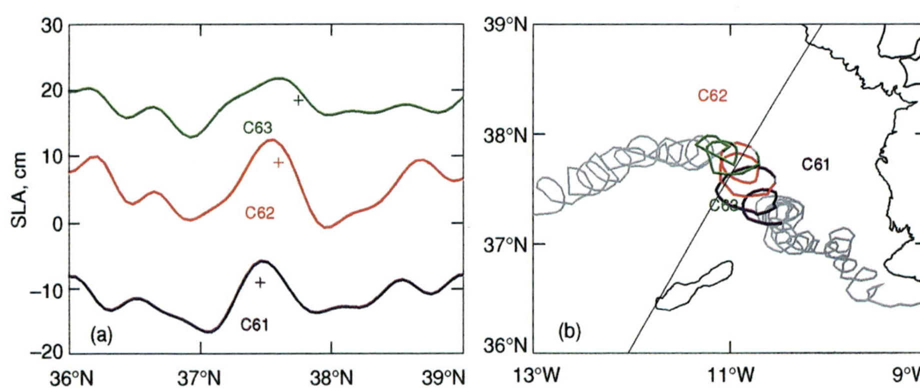

Figure 10. Same as Figure 8, but from T/P ground track 137 for cycles 61 (blue), 62 (red), and 63 (green) (11 May–8 June 1994) and the trajectory of RAFOS am129.

The cross-track component of the geostrophic velocity anomaly was computed from the SLA profiles. At a half-radius distance, surface geostrophic velocity anomalies were 17 cm s^{-1} for the meddy followed by float am103b (cycle 53), 23 cm s^{-1} for the meddy tracked by float mdl37 (cycle 49), and 24 cm s^{-1} for the meddy followed by float am129 (cycle 62).

The SLA* and the diameter of the meddy portrayed in Figure 4 were 8 cm and 78 km, respectively. The geostrophic velocity anomaly was 22 cm s^{-1}, which is close to the value (18 cm s^{-1}) obtained with surface drifter 694 at almost the same distance from the meddy center.

The meddy that stayed more or less stationary near the northwest edge of the Estremadura Promontory (float am114) was located in the middle of the diamond-shaped cell formed by T/P tracks (Figure 2) and, therefore, was not detected by the altimeter.

Figures 4, 8, 9, and 10 demonstrate that whenever meddies were overflown by the T/P satellite, a positive SLA* was consistently recorded by the altimeter.

4. Discussion

Analysis of RAFOS trajectories shows that during 49 days in early spring 1994, four meddies were present simultaneously in an area of only 330 km × 350 km off the southwest coast of Portugal. A sequence of 6 IR images during the same period showed that all meddies had a surface signature detectable in the SST field at some stage in their lifetimes. In addition, the meddies always produced a positive SLA*. The SST signature of a meddy frequently was associated with a cyclonic feature in its vicinity; some cyclones appeared to be related to counterclockwise rotation of RAFOS floats. A sequence of six SST images (Figures 6 and 7) illustrates the development of the surface expression of cyclones in the vicinity of meddies: (1) cyclone C4 near the meddy tracked by float am114; (2) cyclones C1, C2, and C3 near the meddy tracked by float am129. Interpretation of cyclones and anticyclones identified in the SST field as manifestations of subsurface vortices with similar shapes and rotations provides clues to the movement of meddies. Figure 11 presents the sequence of positions of the ellipses fitted to the signatures of the meddies and neighboring cyclones on the surface thermal field.

Two examples of eddy-eddy interaction are described. The meddy tracked by float am129 (with SST signature A1) moved away from Cape St. Vincent, towards the northwest. On 14 March (Figures 6 and 11) cyclone C1 was present near the southern edge of A1. On 24 March, when the meddy had stopped its northwestward movement, the cyclone at its southern edge was still identifiable in the IR image. It is tempting to speculate that retardation of the meddy might be related to the presence of an eddy (cyclonic or anticyclonic) at the northern edge of the meddy. During 24 March–8 April this meddy moved towards the west. Throughout the period, cyclone C2 was already present near the southwest edge of the meddy, forming a westward-moving dipole.

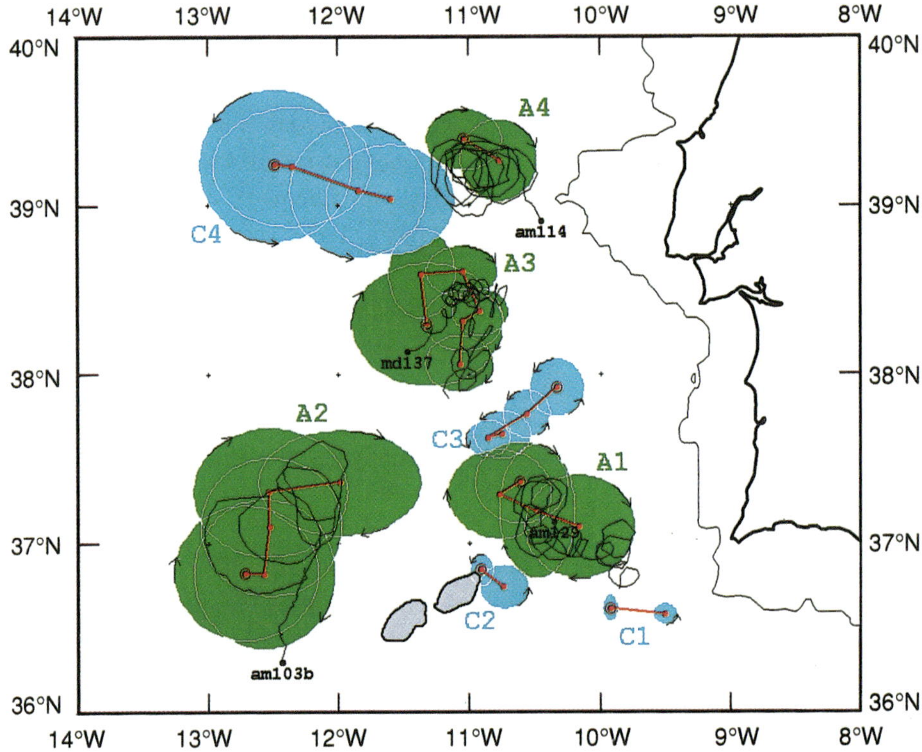

Figure 11. Evolution of the eddy-like surface features identified in the SST fields above, or in the vicinity of, meddies: anticyclones are colored in green and cyclones in cyan. The centers of the features are marked with red dots and are sequentially connected with red lines; the last position is indicated with a small dot and a concentric circumference. Superimposed are trajectories of RAFOS floats am103b, am129, md137, and am114, which tracked meddies during the study period (14 March–2 May 1994).

Meddy-cyclone interactions can be explained using classical results on vortex interactions, which have the following features (Lamb 1932): (1) two vortices separated by some distance will rotate around a common center, lying on the line passing through their centers, with an angular velocity directly proportional to the sum of their strengths and inversely proportional to their separating distance; (2) the common rotation center is located between the centers of the vortices if they have the same rotation sense, but outside if they are counter-rotating; (3) in particular, if the two vortices are counter-rotating and have equal strength, the common rotation center will be at infinity and the pair will move along straight lines perpendicular to the axis connecting their centers. On 12 April

(Figures 7 and 11) cyclone C2 appears to have moved a little to the northwest, probably contributing to a slight change in the direction of displacement of the meddy associated with A1. But the meddy stalled during 12–28 April, presumably because cyclone C3 blocked its northward progression. The 2 May SST image indicated that the meddy restarted its movement, probably under the interaction of the cyclone located at its southeast edge (not shown in Figure 11 due to its superimposition with the meddy signature at earlier dates; see Figure 7, bottom).

In the second case, the meddy tracked by float am114 (A4) probably formed near the Estremadura Promontory and remained stationary for several months adjacent to the slope. Bower et al. (1997) suggested that this unusual position could have important implications for the continuity of the Mediterranean Undercurrent around the Estremadura Promontory. On 14 March this meddy was near the slope, while meddy Pinball (A3) moved to the northeast; between these two meddies was a patch of colder water. We hypothesize that cyclone C4 was already present at depth under the colder water and became visible in the 8 April SST image near the southwest edge of the Promontory meddy revealed by float am114 (A4). This would explain the movement of meddy Pinball to the northeast where Pinball was close enough to the Promontory meddy to stop its movement (see Figures 6, 7, and 11). Then, when cyclone C4 moved far enough to the west, the interaction between the three eddies stopped and Pinball started to move southwestwards. At this stage, another meddy was probably forming near the Promontory (see float am135b on 28 April and 2 May, Figure 7) to disrupt the interaction between the three pre-existing eddies.

5. Conclusions

Analysis of the relationships between anticyclonic trajectories at mid-depths and contemporaneous SST fields has shown that, generally, SST circular patterns correspond to float loops, although some RAFOS float trajectories crossed surface thermal structures and others evolved underneath homogeneous SST fields. These results lead to the conclusion that meddies only have a surface thermal signature when there are SST gradients. During the 49-day period covered by the present research, 14 March–2 May 1994, SST gradients were associated with the northward flow of warm water between the cold water on the shelf and offshore waters that occur during winter and early spring (Frouin et al. 1990), and with the offshore progression of cool jets and filaments resulting from coastal upwelling (Fiúza 1983, 1996). Typically, there is a better expression of the meddies on the surface thermal field during the winter/spring northward surface current and during the relaxation phase of coastal upwelling. A positive sea level anomaly of about 10 cm, consistent with anticyclonic rotation, existed whenever T/P flew directly above a meddy. The main limitations of T/P data for identifying and tracking meddies are the large separation between satellite tracks and the low repeatability of the orbits. Furthermore, the high meddy translation speeds lead to an underestimation of the number of meddies in each satellite cycle. Finally, there is the problem of distinguishing positive sea level anomalies associated with meddies from those related to surface eddies.

Evidence was also found that important aspects of meddy dynamics, such as their association with submesoscale cyclones assuming a dipolar pattern, are revealed in surface thermal imagery. Analysis of a sequence of IR images illustrated that irregular speeds of meddies, a feature commonly observed with floats, could be attributed to vortex interaction. This idea, first put forward by Armi et al. (1989), is strongly supported by our results.

In conclusion, although it is not possible to unambiguously identify meddies from only remotely sensed data, satellite infrared and altimetry data can be very useful in the study of MW eddies. The combined analysis of IR and altimetry data, and knowledge of meddy generation sites and their preferred trajectories in the early stages of their lifetimes, will hopefully provide the background information needed to establish a satellite-based system for monitoring meddies and provide estimates of meddy population and generation rates.

Acknowledgments. This work was supported mainly by the European Union MAST-3 CANIGO Project (Contract no. MAS3-CT96-0060). The establishment and development of the NOAA/HRPT receiving station at the Institute of Oceanography in Lisbon was funded under the PO-SATOCEAN Project (NATO Science for Stability Programme) and, in part, under the European Union MAST-2 MORENA Project (Contract no. MAS2-CT93-0065). The SLA products were supplied by the CLS Space Oceanography Division, Toulouse, France (AVISO/Altimetry), with financial support from the CEO (Centre for Earth Observation) programme and the Midi-Pyrénées regional council. The AMUSE Project was funded by the National Science Foundation through grant OCE-9101033 to the Woods Hole Oceanographic Institution, grant OCE-9100724 to the Scripps Institution of Oceanography, and by the Luso-American Foundation for Development through grant 54/93 to the University of Lisbon. Data from float mdl37 were retrieved from the WOCE Subsurface Float Data Assembly Center at the Woods Hole Oceanographic Institution. We thank the two anonymous reviewers for their thoughtful and constructive comments on this paper.

References

Arhan, M., A., Colin de Verdière, and L. Mémery, The eastern boundary of the subtropical North Atlantic, *J. Phys. Oceanogr.*, *24*, 1295–1316, 1994.

Armi, L., and H. Stommel, Four views of a portion of the North Atlantic subtropical gyre, *J. Phys. Oceanogr.*, *13*, 828–857, 1983.

Armi, L., D. Hebert, N. Oakley, J. Price, T. Rossby, and B. Ruddick, Two years in the life of a Mediterranean salt lens, *J. Phys. Oceanogr.*, *19*, 354–370, 1989.

AVISO, *Archiving, Validation, and Interpretation of Satellite Oceanographic Data, AVISO user handbook: Sea level anomalies (SLA),* 2nd ed., Handb. AVI-NT-011-312.CN, Toulouse, France, 1997.

Barton, I. J., Satellite-derived sea surface temperatures: current status, *J. Geophys. Res.*, *100*, 8777–8790, 1995.

Bower, A., L. Armi, and I. Ambar, Lagrangian observations of meddy formation during A Mediterranean Undercurrent Seeding Experiment, *J. Phys. Oceanogr.*, *27*, 2445–2575, 1997.

Cheney, R., and J. Marsh, Seasat altimeter observations of dynamic topography in the Gulf Stream region, *J. Geophys. Res.*, *86*, 473–483, 1981.

D'Asaro, E., Generation of submesoscale vortices: A new mechanism, *J. Geophys. Res.*, *93*, 6685–6693, 1988.

DeRycke, R., and P. Rao, Eddies along a Gulf Stream boundary viewed from a very high resolution radiometer, *J. Phys. Oceanogr.*, *3*, 490–493, 1973.

Dickson, R., and D. Hughes, Satellite evidence of mesoscale eddy activity over the Biscay abyssal plain, *Oceanol. Acta*, *4*, 43–46, 1981.

Emery, W., A. Thomas, M. Collins, W. Crawford, and D. Mackas, An objective method for computing advective surface velocities from sequential infrared satellite images, *J. Geophys. Res.*, *91*, 12865–12878, 1986.

Fiúza, A., Upwelling patterns off Portugal, In *Coastal Upwelling: Its Sediment Record,* edited by E. Suess and J. Thiede, Plenum, New York, 85–98, 1983.

Fiúza, A., Mesoscale and submesoscale shelf-ocean exchange processes off western Iberia, *MORENA Scientific and Technical Report*, *39*, Instituto de Oceanografia, Universidade de Lisboa, Lisbon, 36 pp, 1996.

Frouin, R., A. Fiúza, I. Ambar, and T. Boyd, Observations of a poleward current off the coasts of Portugal and Spain during winter, *J. Geophys. Res.*, *95*, 679–691, 1990.

Greenslade, D., D. Chelton, and M. Schlax, The midlatitude resolution capability of sea level fields constructed from single and multiple altimeter datasets, *J. Atmos. Oceanic Tech.*, *14*, 849–870, 1997.

Huang, N., C. Leitao, and C. Parra, Large-scale Gulf Stream frontal study using GEOS 3 radar altimeter data, *J. Geophys. Res.*, *83*, 4673–4682, 1978.

Hunt, H., C. Wooding, C. C. L., and A. Bower, A Mediterranean Undercurrent Seeding Experiment (AMUSE), Part II: RAFOS float data report May 1993–March 1995, *Technical report WHOI-98-14,* Woods Hole Oceanographic Institution, 1998.

Kahru, M., B. Häkansson, and O. Rud, Distributions of the sea-surface temperature fronts in the Baltic Sea as derived from satellite imagery, *Cont. Shelf Res.*, *15*, 663–679, 1995.

Käse, R., and W. Zenk, Reconstructed Mediterranean salt lens trajectories, *J. Phys. Oceanogr.*, *17*, 158–163, 1987.

Lamb, H., *Hydrodynamics,* 6th ed., 738 pp, Dover, New York, 1932.

Le Traon, P., and F. Ogor, ERS-1/2 orbit improvement using TOPEX/Poseidon: The 2-cm challenge data, *J. Geophys. Res.*, *103*, 8045–8057, 1998.

Le Traon, P., P. Gaspar, F. Bouyssel, and H. Makhmara, Using TOPEX/Poseidon data to enhance ERS-1 data, *J. Atmos. Oceanic Tech.*, *12*, 161–170, 1995.

Levitus, S., R. Burgett, and T. Boyer, World Ocean Atlas 1994, 3: Salinity, NOAA/NEDIS Atlas 3, U.S. Department of Commerce, Washington, D.C., 1994.

Marshall, J., Submarine salt lenses, *Nature*, *333*, 594–596, 1988.

Martins, C. S., Estudo da Circulação Oceânica Superficial no Atlântico Nordeste Utilizando Bóias Derivantes com Telemetria por Satélite, Ph.D. thesis, Instituto de Oceanografia, Faculdade de Ciências da Universidade de Lisboa, Lisbon, 1997.

Mazé, J., M. Arhan, and H. Mercier, Volume budget of the eastern boundary layer off the Iberian Peninsula, *Deep-Sea Res.*, *44*, 1543–1574, 1997.

McDowell, S., and H. Rossby, Mediterranean Water: An intense mesoscale eddy off the Bahamas, *Science*, *202*, 1085–1087, 1978.

Needler, G., and R. Heath, Diffusion coefficients calculated from the Mediterranean salinity anomaly in the North Atlantic Ocean, *J. Phys. Oceanogr.*, *5*, 173–182, 1975.

Pingree, R., The droguing of meddy Pinball and seeding with ALACE floats, *J. Mar. Biol. Assoc. U.K.*, *75*, 235–252, 1995.

Pingree, R., and B. Le Cann, Three anticyclonic Slope Water Oceanic Eddies (SWODDIES) in the southern Bay of Biscay in 1990, *Deep-Sea Res.*, *39*, 1147–1175, 1992.

Pingree, R., and B. Le Cann, A shallow meddy (a smeddy) from the secondary mediterranean salinity maximum, *J. Geophys. Res.*, *98*, 20169–20185, 1993.

Prater, M., and T. Sanford, A meddy off Cape St. Vincent, Part I: Description, *J. Phys. Oceanogr.*, *24*, 1572–1586, 1994.

Rao, P., A. Strong, and R. Koffler, Gulf Stream meanders and eddies as seen in satellite infrared imagery, *J. Phys. Oceanogr.*, *3*, 237–239, 1971.

Richardson, P., D. Walsh, L. Armi, M. Schröder, and J. Price, Tracking three meddies with SOFAR floats, *J. Phys. Oceanogr.*, *19*, 371–383, 1989.

Richardson, P., A. Bower, and W. Zenk, A census of meddies tracked by floats, *Prog. Oceanogr.*, in press, 1999.

Robinson, A., Overview and summary of eddy science, In *Eddies in Marine Science,* edited by A. Robinson, Springer-Verlag, Berlin, 3–15, 1983.

Rossby, T., D. Dorson, and J. Fontaine, The RAFOS system, *J. Atmos. Oceanic Tech.*, *3*, 672–679, 1986.

Schultz Tokos, K., H.-H. Hinrichsen, and W. Zenk, Merging and migration of two meddies, *J. Phys. Oceanogr.*, *24*, 2129–2141, 1994.

Simpson, J., On the accurate detection and enhancement of oceanic features observed in satellite data, *Remote Sensing Environ.*, *33*, 17–33, 1990.

Stammer, D., H.-H. Hinrichsen, and R. Käse, Can meddies be detected by satellite altimetry?, *J. Geophys. Res.*, *96*, 7005–7014, 1991.

Stumpf, H., and P. Rao, Evolution of Gulf Stream eddies as seen in satellite infrared imagery, *J. Phys. Oceanogr.*, *5*, 388–393, 1975.

Wunsch, C., Low frequency variability of the sea, *Evolution of Physical Oceanography*, edited by B. Warren and C. Wunsch, MIT Press, Cambridge, MA, 342–374, 1981.

Paulo B. Oliveira, Instituto de Oceanografia, Faculdade de Ciências da Universidade de Lisboa, Campo Grande, 1749-016 Lisboa, Portugal. (email, pboliv@fc.ul.pt; fax, +351-21-750-0009)

Satellites, Oceanography and Society
edited by David Halpern

Chapter 8

Why care about El Niño and La Niña?

Michael H. Glantz

Environmental and Societal Impacts Group, National Center for Atmospheric Research, Boulder, Colorado

Abstract. At the time of writing this chapter in February 2000, many aspects of global weather were under the influence of a cold sea surface temperature event along the equator in the eastern Pacific Ocean. This type of event is known as La Niña. The cold event began in mid-1998, following on the heels of the most intense and damaging warm event, or El Niño, of the twentieth century. El Niño and La Niña are important applications of satellite data, as described elsewhere in this book. Episodes of El Niño and La Niña occur at approximately 3–7 year intervals to disrupt human activities worldwide in both positive and negative ways.

The structure, content, and approach taken for this chapter reflect my desire to produce a concise overview of many attributes regarding El Niño and La Niña, which could be reproduced for presentation to various audiences, from university students to researchers in other disciplines, to policy makers, and to the public. Each page provides a brief stand-alone explanation of characteristics of El Niño and La Niña.

The chapter begins with a brief description of how the public learned about "El Niño" and "La Niña" from the media, which is credited with rapidly informing the public about the impacts of El Niño and La Niña. It seems that nearly everyone has heard of "El Niño" and "La Niña," but very few know what El Niño and La Niña are, what they do, and why societies should care about them. Since early 1997, oceanographers and meteorologists readily learned the difficulties inherent in explaining the complex ocean-atmosphere interaction process of El Niño or La Niña in a few words or in 15-second interviews. The second part is a succinct description of the El Niño and La Niña phenomena. Then four El Niño and La Niña impacts are shown: marine living resources, excessive rainfall, drought-related wildfires, and hurricanes. The chapter concludes with some general lessons about El Niño and La Niña.

1. El Niño, La Niña, and the Media

Before the 1982–1983 El Niño, mention of El Niño in popular media headlines was virtually nonexistent. During the past decade, the public has become familiar with the term El Niño through its use in the media, especially with regard to the major event in 1997–1998. Headlines reporting on the 1997–1998 E1 Niño, which have been taken from the print and electronic media, appeared in English, Russian, Spanish, Swahili, and (not shown here) in Portuguese, Malay, Filipino, and Chinese, representing the wide range of geographic and economic interests in the phenomenon. The headlines also depict interest in El Niño and, for example, the oceans, ecosystems, energy, water, food, commodities, drought, and politics. The Russian headline is interesting because it is from a country that is not directly affected by El Niño.

The 1997–1998 El Niño event made El Niño a household word, as suggested when the cover and feature story of the 6 October 1997 *US News and World Report* (a weekly newsmagazine) was "The Power of El Niño, Our Century's Biggest Weather Event is Underway." During the 1997–1998 event, the El Niño theme was widely used in advertising for the first time, mostly in humorous ways. The advertisements shown below, which appeared in different issues of the *Denver Post* (a daily newspaper) in November 1997, are representative of favorable media "hype," which educates the public about El Niño and its potential climate-related impacts. Advertisements tend to be tailored to El Niño impacts expected to occur within a region. Labels for bottled water, alcoholic drinks, and beachwear trademarks, for example, have contained the words "El Niño."

For the first time ever, numerous cartoons used El Niño in their punch lines. Most of the texts make fun of the phenomenon. For example, when its adverse impacts failed to occur in California in late 1997, despite numerous forecasts, the *Los Angeles Times* (12 July 1998) referred to it as "El Scapegoat." Dialogue in cartoons suggests the variety of ways that the public tends to perceive the phenomenon. In 1999, La Niña began to appear in cartoons as well.

(a)

(b)

"El Nino!"

(c)

Cartoons: (a) Walt Handelsman, reprinted with permission of Tribune Media Services; (b) Bil Keane, reprinted with special permission of King Features Syndicate; and (c) Ed Stein, reprinted with permission of *Rocky Mountain News*.

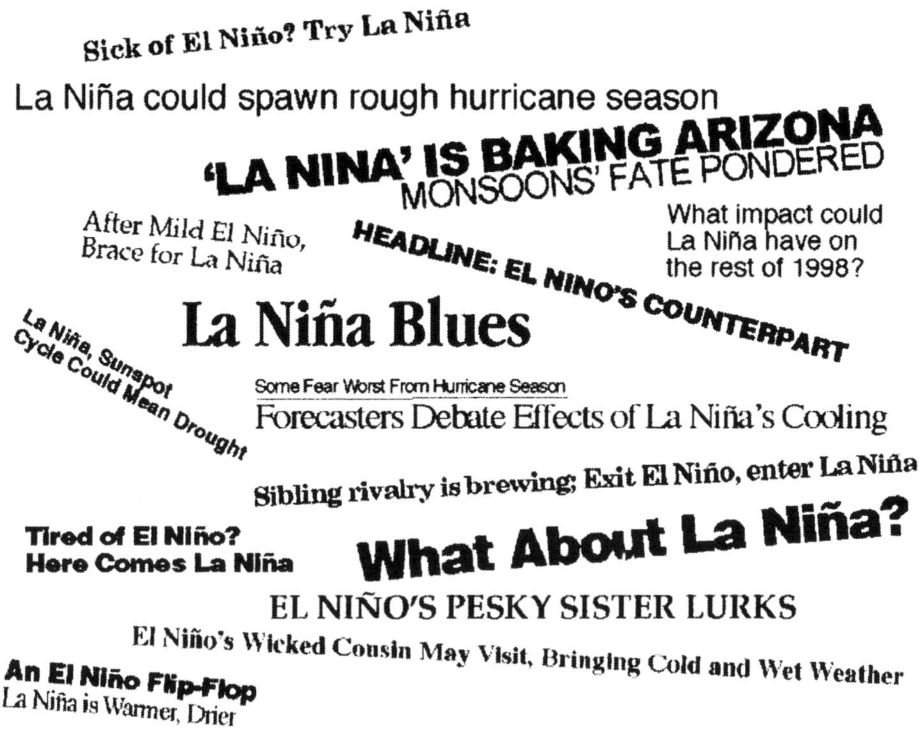

El Niño and La Niña are, respectively, the warm and cold phases of the oscillation of sea surface temperature in the eastern tropical Pacific. Following the collapse of the 1997–1998 El Niño event in May 1998, a cold event began to develop. Numerous headlines about La Niña appeared, as shown above; however, the headlines were much less spectacular than those for El Niño because, perhaps, scientists, during the past two decades, had focused their research on warm events and much less on cold events. El Niño has received the lion's share of attention, because it had been linked to disasters ever since 1892 (Carrillo, C., Disertacion sobre las corrientes oceanicas y estudios de la Corriente Peruana de Humboldt, *Bol. Sociedad Geografico Lima, 11*, 84, 1892). Also, since the early 1970s, there have been twice as many El Niños than La Niñas. However, scientists have come to realize that La Niña is as important to understand as El Niño, and that La Niña, too, is associated with an increase in disasters worldwide. For example, during El Niño there have been relatively few hurricanes in the tropical Atlantic and the Caribbean, whereas during La Niña, the number of tropical storms and hurricanes in the Atlantic is above average. Also, some countries (e.g., the Philippines, Indonesia, Malaysia) that suffer from drought during El Niño are often affected by excessive rainfall and flooding during La Niña.

2. What are El Niño and La Niña?

El Niño 'el nē' nyō *noun* [Spanish] \\ 1: The Christ Child 2: the name allegedly given by Peruvian sailors in the 1800s to a seasonal, warm southward-moving current along the Peruvian coast <la corriente del niño> 3: name given to the occasional return of unusually warm water in the normally cold water [upwelling] region along the Peruvian coast, disrupting local fish and bird populations 4: name given to a Pacific basin-wide increase in both sea surface temperatures in the central and/or eastern equatorial Pacific Ocean and in sea level atmospheric pressure in the western Pacific (Southern Oscillation) 5: used interchangeably with ENSO (El Niño-Southern Oscillation), which describes the basin-wide changes in air-sea interaction in the equatorial Pacific region 6: ENSO warm event *synonym* warm event *antonym* see La Niña \\ [Spanish] \\ the young girl; cold event; ENSO cold event; non-El Niño year; anti-El Niño or anti-ENSO (pejorative); El Viejo \\ 'el vyā hō\\ noun [Spanish] \\ the old man

El Niño has more than a single meaning. It encompasses both a localized coastal ocean warming off the coasts of Ecuador, Peru, and Chile and the much broader basin-wide event in the equatorial Pacific. Researchers use different quantitative measures to identify conditions that they define as El Niño and La Niña events. The dictionary-like definition of El Niño shown on this page (Glantz, M. H., *Currents of Change: El Niño's Impact on Climate and Society*, Cambridge Univ. Press, Cambridge, England, 194 pp, 1996) encompasses a large range of meanings and attributions.

In the equatorial zone of the Pacific Ocean the sea surface temperature is typically 29°C in the west and 23°C in the east, as illustrated on this page (upper panel) for January 1997 (redrawn from the *Climate Diagnostics Bulletin* distributed by the National Oceanic and Atmospheric Administration National Centers for Environmental Prediction, Camp Springs, Maryland), three months before the start of the 1997 El Niño. The 6–7°C east-west difference in sea surface temperature is not produced directly from the sun; the temperature difference represents a balance between the westward-blowing wind near the sea surface and the density (or temperature) and current in the upper 300 m of the ocean.

Every 3–7 years when an El Niño occurs, the sea surface temperatures in the western and eastern equatorial Pacific will drop 1°C and rise 2–3°C, respectively. In the 1997–1998 El Niño, the sea surface temperature in the eastern equatorial Pacific increased 5–6°C, creating uniform sea surface temperature along the equator across the entire width of the Pacific, nearly one-half the circumference of the Earth. At the peak of the 1997–1998 El Niño in December 1997, the sea surface temperature anomaly in the eastern equatorial Pacific, as illustrated in the lower panel on this page (redrawn from the *Climate Diagnostics Bulletin*), reached 5°C over a width approximately the distance between New York and Seattle.

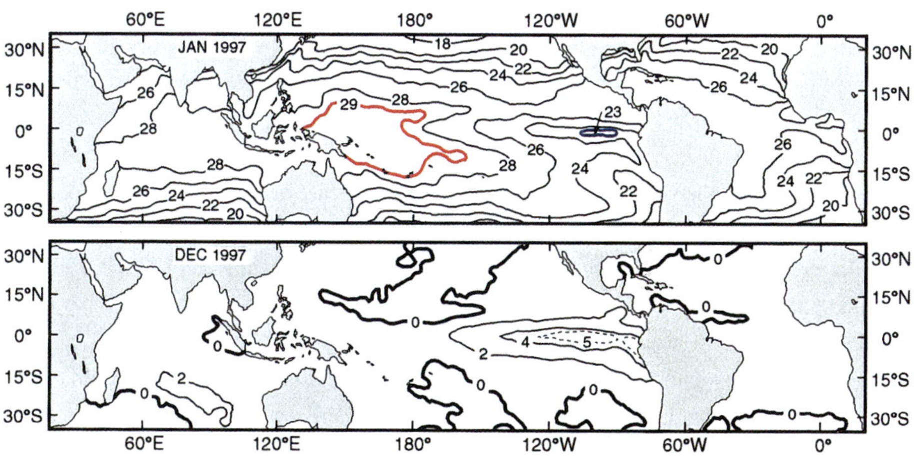

In La Niña conditions (upper diagram on the facing page; redrawn from http://www.pmel.noaa.gov/toga-tao/pmel-graphics/web-graphics.html), the easterly tradewind and westward-flowing South Equatorial Current in the Pacific equatorial zone are stronger than usual. In the western tropical Pacific, the warm water with sea surface temperature greater than 29°C is moved westward by the current, creating a 150- to 200-m thick warm water layer in the western tropical Pacific. In the eastern equatorial Pacific, sea surface temperature is low (<23°C) and the thickness of the upper layer is 25–50 m. The "thermocline" (see blue subsurface layer in the diagram), which is a 100-m thick layer separating the warm upper ocean from the cold deep ocean, slopes upward from west to east. As a consequence of the east-west slope of the thermocline, sea level along the equator is about 0.5 m higher in the west than in the east.

The easterly tradewind is the surface expression of the equatorial wind pattern. In the tropics where air occurs over sea surface temperatures greater than 27°C, the surface layer of the atmosphere, which is fully laden with moisture, rises to form rain-bearing clouds, as depicted in the diagrams. At 10-km altitude, the air flows eastward to sink over the colder water of the eastern equatorial Pacific. The descending air is dry when it reaches the relatively cold ocean in the eastern region. Therefore, in the tropical Pacific, La Niña represents wet weather over northern Australia and Indonesia and arid conditions along the coasts of Ecuador and Peru.

An El Niño begins when the strength of the easterly surface wind is reduced for several weeks; then, the thermocline in the west will rise, the thermocline in the east will drop, and warm water will flow from west to east. After three or four months of reduced easterly wind and, sometimes, the wind reverses direction and blows toward the east (as depicted in the lower diagram), thermocline depths in the west and east become 100–150 m and 75–100 m, respectively. In the eastern equatorial zone, a 75-m layer of warm water occurs above the thermocline. Accompanying the eastward flow of warm water and high sea surface temperature are clouds and rainfall. Therefore, in the tropical Pacific, El Niño represents less rainfall in the western Pacific and excessive rainfall in the eastern Pacific.

The worldwide weather influence of El Niño and La Niña is produced by a change in the latitude of the position of the mid-latitude jet streams, which represent the overall coupling between ocean and atmosphere, such as the sea surface temperature contrast between equator and pole. The eastward-flowing jet stream is the main carrier of local weather for much of the globe. The latitude of the position of the jet stream is related to the location (eastern or western equatorial Pacific) of the warm water with sea surface temperatures above 27°C. For example, during La Niña, the jet stream enters North America between Portland, Oregon, and Seattle, Washington, and in El Niño the jet stream is over southern California (see Glantz, M. H., *Currents of Change: The Impacts of El Niño and La Niña on Climate and Society*, Cambridge University Press, Cambridge, United Kingdom, in press, 2000).

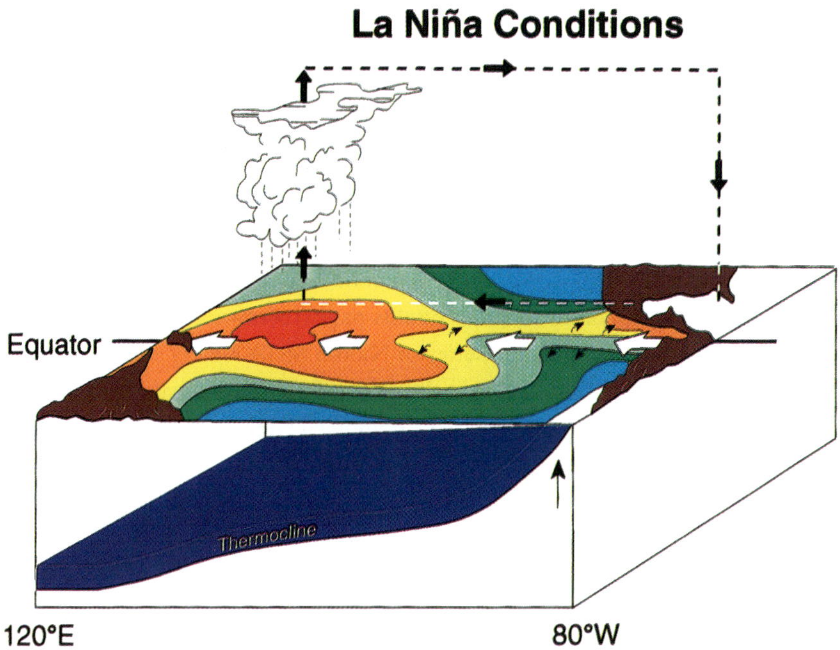
La Niña Conditions
Equator
Thermocline
120°E
80°W

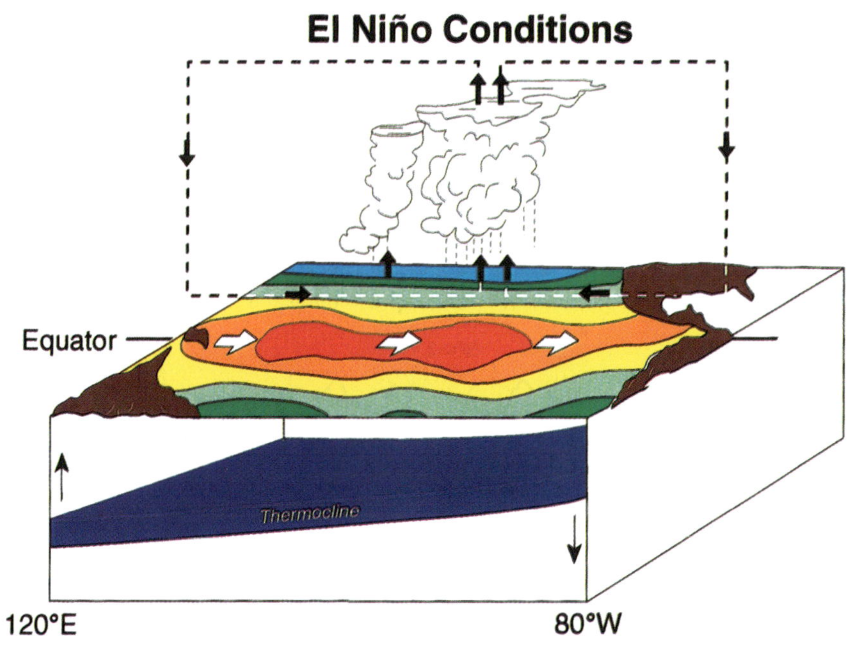
El Niño Conditions
Equator
Thermocline
120°E
80°W

Some years ago, scientists divided the tropical Pacific into four regions (Niño 1, Niño 2, Niño 3, Niño 4) to study sea surface temperature change. The difference between the monthly average sea surface temperature and the long-term mean monthly sea surface temperature, named sea surface temperature anomaly (SSTA), is a measure of the strength of El Niño or La Niña. Most recently, scientists have identified Niño 3.4 as a region that provides the most useful information on sea surface temperature changes related to the onset of El Niño. Australians have apparently shown more interest in SST changes in Niño 4, while Peruvians are most concerned about SST changes off their coast in regions called Niño 1 and Niño 2. In the diagram on this page (redrawn from the *Climate Diagnostics Bulletin* (see page 7)), the time series of Niño-3 SSTA clearly shows the El Niños of 1982–1983, 1986–1987, 1991–1994, and 1997–1998. The extended duration of the 1991–1994 El Niño remains unexplained. The combined Niño-1 and Niño-2 SSTA shows that the El Niños of 1982–1983, 1986–1987, and 1997–1998 had a double-peak structure.

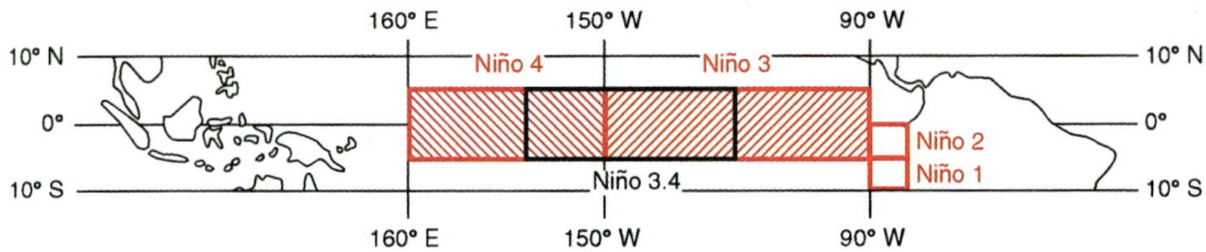

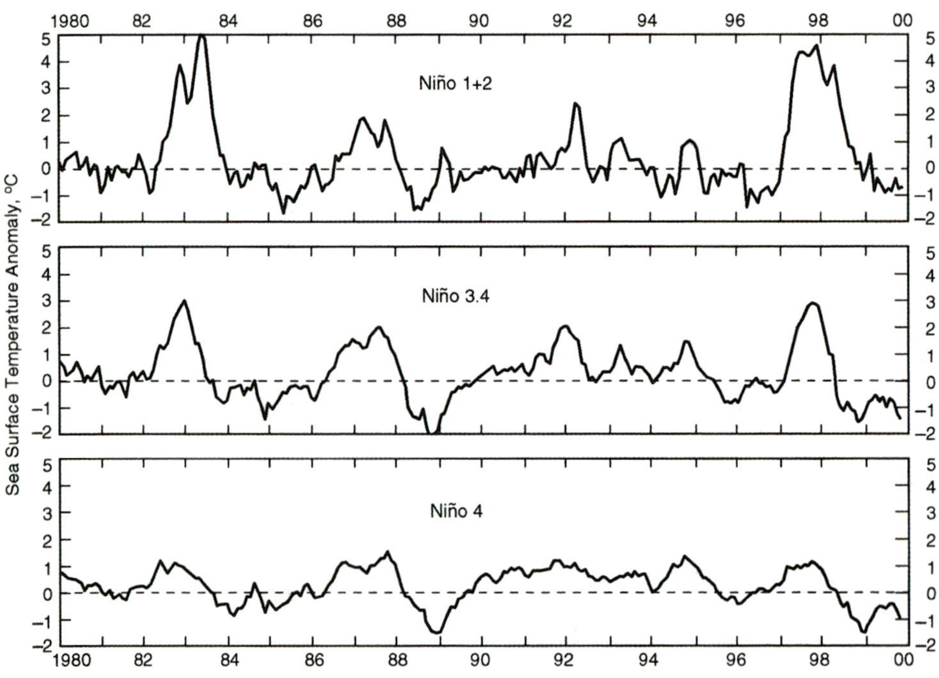

How we know what we know about the El Niño and La Niña episodes of 1997–2000 is the result of an evolutionary development, begun in the late 1970s, of an integrated ocean observing system: subsurface ocean temperature measurements from ships and moored buoys; subsurface ocean current measurements from buoys moored to the ocean bottom at 4-km depth; sea surface temperature observations from drifting buoys and infrared radiometer measurements from satellites; sea surface height from tide gauges and satellite radar measurements; surface winds from moored buoys and scatterometer and microwave radiometer measurements from satellites. Subsurface oceanographic data are sparsely sampled and are assimilated into numerical ocean models for dynamical interpolation. A schematic diagram of the in-situ network of the observing system is shown below. Many datasets are available in almost-real time on the Internet. Two widely distributed bulletins of monthly oceanographic and meteorological conditions in the Pacific Ocean are the *Climate Diagnostics Bulletin* (available from the National Oceanic and Atmospheric Administration National Centers for Environmental Prediction, Camp Springs, MD 20746, U.S.A.) and the *Monthly Ocean Report* (available from Climate and Marine Department, Japan Meteorological Agency, 1-3-4 Otemachi, Chiyoda-ku, Tokyo 100-8122, Japan).

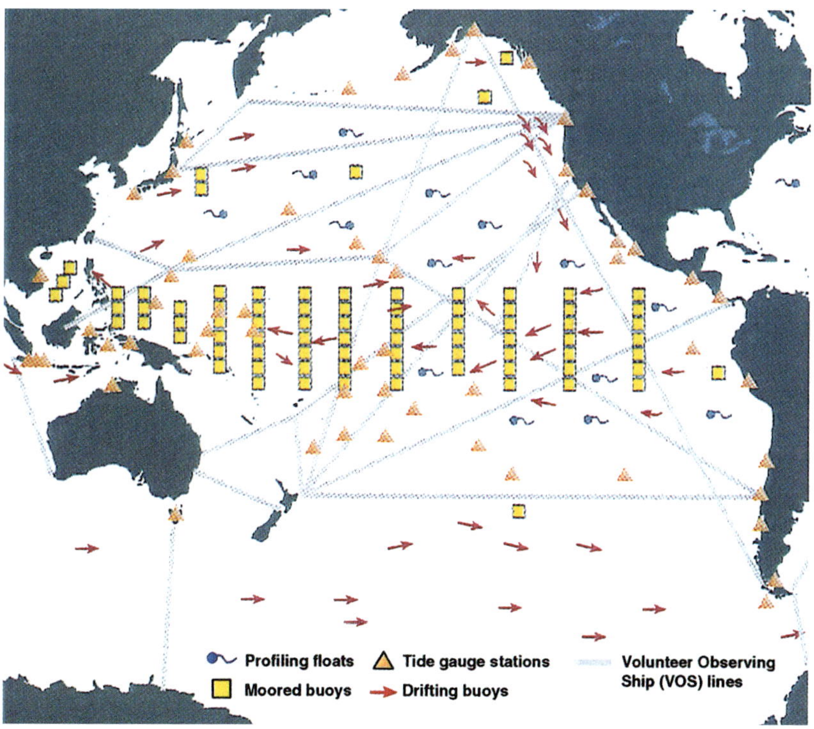

Diagram courtesy of the National Oceanic and Atmospheric Administration Office of Global Programs, Silver Spring, Maryland.

3. El Niño and La Niña Impacts

In 1970, Peru ranked as the number-one fishing nation in the world. The primary fish stock of commercial interest off the coast of Peru is the anchovy, illustrated in the upper panel. The anchovy is considered an industrial fish and is not used for human consumption. Anchovies are converted into fishmeal, which is used as a high-protein feed supplement for chickens and cattle and in fish farms. The anchovy fishery (lower diagram) collapsed in the 1972–1973 El Niño by overfishing and a failure of the fish to reproduce in abundance in a warm water event. Catches dropped sharply in the 1997–1998 El Niño in part as a result of oceanic conditions and government-imposed closed fishing seasons (*vedas*).

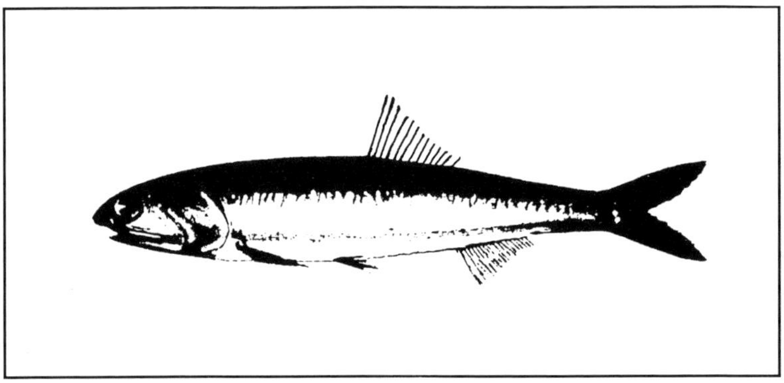

Anchovy, adult actual size 17 cm. Photo courtesy Instituto del Mar del Peru.

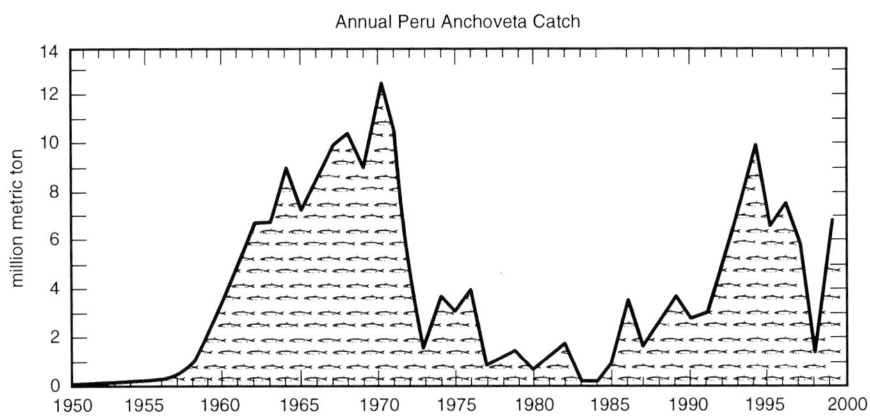

Colony of guanayes, Pescadores Island, 1919. Neg. No. 262158. Photograph by R. C. Murphy. Courtesy Department of Library Services, American Museum of Natural History, New York.

The photograph on this page shows the high density of guano birds that inhabited Peru's various rocky islands and arid coastline. Guano birds—cormorants, pelicans, and gannets—produce excrement (in Spanish, *guano*). Guano birds' main food source has been the anchovy in the coastal waters of Peru. Guano has been a highly valued rich fertilizer used in agriculture, and at the turn of the twentieth century armed guards were posted to keep natural predators away from the birds. A census of the guano bird population in 1950 suggested that there were more than 50 million birds. When El Niño occurs and warm, nutrient-poor water invades Peru's coastal region, the anchovy fishery shifts to deeper water, migrates southward in search of colder water, or fails to reproduce. As a result, guano birds have great difficulty to find their main food source. In the past, they starved to death by the millions, washing up on Peruvian beaches. Following the 1982–1983 El Niño, the number of guano birds was estimated at 400,000. From the perspective of the anchovy, the fishery has three predators: guano birds (constant), fishermen (seasonal), and warm events (every 3–7 years).

Neg. No. 133830. Copied by H. S. Rice. Courtesy Department of Library Services, American Museum of Natural History, New York.

This photograph, taken in the early part of the twentieth century, shows a mountain of guano on Peru's Chincha Island. Guano has been deposited by the birds along the rocky coast of Peru for thousands of years. It was mined as if it were an ore of a precious metal like gold or silver. The size of the mountain can be judged from the height of the miners in the center of the photo. Today, that guano mountain has been mined out. In the first decade of the 1900s, the Peruvian government became concerned about the rapid depletion of guano and established the Guano Administration Company to oversee the entire industry. El Niño's impact on sea birds and, therefore, the guano industry, was a major concern.

The East African country of Kenya was among the most adversely affected countries with respect to the impacts of the 1997–1998 El Niño. Kenya's National Weather Service announced in late 1997, several months in advance of its rainy season, that the El Niño impact would be heavy rains. Kenyans got much more than just heavy rains. They were plagued by severe flooding in cities and in rural areas, and by a major outbreak of Rift Valley Fever, which was blamed for the death of more than a thousand people. Kenya's infrastructure (roads, railroads, and bridges) was greatly damaged, as illustrated by the photos on this page. In previous El Niño events, the impact on Kenyan rainfall varied from drought conditions in 1992 to heavy rains in 1982–1983. The 1997–1998 devastating rains in Kenya have also been linked to the very warm sea surface temperatures in the western Indian Ocean.

Photographs courtesy of Nation Newspapers Ltd., P.O. Box 49010, Nairobi, Kenya.

Satellite imagery captured the location and extent of numerous forest fires in Indonesia and Malaysia (Borneo), which were directly and indirectly related to the 1997–1998 El Niño, when an estimated 3–5 million hectares (1.0 ha = 2.2 acres) of rainforest were burned. The reduced rainfall during El Niño enables fires to burn larger areas. Equally destructive fires in the same area were recorded by satellite during the 1982–1983 El Niño. During both El Niños, some fires were associated with traditional land-clearing practice to cultivate the land for food production. However, some fires were intentionally set by companies to expand palm oil and other plantations. Companies had sought to use El Niño-related regional drought as an excuse for the extensive fires. Haze from the fires was blamed for pollution-related health problems in Singapore and Kuala Lumpur (Malaysia), two ship collisions, and a deadly plane crash. In response to an international outcry against the fires and haze, the Indonesian Government brought legal action against more than 30 companies for their complicity in hiring people to torch the rainforest.

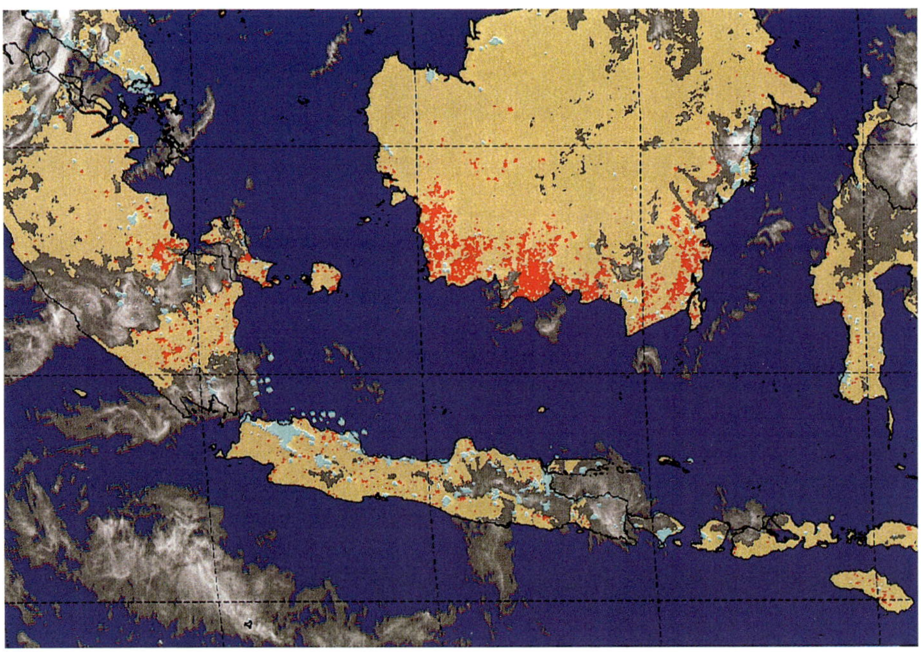

Locations of fires detected the night of 21 September 1997 in Indonesia and Malaysia (Borneo) using data from the U.S. Air Force Defense Meteorological Satellite Program (DMSP) Operational Linescan System (OLS). The OLS has a unique capability to perform low-light imaging of the Earth at night. Data processing by the National Oceanic and Atmospheric Administration National Geophysical Data Center, Boulder, Colorado.

In the Atlantic Ocean, fewer hurricanes occur during El Niño because the tops of large, high-cloud systems are sheared off by strong upper-level winds associated with the warming of sea surface temperatures in the central and eastern tropical Pacific Ocean. Conversely, during La Niña, there is an above-average number of hurricanes in the Atlantic. The satellite image (below) shows five tropical storms in the Atlantic on 23 August 1995, during La Niña. This is the largest number in decades of tropical storms recorded on a single day.

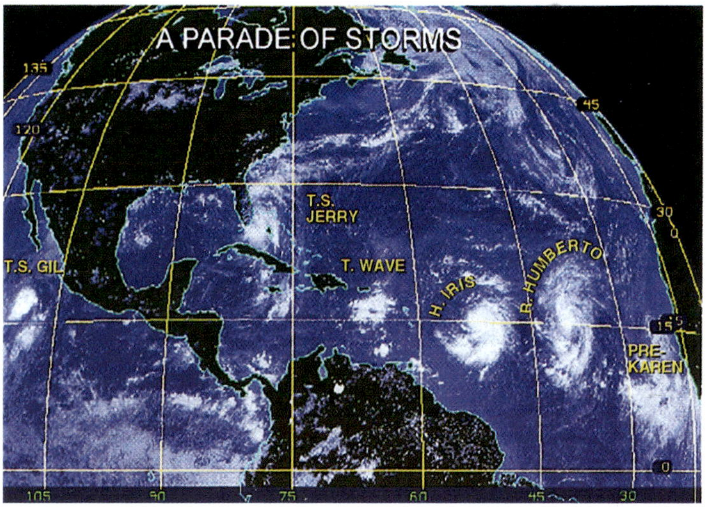

Photograph courtesy of National Oceanic and Atmospheric Administration National Hurricane Center, Miami, Florida.

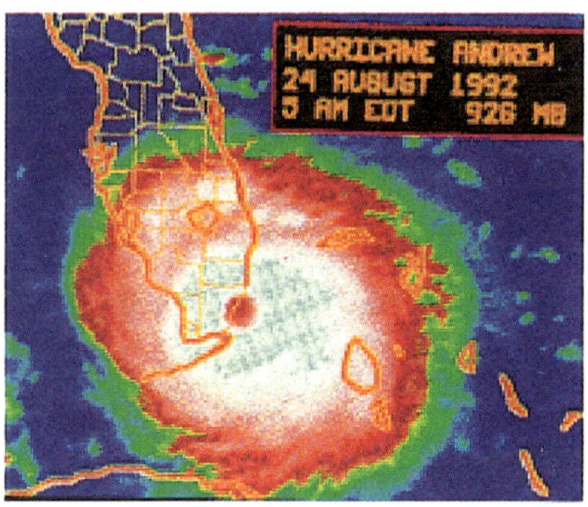

Fewer hurricanes in the Atlantic during an El Niño episode, however, does not mean that there cannot be an intense one. For example, although there were relatively few hurricanes and tropical storms during the 1991–1992 El Niño, in August 1992, Hurricane Andrew, with its huge area of high winds (see diagram above), became the most expensive hurricane to make landfall in North America until that date.

4. El Niño/La Niña Lessons

Seven Things To Know About El Niño

1. **El Niño does not represent unusual behavior of the global climate.**
2. **El Niño is part of a cycle.**
3. **Every weather anomaly throughout the world that occurs during El Niño is not caused by El Niño.**
4. **El Niño has a positive side.**
5. **There will continue to be surprises associated with El Niño events.**
6. **The impact of global warming on El Niño is not known.**
7. **Forecasting El Niño is different than forecasting the impacts of El Niño.**

4.1 El Niño does not represent unusual behavior of the global climate

El Niño and La Niña events are part of the global climate system. They are anomalous in that they are the extremes of sea surface temperature changes (warming and cooling, respectively) in the tropical Pacific Ocean, and not just any change from average (or expected) conditions at any given point in time. They have occurred for millennia and can be expected to do so well into the future.

4.2 El Niño is part of a cycle

While considerable attention has been focused on El Niño, much less has been focused on La Niña. In addition, there are lengthy periods when sea surface temperatures in the tropical Pacific are around average. Needless to say, extreme meteorological events occur during these "normal" years as well.

4.3 Every weather anomaly throughout the world that occurs during El Niño is not caused by El Niño

There is a tendency to blame just about everything that happens during an El Niño episode (which can last 12-18 months) on that particular El Niño. However, only some parts of the globe are directly influenced by El Niño-spawned regional climate anomalies, and even those areas are not necessarily influenced in the same way by different El Niño events. Every year, extreme record-setting weather events are occurring at various locations around the globe. The linkages between El Niño and regional climate anomalies have been identified through: (1) observations of direct linkages; (2) statistical measures; and (3) wishful thinking.

4.4 El Niño has a positive side

While most studies have focused on the negative aspects of El Niño, El Niño has brought benefits to some sectors and to some regions. One of the most obvious examples of a benefit would be the reduced number of hurricanes that can be expected to form in the Atlantic Ocean during El Niño. In various locations, there are shifts in the availability and abundance of living marine resources, available to those with the resources to capture them.

4.5 There will continue to be surprises associated with El Niño events

Scientists have concentrated their research efforts on El Niño as a Pacific basin-wide phenomenon since the 1970s. They have not yet witnessed all the ways El Niños can form, nor have they witnessed all of the combinations of ways that El Niño can simultaneously affect societies and ecosystems worldwide. Thus, each succeeding El Niño will likely surprise scientists and the public with respect to its timing of onset, frequency, intensity, or the severity of its impacts.

4.6 The impact of global warming on El Niño is not known

The scientific community is unable at this time to say with any degree of reliability or confidence what the impacts of global warming will be on El Niño.

4.7 Forecasting El Niño is different than forecasting impacts of El Niño

A close review of forecasts and projections of the 1997–1998 El Niño shows that the onset of the event and the intensity that it reached, an intensity that gave it the title of "El Niño of the Century," were not correctly forecasted. Impacts of this event were correctly forecasted several months in advance, such as heavy rains in southern California, floods in northern Peru, and droughts throughout Southeast Asia.

Acknowledgments. I am beholden to people who have provided guidance to my work on this chapter. I would like to thank Neville Nicholls (Bureau of Meteorology Research Centre, Australia), Joseph Tribbia, Peter Gent, and Kevin Trenberth (NCAR), and Antonio Busalacchi (NASA/GSFC), as well as the two anonymous reviewers of an earlier draft of this chapter. Their combined wisdom has once again helped me to walk the proverbial mine field of describing scientific information to the public and policy makers in user-friendly terms. Any shortcomings in my attempt to do so successfully, however, rests on my shoulder. In addition, I would like to express my sincere appreciation to D. Jan Stewart, ESIG Program Development Office at NCAR, for her continued support. Special thanks also go to Justin Kitsutaka, who provided considerable graphic support for this chapter. The National Center for Atmospheric Research is sponsored by the National Science Foundation.

Michael H. Glantz, Environmental and Societal Impacts Group, National Center for Atmospheric Research, P. O. Box 3000, Boulder, CO 80307, U.S.A. (email, glantz@ucar.edu; fax, +1-303-497-8125)

Satellites, Oceanography and Society
edited by David Halpern

Chapter 9

Satellites, society, and the Peruvian fisheries during the 1997–1998 El Niño

Mary-Elena Carr
Jet Propulsion Laboratory, California Institute of Technology, Pasadena, California

Kenneth Broad
International Research Institute for Climate Prediction, Columbia University, Palisades, New York

Abstract. The evolution of oceanographic conditions off Peru during 1996–1998, including the 1997–1998 El Niño, is studied with in-situ data from coastal tide gauges and with satellite data from the Advanced Very-High Resolution Radiometer (AVHRR), European Remote-sensing Satellite (ERS), National Aeronautics and Space Administration (NASA) Scatterometer (NSCAT), Ocean Color and Temperature Scanner (OCTS), and Sea-viewing Wide Field-of-view Sensor (SeaWiFS). Maximum anomalies of in-situ sea level (>22 cm) and sea surface temperature (SST) (>4°C) occurred near Callao at 12°S in June–July 1997 and December 1997–January 1998. Scalar wind speed at the coast increased between May 1997 and June 1998, indicating the anomalous oceanographic conditions do not result from cessation of coastal upwelling. Monthly unrestricted catch of small pelagic fish surpassed 1 million tons between December 1996 and June 1997, as catchability increased during the onset of warm conditions. Satellite data contributed to a recognition that the anomalous conditions in April–June 1997, which facilitated unseasonably high catches, were part of a large-scale perturbation. This spurred the implementation of regulatory mechanisms to protect the stock, despite strong opposition from the fishing industry. However, later in the event, misinterpretation of satellite data led to premature claims that El Niño was ending and subsequent poor decision-making and confusion by different actors in society. The prediction of return to normal conditions was premature, as the second peak of the El Niño arrived in December 1997–January 1998. Observations and numerical model simulations from a planktonic ecosystem model are compared with variations of the Peruvian catch of small pelagic fish to quantify the impact of El Niño on pelagic fish catch. The two highest correlation coefficients, r, computed between monthly fish catch and several biological and physical variables were associated with cross-shelf SST difference (r = –0.55) and modeled food available for fish (r = 0.50).

1. Introduction

Coastal waters of Peru (between about 3°S and 18°S) are characterized by large phytoplankton biomass and very productive fisheries. Elevated biological productivity results from coastal upwelling which brings high nutrient concentrations to the surface. The average wind velocity, sea surface temperature (SST), and near-surface chlorophyll-*a* for December 1996 (Figure la) show the typical equatorward alongshore winds, which lead to offshore Ekman transport at the coast and replacement by cold, nutrient-rich waters from depth, thus facilitating phytoplankton growth. The dominant southeasterly winds lead to upwelling throughout the year, though they are strongest in April–October (Bakun and Nelson 1991). The Peru-Chile Current, the eastern boundary current of the South Pacific, flows towards the equator from approximately 40°S (Strub et al. 1998). The eastern boundary current regions, where coastal upwelling occurs, are extremely productive: although their area makes up only 0.1% of the world ocean, they account for 5% of global primary production and 17% of global fish catch (Pauly and Christensen 1995).

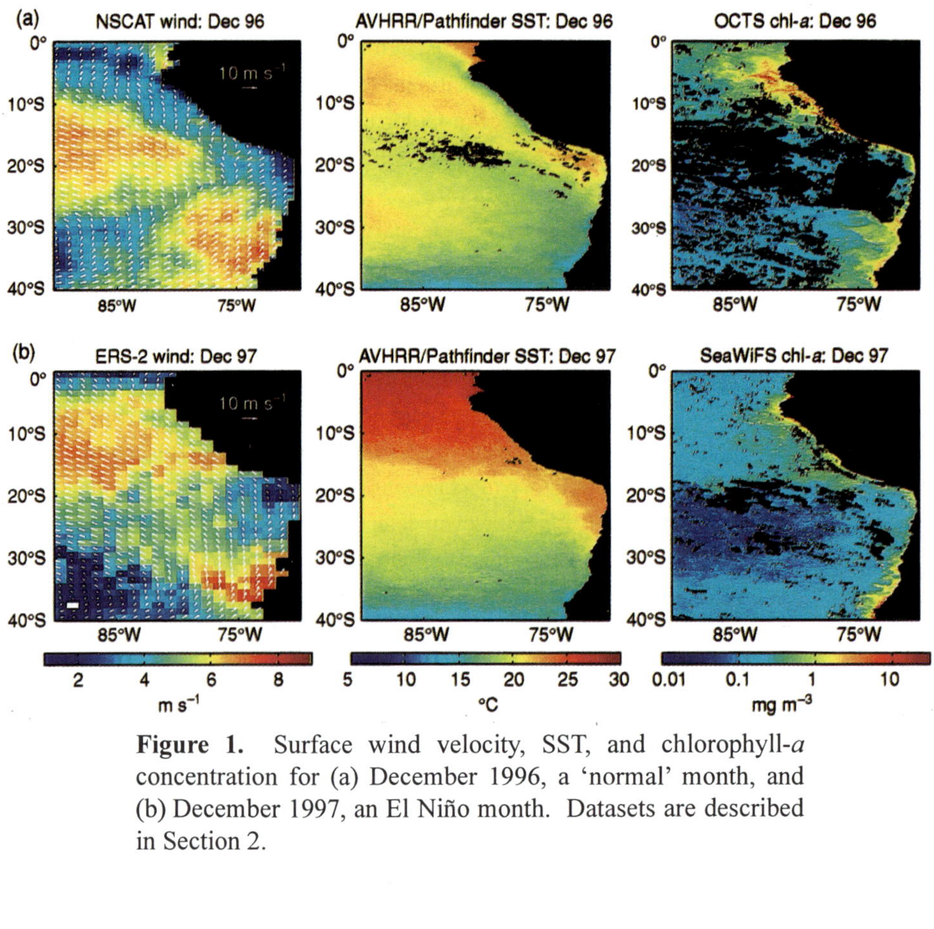

Figure 1. Surface wind velocity, SST, and chlorophyll-*a* concentration for (a) December 1996, a 'normal' month, and (b) December 1997, an El Niño month. Datasets are described in Section 2.

Among these regions, the Peruvian coast is the most productive due to a combination of a wider shelf, upwelling all year, and proximity to the equator because Ekman transport is inversely related to latitude (Barber and Smith 1981; Bakun 1996). Phytoplankton biomass remains high throughout the year (Rojas de Mendiola 1981; Chávez 1995). Generally higher values of chlorophyll-*a* concentration are found inshore in December–June, while the highest concentrations in the 100- to 300-km offshore region occur in July–September (Rojas de Mendiola 1981; Chávez 1995; Walsh et al. 1980).

During an El Niño, there are changes in the current system, sea level rises, mixed layer and thermocline depths increase, SST rises, and sea surface salinity increases (Blanco et al. 1999). These environmental factors tend to reduce the nutrient supply to the euphotic layer, with important consequences for planktonic species composition and production. The effect of El Niño on the survival of higher trophic levels (i.e., fish) is directly influenced by the altered oceanographic conditions (especially high temperatures) and indirectly through reduced planktonic production. While the collapse of the Peruvian anchovy fishery in 1973 following the event of 1972 brought El Niño to the world's attention, fishermen from northern Peru and southern Ecuador were aware of this phenomenon for centuries and named the warm water current that appeared every few years around Christmas time 'El Niño,' or the 'child,' after the baby Jesus.

The Peruvian industrial fishery targets small pelagics (primarily anchovy) to produce fishmeal for animal and aquaculture feed, and had a meteoric rise beginning in 1957–1958 and continuing through the 1960s to peak in 1970 and crash in 1973 (Figure 2). The yield of 12 million tons (1 ton = 1000 kg) in 1970 (made up of a single species, anchoveta) accounted for one-sixth of global fish catch between 1963 and 1972 (Bakun 1996). The late 1980s and early 1990s were characterized by a recovery to high levels of catch (Csirke et al. 1996). In 1996, the last year unaffected by El Niño, the industrial fishery had surpassed 8.5 million tons of small pelagics (Figure 2), contributing over \$1.5 billion (about 4%) to Peru's Gross National Product and employing about 60,000 fleet and plant workers and others in associated industries including netmaking, shipbuilding, and engine repair (Broad 1999). The high catches in the early 1990s leading up to the 1997–1998 event were accompanied by massive financial investments in new boats, and in fishmeal and canning plants. This contributed to the immense political pressure by the industry on regulators to allow continued industrial fishing, even as the El Niño event became increasingly evident in April–June 1997. The artisanal fishery sector, a second component of Peruvian fisheries, supplies fresh fish for local, national, and international markets, and consists of about 50,000 fishermen and divers, who also have financial loans for a variety of fishing gear. Government regulators, members of the financial sector, and fisherman have strong interests in the fishing sector (Broad et al. 2000; Pfaff et al. 1999). The El Niño variability in Peruvian fish catch affects worldwide commodities such as soy meal, a substitute feed (Barber 1988). The 1997–1998 El Niño arrived in Peruvian coastal waters

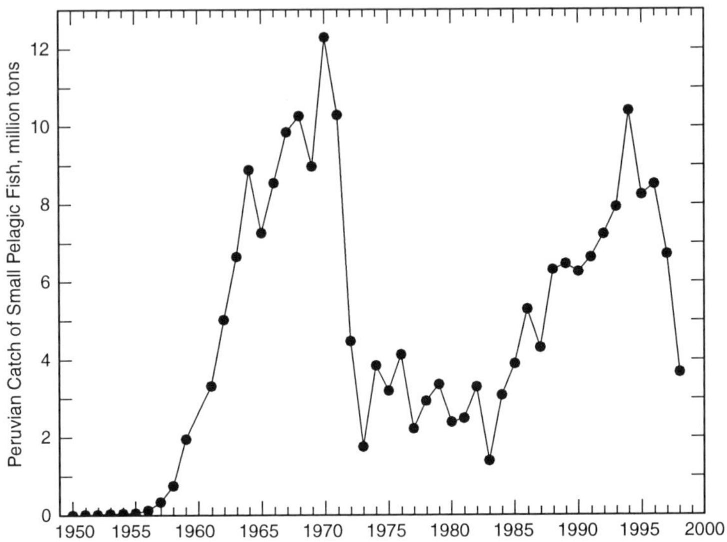

Figure 2. Annual catch of small pelagic fish in Peru. Data are obtained from the Fishmeal Exporter Organization (1998).

within the context of high, but regulated, fishing pressure. The anchoveta population was increasing towards the largest of the three stock levels proposed by Csirke (1996) as indicated by the 1994–1996 annual catches in excess of 8 million tons (Figure 2).

Monthly anomalies of average sea level from Ecuador to Chile, computed from five tide gauges (La Libertad (2°S), Callao (12°S), Antofagasta (23°S), Caldera (27°S), and Valparaiso (33°S)), and monthly anomalies of the National Centers for Environmental Prediction (NCEP) SST (Reynolds and Smith 1994) show that the most distinctive events from 1981 to 1998 were the 1982–1983 and 1997–1998 El Niños (Figure 3). The strongest anomalies occurred within 20° of the equator, though anomalous sea levels extended poleward of 30°S. The 1982–1983 El Niño off Peru was characterized by monthly averaged sea level anomalies of almost 30 cm and SST anomalies exceeding 6°C off Peru. Sea level and SST anomalies during the 1997–1998 event were comparable to those observed in 1982–1983. Anomalies associated with the 1987 and 1992 events were less than 20 cm and 4°C. The 1982–1983 event, sometimes referred to as the 'El Niño of the century' or the 'extraordinary El Niño' (Quinn et al. 1987; Glantz 1996), led to a 'tropicalization' of the ecosystem: warm water coastal species (such as shrimp or scallops) were found further south and had higher growth rates; open ocean fish (such as yellowfin tuna, mackerel, shark, or dolphinfish) were found further inshore; and the usual biota (anchovy and sardine) migrated southward or to greater depth and their reproduction was compromised (Barber and Chávez 1986; Arntz and Tarazona 1990).

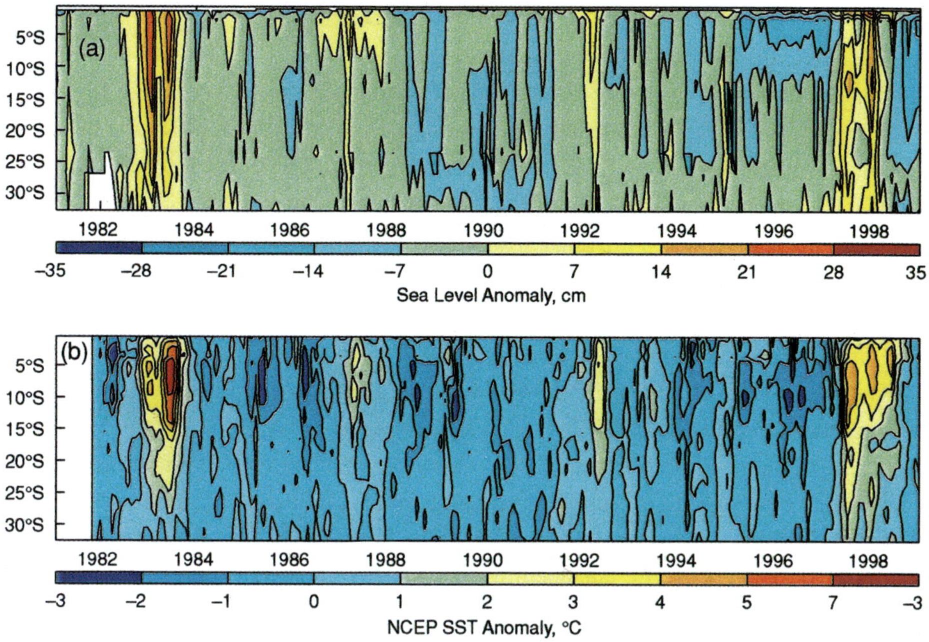

Figure 3. (a) Sea level anomaly from five tide gauge stations (see text for locations) and (b) $1^\circ \times 1^\circ$ NCEP SST anomaly adjacent to the west coast of South America.

Although El Niño influences climate patterns around the globe (Ropelewski and Halpert 1987; Glantz 1996), the Peruvian coast remains one of the areas most consistently and directly impacted by this recurrent event. This cyclical climate event has a range of ecological effects. Warmer waters bring in valuable tropical species of fish and shellfish for the artisanal fisheries. The industrial fisheries, which use purse seine gear to catch small pelagics, are adversely affected because the fish migrate deeper and move further south. In some cases larval survival is reduced in the small pelagics, which fail to reproduce. Heavy rainfall damages the infrastructure of artisanal and industrial fisheries, and higher sea levels cause more port closures due to increased wave height during storms.

Environmental changes associated with El Niño trigger socioeconomic and political reactions in Peru that alter aspects of society (Glantz 1996; Broad 1999). For instance, El Niño contributed heavily to the anchovy collapse in 1973 and, coupled with political change in Peru, led to a nationalization of the Peruvian fishing industry, resulting in massive layoffs and a restructuring (Glantz 1981). At that time, the influential fishermen's labor union fought against nationalization and, in 1976, against de-nationalization. During 1997–1998, the fishermen's labor union was virtually powerless, and unable to secure government aid beyond some provision of foodstuffs (Broad 1999).

2. Data and Methods

The 1982–1983 El Niño was undoubtedly the best studied El Niño until that time and the impacts of the event received worldwide attention. Advanced Very-High Resolution Radiometer (AVHRR) SST data are available starting in November 1981, and the coverage was adequate along the west coast of South America. The Coastal Zone Color Scanner (CZCS) began measuring phytoplankton pigment concentration in 1978, but the coverage along the South American west coast was abysmal, requiring averaging over several months in selected areas. Thomas et al. (1994), using 8-month averages centered in January and July, found anomalously low CZCS pigment concentration off the Peruvian coast in 1983 and 1984.

The 1997–1998 El Niño off Peru was very well-sampled from a suite of spaceborne sensors. In addition to AVHRR SST, sea surface height (SSH) variations were measured by the Topography Experiment (TOPEX)/Poseidon altimeter. These observations were important to quantify the evolution of oceanographic conditions and were used as input for predictive models (Barnston et al. 1999). The Japanese Advanced Earth Observation Satellite (ADEOS) platform, launched in August 1996, carried the National Aeronautics and Space Administration (NASA) Scatterometer (NSCAT), which measured surface wind vector over the ocean, and the National Space Development Agency of Japan (NASDA) Ocean Color and Temperature Scanner (OCTS), which measured near-surface chlorophyll-*a* concentration, until the premature failure of the satellite in June 1997. There were no measurements of pigment concentration until SeaStar was launched in August 1997, with the Sea-viewing Wide Field-of-view Sensor (SeaWiFS). From 1991 to present, the European Remote-sensing Satellite (ERS-1 and ERS-2) made measurements of surface wind vector.

During 1997–1998, the Internet made satellite imagery of SST and SSH available to both the government and the private sector in Peru. The government oceanographic agency, Instituto del Mar de Peru (IMARPE), which makes recommendations to the Ministry of Fisheries (Ministerio de Pesca, MIPE), regularly monitored a variety of Internet websites. Near-real-time observations, forecasts, and historical analogs were used to interpret and speculate on the evolution of the 1997–1998 El Niño event in order to anticipate the state of the fish stocks, contribute to decisions such as setting quotas and fishing bans, and designing the itinerary for in-situ oceanographic measurements from ships. An important benefit of satellite data is that it provides a broad oceanographic context for what may appear to be local conditions.

In this study we use a time series of satellite-derived wind speed, SST, and near-surface chlorophyll-*a* concentration to describe the evolution of oceanographic conditions from a 'normal' (or slightly cold) year, 1996, to the warm event of 1997–1998 and the subsequent return to normal conditions. We then compare the observed fish catch to the evolving environmental conditions.

Sea level data for five tide gauge stations were obtained from the University of Hawaii Sea Level Center, which adjusted the data for the inverse barometer effect and estimated the monthly anomaly by subtracting from each month the monthly average for the 1975–1995 interval. Two SST data products are used. The NCEP 1° × 1° monthly mean SST dataset, which is an optimally interpolated combination of AVHRR and ship and buoy observations (Reynolds and Smith 1994), was used for the 1982–1998 interval. NCEP SST anomalies are estimated as in sea level, except the reference-mean interval was 1982–1998. The 9-km resolution monthly mean AVHRR/Pathfinder SST (Kearns et al. 2000) included daytime and nighttime data. The reference-mean interval to compute AVHRR/Pathfinder SST anomalies was 1984–1993. Satellite wind measurements were acquired from three different sensors: ERS-1 (January–September 1996), NSCAT (October 1996–June 1997), and ERS-2 (July 1997–September 1998). Horizontal resolutions of the gridded ERS monthly and NSCAT 12-h mean wind speeds were 1° × 1° and 0.5° × 0.5°, respectively. Chlorophyll-*a* concentrations were measured by OCTS (November 1996–June 1997) and by SeaWiFS (September 1997–September 1998); in both cases, 9-km resolution data were used.

Monthly fish catch data for the Peruvian coast were provided by the Fishmeal Exporters Organization (1998). Data on the societal uses of climate information come from the International Research Institute for Climate Prediction, Columbia University.

3. Results

3.1 The 1997–1998 El Niño off Peru

The sea level anomaly at Callao at 12°S (Figure 4a) shows that the 1997–1998 El Niño event was comparable in magnitude to the 1982–1983 event. In both El Ninõs, an initial peak value (surpassing 22 cm) was followed by a 1–2 month interval of reduced anomaly (about 10 cm), and then the anomaly reached a second peak greater than 20 cm. In the 1982–1983 event the first peak was larger than the second and in 1997–1998 the second peak was greater than the first. The timing of events differed relative to the annual cycle: in 1982–1983 the peaks occurred in November 1982 and March 1983, while in 1997–1998 the peaks were in June 1997 and December 1997. The 1997 event started in March–April when upwelling favorable winds are maximum at 12°S. Sea level anomaly became negative in May 1998 (Figure 4b), perhaps reflecting La Niña conditions, and remained at approximately –15 cm from August to December 1998. The SST anomaly underwent a comparable evolution (Figure 4a). At the end of both El Niños the return to normal SST conditions was several months slower than for sea level (Figure 4a). During the 1982–1983 and 1997–1998 events, SST anomaly peaks coincided with those of sea level. The second peak in SST was stronger in 1982–1983 and the first peak was slightly stronger in 1997; in each event the maximum peak was about 5°C. The relaxation

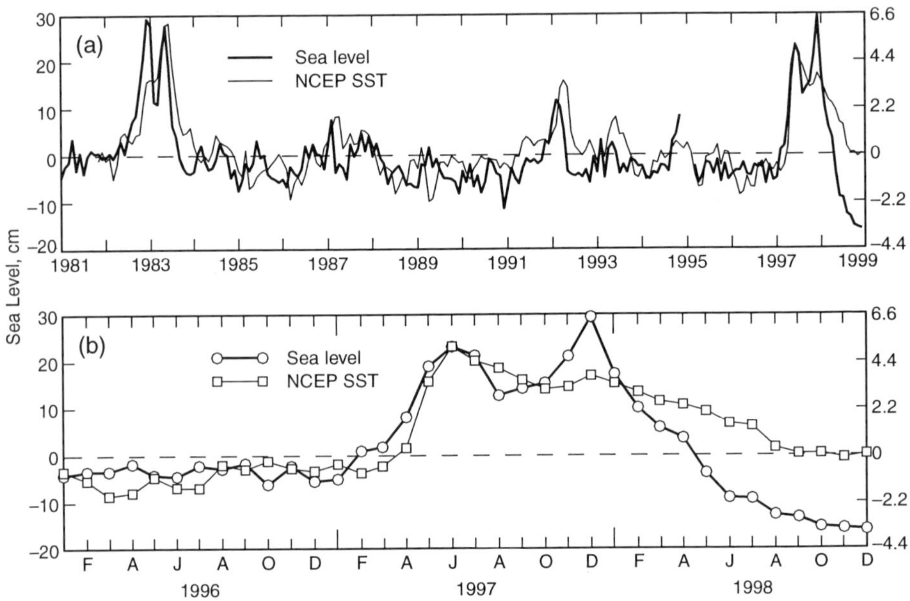

Figure 4. Monthly average anomalies of sea level at Callao (12.5°S) and NCEP 1° × 1° SST adjacent to Callao.

between the two SST peaks in the 1997–1998 event was very weak because of the relatively small size of the second peak. The SST anomaly became zero in September 1998 and SST remained 'normal' until the end of the year.

In December 1997 (Figure 4b) the sea level and SST anomalies were near peak values, while in December 1996 the anomalies were small (Figure 4b). The effect of coastal upwelling in December 1996 (Figure 1a) is evident in the band of cold water along the coast, which is at about 17°C between 5°S and 12°S and in a much narrower, slightly warmer band along southern Peru and northern Chile (14–23°S). The response to normal upwelling is phytoplankton growth: a broad band of high chlorophyll-*a* concentration (> 5 mg m^{-3}) extended 500 km offshore along the northern and central Peruvian coast in December 1996. Although the zonal width of the alongshore band decreased in offshore extent towards the south, reaching the minimum width in northern Chile, enhanced chlorophyll-*a* concentrations continued to occur along the coast. However the situation is dramatically different the following year in December 1997 (Figure 1b). Uniformly warm water (> 22°C) extended to 13°S and along the coast into north-central Chile. Chlorophyll-*a* concentrations were reduced and the width of the productive region had shrunk; off Peru, chlorophyll-*a* values greater than 1 mg m^{-3} were restricted to within 50 km of the coast.

In December 1997, SST anomalies greater than 2°C extended to 12°S, with anomalies surpassing 5.5°C along the coast of Ecuador and northern Peru (Figure 5). Anomalies were weak (about 1°C) in southern Peru, but a secondary area with anomalies reaching 3°C appeared in north-central Chile.

The major fishing region of the productive north-central Peruvian waters is located near Chimbote, at 9.5°S. Along a 500-km onshore-offshore transect at 9.5°S, the time series of scalar wind speed, SST, difference in onshore-SST and offshore-SST, and chlorophyll-*a* pigment concentration (Figure 6) reflect both the seasonal cycle (most clear in 1996) and the changes associated with the El Niño and subsequent La Niña. During 1996, the wind speed generally decreased approaching the coast, was maximum in August–November, and was minimum near the coast in January–March (Figure 6a). The seasonal cycle in 1997 was interrupted in June when the wind offshore increased; enhanced wind speeds reached the coast in October. The onshore-offshore gradient in

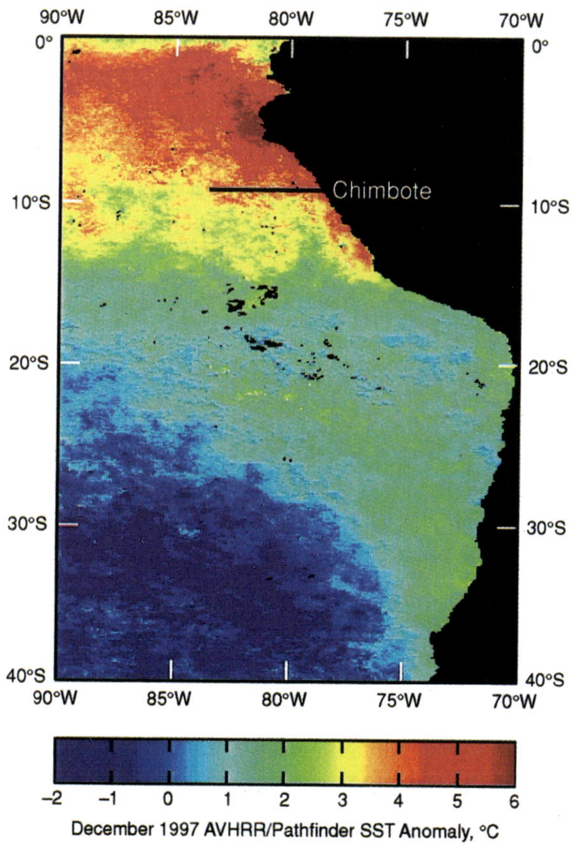

Figure 5. AVHRR/Pathfinder SST anomaly in December 1997. The transect off Chimbote is shown in black.

wind speed changed during the El Niño, with uniform wind speed extended to the coast, alternating with time between high values (>6 m s^{-1}) in October–December 1997 and June–July 1998, and moderate values (>5 m s^{-1}) in January–June 1998. Weaker coastal wind speed occurred in July 1998. Caution is advised in the interpretation of the time series because there were two sensor shifts, one in July 1997.

In 1996 the SST along the Chimbote transect reflected the seasonal heating cycle: cool between July and November and warm between January and May (Figure 6b). The near-shore region was always colder due to coastal upwelling. In 1997, the onset of the cool period never took place and June temperatures were maintained until August. Within 100 km of the coast, SST remained below 22°C between August and November 1997, which was about 6°C higher than in the previous year. In October 1997, a pocket of cool water, extending 100 km offshore, was associated with enhanced coastal wind. Temperatures over 26°C, normally found beyond 400 km offshore, extended to the coast in January 1998. The cooling period started in July 1998.

The onshore-offshore SST difference (Figure 6c), defined as the difference between the SST at 700 km offshore and at each location along the transect, highlights the occurrence of coastal upwelling and removes the seasonal heating cycle. Low values indicate cooler water than offshore. In 1996, large negative values (<–4°C) extended 200 km from the coast between February and June, a few months prior to the wind maximum along the coast. Although the SST difference was less between September and December 1996, it was consistently negative (cooler water) within 500 km of the coast. In June 1997, the SST difference was reduced and the value at 200 km offshore in the previous year had reached the coast. The October 1997 wind maximum enhanced the onshore-offshore SST difference, but in January 1998 the maximum onshore-offshore SST difference along the entire transect was 2°C. By April 1998 the coastal region was 3°C cooler than offshore. The SST difference within 200 km of the coast continued to increase until August 1998.

The chlorophyll-*a* concentration along the transect was unfortunately not measured for much of the 1996–1998 period (Figure 6d). An idea of normal conditions is given by the OCTS measurements in late 1996 and early 1997, when uniform chlorophyll-*a* of about 1 mg m^{-3} extended 400 km offshore. In February 1997, chlorophyll-*a* increased above 1.6 mg m^{-3} with onset of upwelling. However, in June 1997, values of 0.5 mg m^{-3} advanced shoreward to within 100 km of the coast. There was no ocean color sensor from July to September 1997, and by October 1997 near-shore concentrations were again over 1 mg m^{-3}. In the near-shore region, very low chlorophyll-*a* was observed in December 1997 and March 1998. The minimum offshore extent of chlorophyll-*a* concentrations greater than 0.5 mg m^{-3} occurred in March 1998 rather than in December 1997, probably responding to decreased alongshore winds (Figure 6a). After March 1998 the near-shore chlorophyll-*a* concentration increased above 1 mg m^{-3} and the offshore extent of high (> 0.5 mg m^{-3}) values increased.

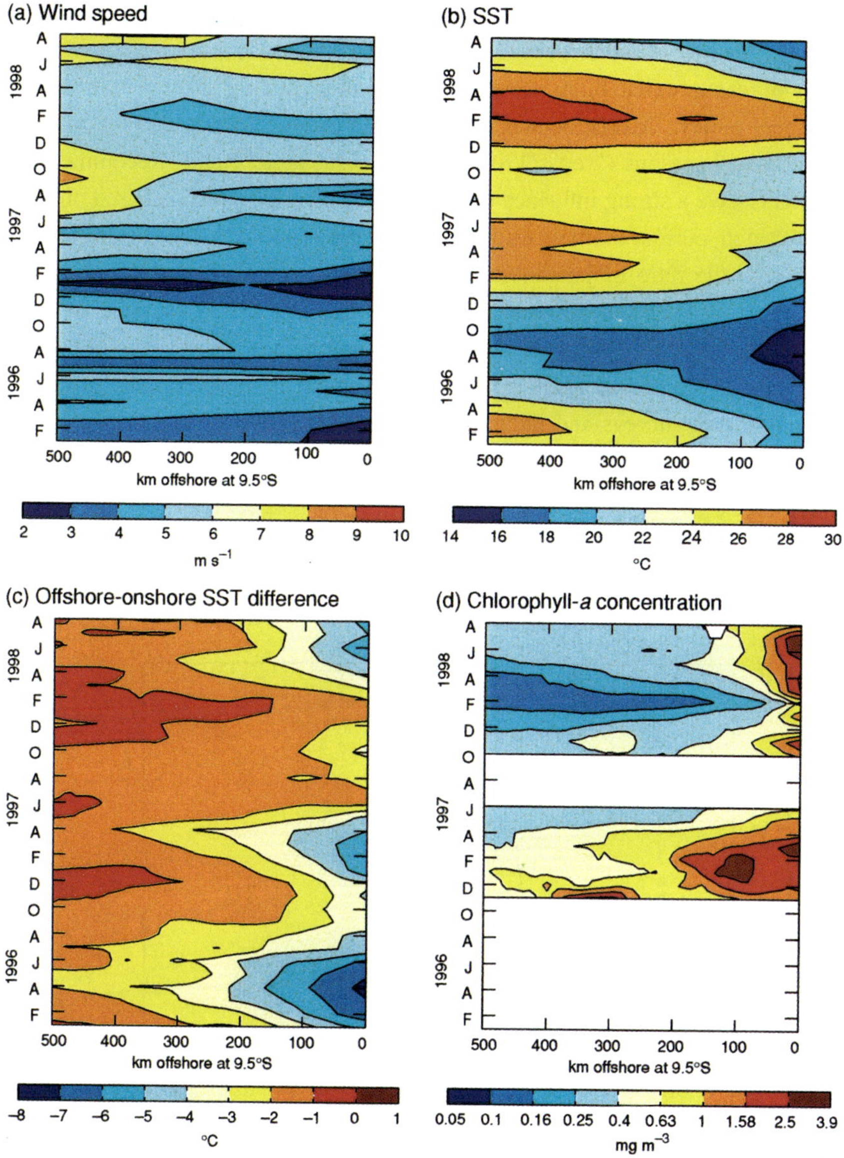

Figure 6. (a) ERS-1, NSCAT, and ERS-2 wind speed, (b) AVHRR/Pathfinder SST, (c) offshore-onshore AVHRR/Pathfinder SST difference, and (d) OCTS and SeaWiFS chlorophyll-*a* concentration along the 9.5°S transect from January 1996 to August 1998. The offshore-onshore SST difference is estimated by subtracting the SST at each point from the SST at 84.7°W (700 km offshore). SST and SST difference are 100-km averages.

3.2 Peruvian fish catch during the 1997–1998 El Niño

The catch of small pelagics for the Peruvian coast (Figure 7) reflects a combination of availability of resource (fish biomass) and its distribution, fishing effort, and the influence of government policy. MIPE dictates the allowable catch ('cuota') and the timing and duration of closed seasons ('vedas') to maximize sustainable yield. Recommendations from IMARPE have a strong influence on the allowable catch, which is based on the previous year's catch, current biomass estimates, and various indices of the health of the current stock (e.g., the ratio of juveniles to adults, fat content, status of the reproductive organs). The allowable catch can be reassessed and changed within a season. The closed seasons take place traditionally twice a year, coinciding with times of spawning (approximately in February–June and again in August–November), but determined in large part on the basis of ship surveys and industry pressure. Closed seasons aim to protect the spawning fish and juveniles (Csirke personal communication 1999), though they also coincide with periods in which the population is more dispersed (Paulik 1981). Vedas are announced for specific species in specific regions at different times, i.e., a fishing ban of a subgroup of the small pelagic stock at some area along the coast. Thus, a 100% closure,

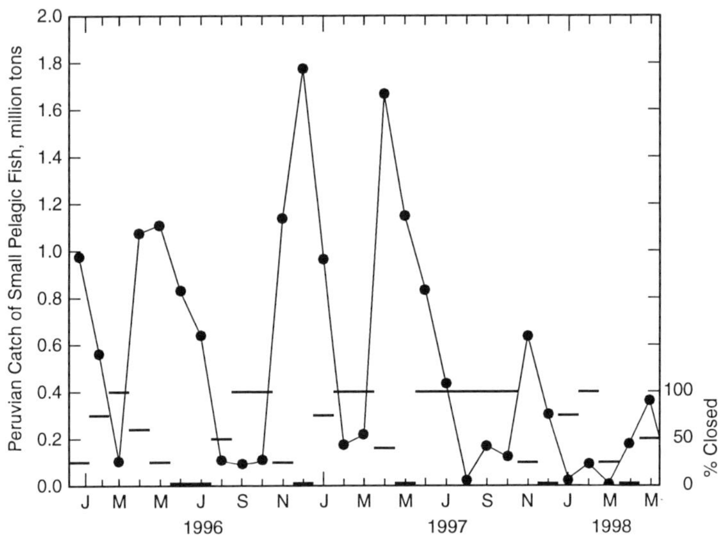

Figure 7. Monthly catch of small pelagic fish in Peru between January 1996 and May 1998. The percentage of time in which the fishery is closed represents the total amount of time in which any subgroup of the stock is banned anywhere along the coast. Consequently there are instances of 100% closure and non-zero catch.

as shown in Figure 7, does not necessarily imply a completely closed fishery and there can be non-zero catch, e.g., June, July, and September 1997 (Figure 7).

In 1996, when 'normal' oceanographic conditions prevailed, the monthly fish catch had strong oscillations, responding dramatically to the closings in March and September–October (Figure 7). Maximum fish catch (1.8 million tons) occurred in December, and high monthly catches, exceeding 1 million tons, continued into 1997. Although anomalous oceanographic conditions were observed as early as April 1997, and the onset of the El Niño was known by May 1997 (Figure 4b), the catch continued to be high in May 1997. Catchability rises (Csirke 1988) as temperature increases. Coastal regions of cold water shrink in size and become traps for the anchovy. Given a way to find the remaining areas of cold water by AVHRR SST data, it becomes easier to harvest the available stock. Additionally, anchovy aggregate in schools of an optimal size which do not appear sensitive to environmental conditions; as biomass decreases, most of the remaining fish may be found in a few schools (Csirke 1988). These two factors lead to high catches in El Niños, precisely when the stock are most vulnerable. The fishery was closed in July 1997, having reached the allowable catch and awaiting the developing El Niño conditions. It reopened briefly in November 1997, but the catch was small, and in early 1998 the catch continued to be below normal. El Niño has multiple impacts on the small pelagic fish population. Anchovy have a fairly narrow preferred temperature range (14.5–21°C) (Jordán 1971), and cease reproduction when the temperature is over 20°C, as no eggs have been found over 19.3°C (Csirke, personal communication 1999). At high temperatures the anchoveta length is reduced (Pauly and Soriano 1989) and egg mortality is increased (Pauly and Soriano 1987). Unlike sardine, which are slightly larger, anchovy are not very strong swimmers and are less likely to migrate southward to escape rising temperatures (Barber and Chávez 1986). Instead, they tend to aggregate in pockets of cool water and to swim deeper, sometimes beyond the reach of the fishing nets. In addition to the direct influence of temperature on the fish, there is a secondary effect of food availability. If phytoplankton concentrations have decreased significantly or if the phytoplankton species composition changes (especially towards small cells) the fish can starve, as they have a marked preference for large cells and mesozooplankton (which feed on large cells) (Walsh et al. 1980; Arntz and Tarazona 1990; James and Findlay 1989). Furthermore, increased predation, even cannibalism, can reduce larval survival (Walsh et al. 1980; Fiedler et al. 1986).

In an attempt to predict the available food for fish, we used a time-dependent planktonic ecosystem model (Moloney and Field 1991; Carr 1998; Carr 2000) forced by upwelling (estimated from scatterometer wind measurements), thermocline depth, and two characteristics of the upwelled water at 80-m depth. Thermocline depth and nitrate concentration and temperature of the upwelled water were estimated from a statistical relationship computed from tide gauge sea level anomaly and these three variables measured by Barber and Kogelschatz (1990) in 1982–1984. Thus, the observed sea level anomaly (Figure 4b) provided estimates of the three variables for 1996–1998. The model is size-based, i.e., rate processes are obtained from an allometric relationship with size, and the model is run with three phytoplankton size classes, three zooplankton size classes, and bacteria. The upper limit of the available carbon for small pelagic fish, which eat either mesozooplankton or net-phytoplankton, is equal to the rate of mesozooplankton grazing. One of the difficulties in comparing the numerical simulation of a planktonic ecosystem model with a long-lived, mobile vertebrate is that the response time of fish is much slower and they are able, to some extent, to 'wait out' poor environmental conditions.

Correlation coefficients were computed between fish catch (only for months when the closed season was less than 100%) and observations of sea level anomaly, SST, onshore-offshore SST difference, and chlorophyll-*a*, and modeled phytoplankton biomass and carbon available for fish. The highest correlation coefficient, r, was with the SST difference ($r = -0.54$) and the next highest correlation coefficient was with the modeled food available for fish ($r = 0.50$). Correlation coefficients of fish catch with SST anomaly was –0.40, with modeled phytoplankton it was 0.36, and with sea level it was –0.32. The lowest correlation coefficient was with chlorophyll-*a* ($r = -0.21$), which had few data and almost no measurements in 1996 (during 'normal' conditions). As expected, the statistical relationship between environmental parameters and fish catch was not strong because we are dealing with catch, not biomass, and because of complex strategies the fish use to survive poor conditions.

4. Summary and Discussion

4.1 Environmental conditions

The coastal tide gauge data, which, together with the satellite SST anomaly, show, perhaps, the clearest signal of anomalous conditions (Figure 4). The double peaks in sea level are consistent with the arrival in May–August 1997 and October 1997–February 1998 of two Kelvin waves originating in the equatorial western Pacific (Chávez et al. 1998; McPhaden 1999). The association between equatorial waves and coastal conditions along South America has been established for the 1982–1983 and 1991–1992 El Niños (Enfield et al. 1987; Shaffer et al. 1997). In the 1997–1998 El Niño, the SST anomaly (Figure 4) had a single peak and a slow return to normal conditions. This may reflect the broader spatial extent ($1° \times 1°$) of the SST data compared to tide gauge data, but is more likely an indication that the region is responding to more than the passage of a coastally trapped wave, which has its maximum expression at the coast. Increased poleward transport during El Niño (Huyer et al. 1991) brings warmer water into the region, and SST anomalies extend to greatest depth at the time of the two sea level peaks (Blanco et al. 1999).

Although the timing of the evolution of the 1997–1998 El Niño off Peru is consistent with an equatorial source (i.e., remote forcing), local forcing may also play a role. Local forcing implies that local winds were significantly decreased or even became favorable for downwelling. NSCAT and then ERS-2 wind speeds increased between June 1997 and August 1998 with respect to ERS-1 wind data in 1996 (Figure 6). There may be some uncertainty in this interpretation, as the time series involved three different scatterometers. The temporal evolution of the wind field along the coast (not shown here) also supports a continuous increase between the NSCAT and ERS-2 records.

The effect of the El Niño, where coastal upwelling brings warm, nutrient-poor water to the surface, is to reduce near-shore chlorophyll-*a* concentration and the width of the coastal band of high pigments (Barber and Chávez 1986). This occurred in 1982–1983 (Thomas et al. 1994) and in 1997–1998 (Figure 6d). Chlorophyll-*a* concentration was also sensitive to local wind changes: though minimum near-shore values were observed in December 1997 and March 1998, values of 0.5 mg m^{-3} are closest to the coast in March 1998, during a period of weakened wind speed.

The best 'predictor' of catch for the 1997–1998 period was the SST difference ($r = -0.54$), which explained 29% of the variance of fish catch. SST difference is an indicator of upwelling intensity and of food availability. The correlation coefficient is negative because greater upwelling intensities lead to larger negative values of SST difference. Mesozooplankton grazing, the model proxy for food available for fish, had the highest correlation with fish catch. Increasing the complexity of the model by including food web dynamics improved correlation coefficients by 14%. Though the total model phytoplankton biomass was only slightly reduced during the El Niño, the largest size-class (net-phytoplankton) was decimated. Mesozooplankton grazing depends on net-phytoplankton and, consequently, is greatly reduced when nutrient supply is reduced. Recovery of oceanographic conditions was rapid after March 1998, which was not reflected in fish catch (nor probably fish biomass).

Any attempt to predict the fish catch on the basis of observed environmental variables is fraught with uncertainties, i.e., mobile vertebrates, with a multi-year life span, have some ability to 'escape' or 'wait out' unfavorable environmental conditions. Using the fish catch instead of biomass (not presently available) introduces additional uncertainty by including the complication of management, with imposed minima due to cessation of harvesting, irrespective of biomass fluctuations. Fishery management decisions (size of allowable catch and closed seasons) make the relationship between environment and catch highly nonlinear by affecting the evolution of the stock itself. It seems remarkable that the correlations between environmental variables and fish catch are as good as they are.

4.2 Societal decision-making

Seasonal-to-interannual forecasts and concurrent satellite and in-situ data play an important role in fisheries-related decisions. The 1997–1998 El Niño in Peru, in which such information was more readily available than ever before, provides an opportunity to study how this information influenced decisions and the constraints of their utility.

Toward the start of the El Niño in April and May 1997, when the catch of small pelagics was large, AVHRR SST images contributed to the realization that the larger catch was not a result of an increase in the total abundance in anchovy, but that the fish were distributed in dense schools extremely close to the coast in the few remaining pockets of cool, nutrient rich water. In recognition of the larger oceanographic context, the Peruvian Fisheries Minister implemented in April 1997 a fishing ban on anchovy to protect the vulnerable stock. The ban was reversed 10 days later due to intense pressure from the politically powerful fishing industry.

Once it was widely held that the 1997–1998 El Niño event was underway, satellite images were used by a range of agents to confirm, deny, and speculate on the progress of the event. However, a lack of understanding of the images led to poor interpretation of information. For instance, the widely publicized statement of a climate trend indicating the breakup of the event in early December 1997 (Kasindorf 1997) that was based on a single satellite snapshot (http://topex-www.jpl.nasa.gov/enso97/el_niño_1997.html) led some fishing firms to believe that the fishing would return to 'normal' by early 1998, and some banks continued to invest capital in the sector.

The industrial sector makes use of real-time AVHRR SST data to identify areas of optimal temperatures for their target species and direct their fleets accordingly. Some of the largest fishing firms purchased satellite-data receiving stations prior to IMARPE's acquisition in late 1997. Large Peruvian banks, which have heavy investments in the fishing industry, also use satellite data and forecasts to make loans based on anticipated conditions. Interviews with executives of various fishing firms and bank loan officers revealed that a fishing firm was waiting to receive about 60 million dollars in bank loans, and began to sell bonds. Word of the impending strong El Niño resulted in the loan being denied and discouraging the investment in bonds. In contrast, other large industrial fishing firms heavily indebted to the banks began to renegotiate debt payments.

In the relaxation period prior to the second peak in January 1998, some Peruvian scientists interpreted a single SSH image taken from the Internet (http://topex-www.jpl.nasa.gov/enso97/el_nino_1997.html) to indicate a breakup of the warm water in the Pacific warm pool. This image was assigned predictive value and thought to signal the demise of El Niño (*El Comercio*, 16 January 1998). Actually, anomalous SST and SSH persisted until August and May, respectively.

Satellite data increase the efficiency of fish harvesting, resulting in increased exploitation of a natural resource. While this strategy may have short-term benefits for a company, it could lead to overfishing and decimation of the stock. The use of satellite data during the 1997–1998 El Niño helped anticipate fluctuations in small pelagic stock and thus, in theory, enhance responsible resource management and industrial decision-making. Such use of information, however, is in large part constrained by understanding the El Niño phenomenon and access to data.

Once the 1997–1998 El Niño event was under way, satellite data were used to confirm, deny, and speculate on the progress of the event. However, sometimes a lack of understanding of the dynamics of the coupled ocean-atmosphere system led to poor usage of satellite data. For instance, the widely publicized breakup of the El Niño in November 1997 was premature because it did not wait for additional TOPEX/Poseidon data, nor did it consider information from AVHRR SST, and it did not take into account the common two-peak structure of El Niño.

Attempts to influence opinion on the evolution of El Niño were often played out in the media and in public meetings, and satellite data were at times selectively interpreted. Given the uncertainty in forecasts, there was major controversy over the prediction of the onset of El Niño (*El Comercio*, 8 May 1997) and its duration (*El Comercio*, 10 November 1997; *El Comercio*, 7 January 1998). At times, statements about the impact of El Niño on Peru based on satellite data contradicted reports based on season-to-interannual climate forecasts, which in turn contradicted local forecasts. This led to confusion. Uncertainty was compounded because the 1997–1998 El Niño differed in its onset, duration, and biological characteristics from the extraordinary event of 1982–1983, which still loomed large in people's memory.

Not everyone in Peru has equal access to the Internet, the primary source of satellite data and seasonal-to-interannual climate forecasts, and those with access to up-to-date environmental data have an advantage. For instance, with advance information on probable changes in fish catch, a firm may buy or sell equipment at an advantage compared to an uninformed company. Similarly, a firm may lay off workers in anticipation of a decline in catch. Increased access to information leads to increased fleet efficiency. Satellite data in 1997–1998 allowed the Peruvian fishing fleets in the north to relocate in the south in anticipation of the southward migration of the anchovy stocks.

4.3 Recommendations

While it is likely that there always will be contradictory environmental information of different precision, mechanisms for standardization of presentation of information would be useful. The case of El Niño and Peru exemplifies the need for better integration of different types of information accompanied by explanation of strengths and limitations. Observations, experimental forecasts, and biological model results, if presented in a coherent and integrated manner, could more holistically contextualize current and future oceanographic conditions. Short-time decisions (e.g., planning sampling cruises) and long-term decisions (e.g., investing in a new fishmeal plant) could then be made with the best available information.

Agencies that produce environmental information should provide clear explanations and warnings about the strengths and limits of the information and aim to ensure equal access to the information. Nonetheless, even with perfect information, differential access to and understanding of information by some groups may limit a society's ability to best adapt to climate variability. However, accessibility alone, without training in the methods of interpretation of the information, is only a partial solution to the challenge.

Acknowledgments. We are grateful to two anonymous reviewers and Blanca Rojas de Mendiola for valuable comments which improved the manuscript. We thank Drs. Tim Liu and Wendy Tang, and the NSCAT Project for the gridded ERS-1 and NSCAT data, CERSAT for the ERS-2 monthly files, the Goddard DAAC and the SeaWiFS Science Project for the SeaWiFS data sets, the NOAA/NASA AVHRR Pathfinder project for providing daily sea surface temperature via the JPL-PODAAC, and the National Space Development Agency of Japan (NASDA) for the OCTS data. NASDA retains ownership of the OCTS (ADEOS) data. NASDA supports the authors in acquiring the satellite data at a marginal cost. The Fishmeal Exporters Organization kindly provided the monthly catch values. The International Research Institute for Climate Prediction, Columbia University, provided data on the societal uses of climate information originating from interviews, focus groups, surveys, participant observation, and archival research. K.B. acknowledges support from the the International Research Institute for Climate Prediction, the Tinker Foundation, the Research Institute for the Study of Man, the NOAA Office of Global Programs, and Columbia University. Funding for M.-E.C. was provided by the NSCAT QuickScience and the NASA Ocean Biogeochemistry Programs. The research described in this paper was carried out by the Jet Propulsion Laboratory, California Institute of Technology, under a contract with the National Aeronautics and Space Administration.

References

Arntz, W. E., and J. Tarazona, Effects of El Niño 1982–83 on benthos, fish, and fisheries off the South American Pacific coast, In *Global Ecological Consequences of the 1982–1983 El Niño-Southern Oscillation*, edited by P. Glyn, Elsevier, Amsterdam, 323–360, 1990.

Bakun, A., Patterns in the ocean: Ocean processes and marine population dynamcs, Report T-037, California Sea Grant College System, La Jolla, California, 323 pp, 1996.

Bakun, A., and C. S. Nelson, The seasonal cycle of wind stress curl in subtropical eastern boundary current regions, *J. Phys. Oceanogr.*, *21*, 1815-1834, 1991.

Barber, R. T., The ocean basin ecosystem, In *Concepts of Ecosystem Ecology*, edited by J. Alberts and L. R. Pomeroy, Springer-Verlag, Berlin, 166–188, 1988.

Barber, R. T., and F. P. Chávez, Ocean variability in relation to living resources during the 1982–83 El Niño, *Nature*, *319*, 279–285, 1986.

Barber, R. T., and J. Kogelschatz, Nutrients and productivity during the 1982/83 El Niño, In *Global Ecological Consequences of the 1982–1983 El Niño-Southern Oscillation*, edited by P. Glyn, Elsevier, Amsterdam, 21–53, 1990.

Barber, R. T., and R. L. Smith, Coastal upwelling ecosystems, In *Analysis of Marine Ecosystems*, edited by A. R. Longhurst, Academic Press, New York, 31–68, 1981.

Barnston, A., M. Glantz, and Y. X. He, Predictive skill of statistical and dynamical climate models in SST forecasts during the 1997–98 El Niño episode and the 1998 La Niña onset, *Bull. Amer. Meteorol. Soc.*, *80*, 217–243, 1999.

Blanco, J., M.-E. Carr, A. Thomas, and P. T. Strub, Oceanographic conditions off northern Chile during the 1996-1998 cold and and warm events, Part 1: Hydrographic conditions, *J. Geophys. Res.*, in press, 2000.

Broad, K., Climate, culture, and values: El Niño 1997–98 and Peruvian fisheries, Ph.D. dissertation, Department of Anthropology, Columbia University, New York, 311 pp, 1999.

Broad, K., A. Pfaff, and M. Glantz, Effective and equitable dissemination of seasonal-to-interannual climate forecasts: Policy implications from El Niño 1997–98 and the Peruvian fishery, *J. Policy Manage.*, submitted, 2000.

Carr, M.-E., A numerical study of the effect of periodic nutrient supply on pathways of carbon in a coastal upwelling regime, *J. Plank. Res.*, *20*, 491–516, 1998.

Carr, M.-E., Simulation of carbon pathways in the planktonic ecosystem off Peru during the 1997–1998 El Niño: Physical forcing versus the phytoplankton size composition in the upwelling source water, *J. Geophys. Res.*, submitted, 2000.

Chávez, F., A comparison of ship and satellite chlorophyll from California and Peru, *J. Geophys. Res.*, *100*, 24845–24862, 1995.

Chávez, F., P. Strutton, and M. McPhaden, Biological-physical coupling in the central equatorial Pacific during the onset of the 1997–1998 El Niño, *Geophys. Res. Lett.*, *25*, 3543–3546, 1998.

Csirke, J., Small shoaling pelagic fish stocks, In *Fish Population Dynamics*, edited by J. Gulland, John Wiley, New York, 271–302, 1988.

Csirke, J., R. Guevara-Carrasco, G. Cárdenas, M. Ñiquen, and A. Chipollini, Situación de los recursos de anchoveta (*Engraulis ringens*) y sardina (*Sardinops sagax*) a principios de 1994 y perspectivas para la pesca en el Perú, con particular referencla a las regiones norte y centro de la costa peruana, *Bol. Inst. Mar, Perú*, *15*, 1–23, 1996.

El Comercio, “Ligero incremento de temperatura de mar eleva calor en el Perú,” *El Comercio*, Lima, 8 May 1997.

El Comercio, “Mar continúa caliente frente a costas de Paita,” *El Comercio*, Lima, 10 November 1997.

El Comercio, “Fenómeno del Niño podría estar llegando a su fin,” *El Comercio*, Lima, 7 January 1998.

Enfield, D., M. Cornejo-Rodriguez, R. Smith, and P. Newberger, The equatorial source of propagating variability along the Peru coast during the 1982–1983 El Niño, *J. Geophys. Res.*, 92, 14335–14346, 1987.

Fiedler, P., R. D. Methot, and R. Hewitt, Effects of the California El Niño 1982–1984 on the northern anchovy, *J. Mar. Res.*, *44*, 317–338, 1986.

Fishmeal Exporter Organization, Proc. Annual Conf. Fishmeal Exporter’s Organization, edited by J. F. Mittaine, Fishmeal Exporter Organization, Paris, 79 pp, 1998.

Glantz, M. H., Considerations of the societal value of an El Niño forecast and the 1972–1973 El Niño, In *Resource Management and Environmental Uncertainty: Lessons from Coastal Upwelling Fisheries*, edited by M. H. Glantz and J. D. Thompson, John Wiley, New York, 449–476, 1981.

Glantz, M. H, *Currents of Change: El Niño's Impact on Climate and Society*, Cambridge University Press, Cambridge, England, 194 pp, 1996.

Huyer, A., M. Knoll, T. Paluskiewicz, and R. L. Smith, The Peru Undercurrent: A study in variability, *Deep-Sea Res.*, *38*, S247–S271, 1991

Huyer, A., R. L. Smith, and T. Paluskiewicz, Coastal upwelling off Peru during normal and El Niño times, *J. Geophys. Res.*, *92*, 14297–14308, 1987.

James, A. G., and K. P. Findlay, Effect of particle size and concentration on feeding behavior, selectivity, and rates of food ingestion by the Cape anchovy *Engraulis capensis*, *Mar. Ecol. Prog. Series*, *50*, 275–294, 1989.

Jordán, R., Distribution of anchoveta (*Engraulis ringens*), *Inv. Pesqu.*, *35*, 113–126, 1971.

Kasindorf, M., El Niño appears to be in retreat, *USA Today*, 10 December 1997.

Kearns, E. J., J. A. Hanafin, R. Evans, P. J. Minnett, and O. Brown, An independent assessment of Pathfinder AVHRR sea surface temperature accuracy using the marine-atmosphere emitted radiance interferometer (M-AERI), *J. Climate*, submitted, 2000.

McPhaden, M.J., Climate oscillations—Genesis and evolution of the 1997–1998 El Niño, *Science*, *283*, 950–954, 1999.

Moloney, C. L., and J. G. Field, The size-based dynamics of plankton food webs: 1. A simulation-model of carbon and nitrogen flows, *J. Plankton Res.*, *13*, 1003–1038, 1991.

Paulik, G. J., Anchovies, birds, and fisherman in the Peru Current, In *Resource Management and Environmental Uncertainty: Lessons from Coastal Upwelling Fisheries*, edited by M. H. Glantz and J. D. Thompson, John Wiley, New York, 35–80, 1981.

Pauly, D., and V. Christensen, Primary production required to sustain global fisheries. *Nature*, *374*, 255–257, 1995.

Pauly, D., and M. Soriano, Monthly spawning stock and egg production of Peruvian anchoveta (*Engraulis ringens*), 1953–1962, In *The Peruvian Anchoveta and Its Upwelling Ecosystem: Three Decades of Change*, edited by D. Pauly and T. Tsukuyama, International Center for Living Aquatic Resources Management (ICLARM), Manila, Philippines, 167–178, 1987.

Pauly, D., and M. Soriano, Production and mortality of anchoveta (*Engraulis ringens*) eggs off Peru, In *The Peruvian Upwelling Ecosystem: Dynamics and Interactions*, edited by D. Pauly, P. Muck, J. Mendo, and T. Tsukuyama, International Center for Living Aquatic Resources Management (ICLARM), Manila, Philippines, 155–167, 1989.

Pfaff, A., K. Broad, and M. Glantz, Who benefits from climate forecasts? *Nature*, *397*, 645–646, 1999.

Philander, G., El Niño and La Niña, *American Scientist*, *77*, 451–459, 1989.

Quinn, W., V. T. Neal, and S. Antunez de Mayolo, El Niño occurrences over the past four and a half centuries, *J. Geophys. Res.*, *92*, 14449–14461, 1987.

Reynolds, R. W., and T. Smith, Improved sea surface temperature analysis using optimum interpolation, *J. Climate*, *7*, 929–948, 1994.

Rojas de Mendiola, B., Seasonal phytoplankton distribution along the Peruvian coast, In *Coastal Upwelling*, edited by F. A. Richards, American Geophysical Union, Washington, D.C., 339–347, 1981.

Ropelewski, C., and M. Halpert, Global and regional scale precipitation patterns associated with the El Niño-Southern Oscillation, *Mon. Wea. Rev.*, *115*, 1606–1626, 1987.

Shaffer, G., O. Pizarro, L. Durfeldt, S. Salinas, and J. Rutllant, Circulation and low-frequency variability near the Chilean coast: Remotely forced fluctuations during the 1991–92 El Niño, *J. Phys. Oceanogr.*, *27*, 217–235, 1997.

Strub, P., J. Mesias, V. Montecino, J. Rutllant, and S. Salinas, Coastal ocean circulation off western South America, In *The Sea*, edited by A. R. Robinson and K. H. Brink, John Wiley, New York, 273–313, 1998.

Thomas, A. C., F. Huang, P. T. Strub, and C. James, Comparison of seasonal and interannual varibility of phytoplankton pigment concentrations in the Peru and California Current systems, *J. Geophys. Res.*, *99*, 7355–7370, 1994.

Walsh, J., T. Whitledge, W. Esaias, R. Smith, S. Huntsman, H. Santander, and B. Rojas de Mendiola, The spawning habitat of the Peruvian anchoveta (*Engraulis ringens*) stocks, *Deep-Sea Res.*, *27*, 1–27, 1980.

Mary-Elena Carr, MS 300-323, Jet Propulsion Laboratory, California Institute of Technology, Pasadena, CA 91109, U.S.A. (email, mec@pacific.jpl.nasa.gov; fax, +1-818-393-6720)

Satellites, Oceanography and Society
edited by David Halpern

Chapter 10

Satellites and fisheries: The Namibian hake, a case study

Ana Gordoa
Centro de Estudios Avanzados de Blanes, Blanes (Girona), Spain

Mercedes Masó
Instituto de Ciències del Mar, Barcelona, Spain

Lizette Voges
Ministry of Fisheries and Marine Resources, Swakopmund, Namibia

Abstract. Remote sensing is an important facet of fishery research and fishing operations because oceanographic conditions strongly influence natural fluctuations of fish stocks. Accordingly, satellite capabilities in fisheries have been long emphasized. Applications of satellite remote-sensing capabilities for fisheries from 1987 to 1998 are reviewed, emphasizing the relationship between sea surface temperature and hake availability in Namibian waters.

1. Introduction

Variations in ocean conditions play an important role in natural fluctuations of fish stocks, including their vulnerability to harvesting (Hela and Laevastu 1963). Satellite ocean remote sensing is considered an important tool in fishery research and management because it provides synoptic oceanic measurements for use in evaluating environmental impacts on the abundance and availability of fish populations (Laurs and Brucks 1985). Satellite remote-sensing applications in fisheries have focused mainly on thermal infrared images to derive sea surface temperature (SST). SST is one of the most easily measured environmental characteristics in the sea, and traditionally the one most often used in different aspects of fisheries oceanography.

There is a close link between climate and fisheries (Cushing 1982), and some knowledge of the effects that environmental parameters have on the life processes of fish is a prerequisite to understanding how and why variability in ocean conditions influences their distribution and abundance. Fish can perceive water temperature changes smaller than 0.1°C (Bull 1952) and temperature can have an impact on fish in many different ways. Temperature affects the rates of metabolic processes and thus modifies their activity level—growth, feeding rates, swimming speed, and spawning time are directly influenced by the temperature of the environment. Indirectly, it may also affect the survival rates of fish populations and, consequently, of fisheries. Laevastu and Hayes (1981) provide an extensive summary of the correlation between temperature and the behavior and occurrence of fish.

Temperature and its variability may also be an indicator of other conditions and changes in the ocean environment that might affect the distribution of fish. Examples are estimation of upwelling intensities, locations of thermal gradients and currents, and identification of surface water types. Temperature contrasts are often boundaries of surface currents, which affect the distribution and accumulation of species. Fish activities influenced by temperature variations are vertical motion, spawning, feeding, and passive transport. Ocean current and temperature are important environmental variables for many species and year-to-year variations affect the seasonal and life-cycle migrations of pelagic and semi-pelagic species. Currents hinder or facilitate fish migrations depending on relative directions. Current and temperature boundaries are usually associated with the distribution of adult fish, in addition to their association with aggregation of fish food. Consequently, the best fishing grounds are frequently located on the boundary region of two currents or in areas of upwelling and divergence. As long as four decades ago, oceanic fronts were considered to be indicators of productive fishing localities (Uda 1959). Other oceanographic features, such as eddies, may also promote the aggregation of fish (Zusser 1958), who rest in the calm center of the eddies and feed in the eddies where plankton and fry accumulate.

Three environmental requirements for predicting whether a fish population is large enough to be profitable were stated by Laevastu and Hayes (1981): (1) optimum water temperature and other environmental factors pertaining to economically significant species; (2) a sufficient number of frequent hydrographical and meteorological observations to locate critical surface isotherms and large surface temperature gradients; (3) changes in hydrographical conditions. Oceanographic features change with the season, and they may also vary on a shorter than seasonal time scale. Prediction of these variations is essential in applied fisheries oceanography. Remote sensing plays an important role in fishery research and fishing operations, since satellites provide a unique view of the ocean, synoptically covering large areas and detecting mesoscale structures through infrared, radar, or color images. Uses of satellite capabilities in fisheries have been emphasized by Montgomery (1981), Gower (1982), Yamanaka (1982), Laurs and Brucks (1985), Fiedler et al. (1985), and Fiúza (1990). The purpose of this chapter is to present an updated review and a new contribution to this relevant topic: the relationship between SST and hake availability in Namibian waters.

2. Remote Sensing and Fisheries

In the last decade, practically all work concerning fisheries and satellites has focused on pelagic species, especially those with migratory patterns, such as tuna, and those that have a great economic impact. As world demand for tuna increases, understanding that most tuna species respond directly to temperature is crucial. Thermal fronts are a good indicator of the location of productive tuna fishing areas. Real-time information of fronts, isotherms, and the location of upwelling zones would lead to decreased search time, lower fuel use, and larger fish catches, thus improving the potential for profit for tuna fishermen, who have long believed that certain types of waters increase fishing success and that some regions are known as "tuna waters" (Alberson 1961). A system to provide information about fronts and isotherm locations to tuna fishermen in the northeast Atlantic is described by Trinanes et al. (1993). Also, in order to reduce the cost of searching for tuna, some cooperative research programs between research institutions and fishermen have been developed (Barbieri et al. 1991).

Satellite-derived SST has been used to locate large and mesoscale ocean features (current boundaries, fronts, eddies). Reddy et al. (1995) analyzed the relationship between southern bluefin (*Thunnus maccoyii*) and albacore (*T. alalunga*) tuna and the occurrence of warm-core eddies and thermal fronts off eastern Tasmania. They found that edges of thermal fronts proved to be good predictors of productive fishing areas. Fiedler and Bernard (1987) showed that the distribution and diet of albacore and skipjack (*Katsuwonus pelamis*) tuna off California were related to mesoscale frontal features visible in satellite SST and phytoplankton pigment imagery. The albacore were caught in the vicinity of a cold, pigment-rich filament, with skipjack caught in warm water. In the coastal region of the Arabian Sea between Bombay and Cochin, Narain et al. (1991) found the highest fish catch was associated with a distinct temperature gradient, which Kumari et al. (1993) reported to be about 1°C in the range between 27° and 29°C. Laurs et al. (1984) and Maul et al. (1984) support the finding that tuna are more abundant near thermal fronts. However, Power and May (1991) did not find any relationship between yellowfin (*T. albacares*) tuna catch per unit effort (CPUE) and SST in the northwestern Gulf of Mexico.

Most tuna species have a preferred temperature range. Data collected from Virginia's recreational fishery showed (Bochenek 1990) that yellowfin tuna were caught at SST from 20° to 30°C, with the majority landed at SST of 24°–26°C. White marlin appear to prefer SST of 23°–27°C. In the Canary Islands, the range of SST in the catches of bigeye (*T. obesus*) tuna was larger than that observed in catches of yellowfin (Ramos et al. 1996).

Tuna are influenced by thermocline and mixed layer depths. In the western Indian Ocean the availability of yellowfin tuna is clearly affected by changes in the depth of the thermocline (Marsac 1996). When the mixed layer is drastically reduced, the CPUE on adult yellowfin tuna is much higher. The skipjack population from the Pacific Ocean showed spatial shifts apparently linked to large zonal displacements of the warm pool that occurs during El Niño events (Lehodey et al. 1997).

Information can be acquired from studying other species of pelagic fish. Kimley and Butler (1988) show the coincidence between the appearance of an assemblage of planktivorous and predatory fishes with an increase in SST and chlorophyll concentrations. Reid et al. (1993) found a positive correlation between satellite SST and herring density in 1989 and 1991 in the region between Scotland and Norway. Tameishi et al. (1994) suggested that the movement of warm streamers has a close relationship with the formation and migration of Japanese sardine (*Sardinops melanosticta*) fishing grounds.

Yanez et al. (1996a) found that SST gradients were significantly related to fishing fleet operations for jack mackerel (*Trachurus murphyi*), anchovy (*Engraulis ringens*), and the common sardine (*Clupea bentincki*). On the contrary, swordfish (*Xiphias gladius*) near the coast of central Chile was associated with temperatures from waters of oceanic origin, rather than with thermal discontinuities formed in the coastal zone (Yanez et al. 1996b). In the coastal fishery of Colima, Mexico, CPUE was related to SST as an indicator of environmental change (Espino et al. 1997). The most influential component corresponded to a lapse of 38 months, suggesting a possible link with El Niño events. However, the time scale of the most important component was 12 months, which is believed to be associated with the seasonal cycle.

In cephalopod fisheries, like the South Atlantic *Illex argentinus* fishery, squid populations (*Martialia hyadesi*) respond to environmental change. The appearance of *M. hyadesi* in the fishery over the last decade has been related to SST anomalies (Gonzalez et al. 1997), suggesting that oceanographic effects probably mediated this species, the squid's prey.

Recurrent outbreaks of disease also have important implications for coastal fisheries. In particular, the rapidly expanding sea urchin fishery on the Atlantic coast of Nova Scotia, Canada, is affected by the high mortality induced by *Paramoeba invadens*. Recent outbreaks of paramoebiasis were associated (Scheibling et al. 1997) with increased proximity to the boundary of warm water masses in the summer/fall, as indicated by satellite SST.

The application of satellite remote-sensing data for demersal fish is practically nonexistent, which is understandable considering the depth at which these species live. Variations in surface conditions may indirectly indicate changes in deeper layers (e.g., SST as an indicator of upwelling) that can affect changes in the fish distribution. Leming and Stuntz (1984) predicted hypoxic areas using satellite SST and chlorophyll data where species like shrimp and finfish were absent.

3. SST Predictor of Availability of Namibian Hake

Hake (family: Merluccidae) fisheries support over one million tons of catch worldwide each year. Although hake may live in a range of habitats, they specifically inhabit ocean fronts in productive upwelling regions that are associated with eastern boundary currents (Pitcher and Alheit 1995). The waters off Namibia are under the influence of the Benguela upwelling system in the southeast Atlantic, and the most valuable fish resource in these waters is hake (*Merlucius capensis* and *M. paradoxus*). This stock, along with the South African population, represents over one-third of the world hake biomass (Pitcher and Alheit 1995). A review of the main aspects of the biology and fisheries of the Namibian hake can be found in Gordoa et al. (1995).

Environmental conditions have been related to large-scale fluctuations in the main pelagic and demersal fisheries of this region (Shannon et al. 1988). However, there is no evidence of a correlation between environmental variables and demersal fishery productivity on a time scale shorter than a year. Changes in fishery productivity (catch rates, CPUE) do not necessarily imply changes in resource abundance: this can result from changes in fishing efficiency or fish availability. By examining short-term fluctuations in productivity, it may possible to detect changes in fish distribution, which may be governed by fish behavior (spawning or feeding migrations) and/or environmental factors.

The first observation of seasonal variability in hake availability as it relates to environmental conditions was deduced from SST satellite infrared images (Macpherson et al. 1991). Biomass in warm summers was anomalously higher than the preceding winter's biomass. The authors hypothesized that anomalous warm conditions could induce hake to concentrate closer to the seabed, making them more susceptible to bottom trawling. The Namibian National Marine Information and Research Center (NatMIRC) has a highly resolved spatial and temporal CPUE dataset based on catch rates from fishing vessel logbooks since 1994. Commercial fishing data are much larger and richer compared to data obtained with research ships. The NatMIRC data set has about 10,000 records per year.

3.1 Relationship between CPUE and SST patterns

The mean monthly CPUE of Namibian hake fisheries followed a clear seasonal pattern during the first three years of testing, 1994–1996, but not during 1997 (Figure 1). The detected seasonality shows maximum catch efficiency during late summer and early autumn. Availability decreases steadily until October, when it reaches its minimum.

The hake fishing grounds are located at the shelf-break between 200- and 500-m isobaths. Spatial analysis (Figure 2) of fishing efficiency showed that it does not change significantly throughout the year. The smaller fishing areas of some periods (mainly summer) is a consequence of less active vessels.

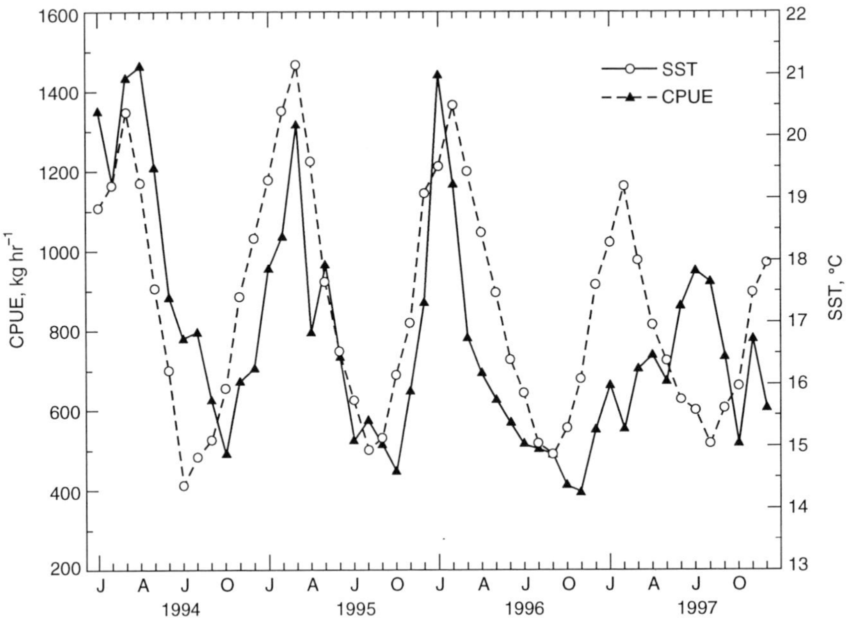

Figure 1. Monthly time series of SST, averaged from 18° to 30°S at the 200-m isobath, and Namibian hake CPUE.

To determine if the origin of the monthly pattern in catch rates is caused by seasonal migratory patterns, as have been observed in other species of hake (Bailey et al. 1982), CPUE was analyzed on a spatial-temporal basis. CPUE was estimated by 32-km × 32-km grid per month, 100-m water-depth intervals, and 1° latitude intervals. Analyses show that the monthly pattern in catch rates occurs at every depth and latitude, i.e., no latitudinal or depth changes in CPUE have been detected in relationship with a CPUE seasonal cycle (Gordoa et al. 2000). Consequently, no migratory patterns could be identified, and the monthly pattern in CPUE is a general feature throughout the whole region. This is true for the years characterized by a clear seasonal CPUE pattern, 1994–1996, and for 1997, when CPUE had a different time pattern.

The SST seasonal cycle along the Namibian coast has been described by Boyd and Agenbag (1985). Nelson and Hutchinqs (1983) and Shannon (1985) summarized the current knowledge of oceanographic processes of the Benguela upwelling system. Namibian waters are characterized by a strong seasonal signal (Figure 3), and SST can be a good indicator of upwelling intensity. Upwelling is intense throughout most of the year but is particularly strong during the winter months, which reinforces the seasonal effect (solar heating) and causes a very definite temperature cycle.

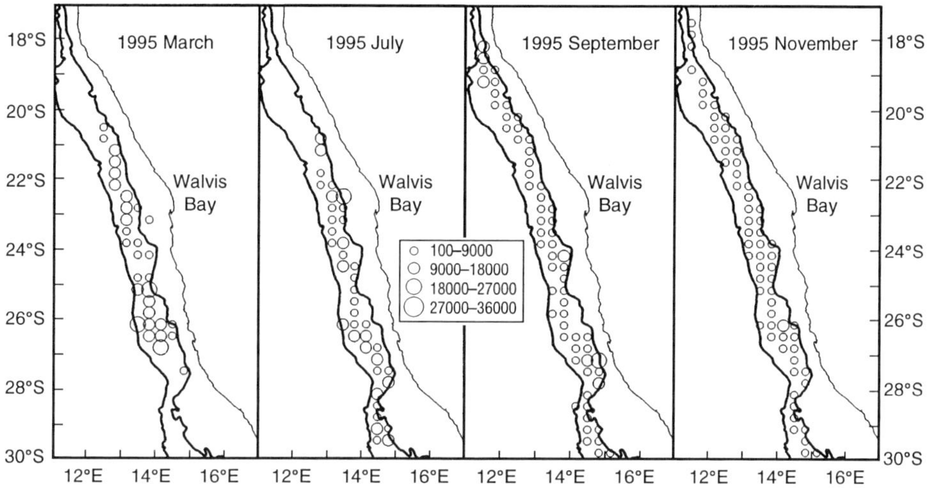

Figure 2. Monthly CPUE (kg day^{-1}) for summer (March), autumn (July), winter (September), and spring (November). Each CPUE is representative of a 32-km × 32-km region. The thick lines are the 200- and 500-m isobaths. Areas without data are areas of no fishing activity, but are not areas of no fish.

Seasonal warming of central and northern Namibia waters occurs during late summer and early autumn due to the intrusion of warm salt water of equatorial origin (Shannon et al. 1987; Boyd et al. 1987). This produces, in addition to the localization of the major upwelling cell around 26°S (Figure 3a), a perceptible SST latitudinal pattern (Figure 4). The seasonal migration to the south of the Benguela-Angola front enlarged the seasonal signal—it occurs during the quiescent upwelling periods. The seasonal migration has an important interannual variability (Shannon et al. 1987; Boyd et al. 1987). During the study period (Figure 4), the summer/autumn of 1995 and, to a lesser extent, 1996, witnessed substantial equatorial water intrusion. When these conditions exist, the northern area is extremely warm; only in the south does upwelling remain active (Figure 3b). Consequently, substantial latitudinal variability of SST occurred. A very different situation was observed in autumn 1997, when upwelling was anomalously active all along the coast, even off central Namibia, and warm equatorial waters were restricted to the northernmost area (Figure 3c).

Although SST had a very clear seasonal pattern, the amplitude of the seasonal signal was different. The 1997 seasonal signal was the weakest (Figure 4). Our results show that the seasonal pattern observed in hake availability from 1994 to 1996 was not altered by the intrusion of low-oxygen water in 1994 (Hamukuaya et al. 1998).

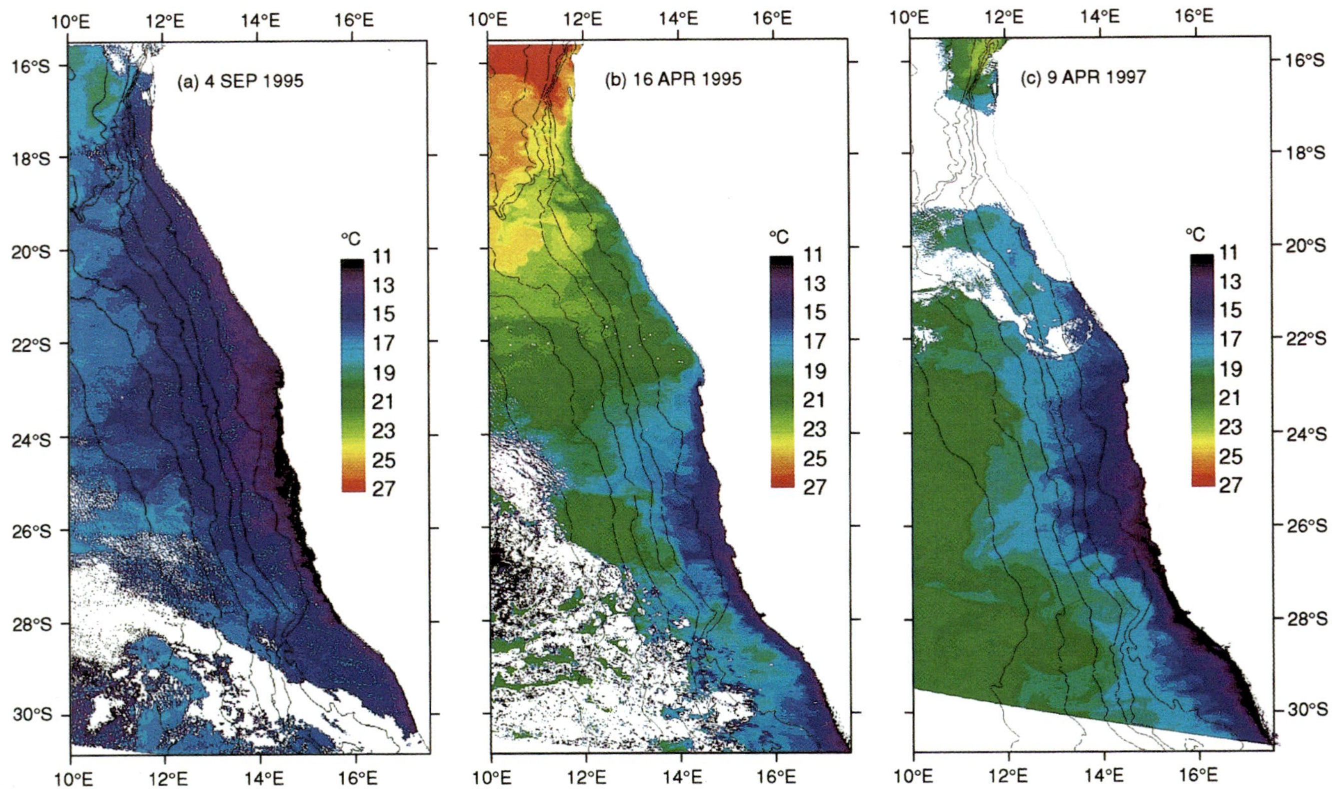

Figure 3. SST of Namibian waters during (a) active upwelling on 4 September 1995, (b) quiescent upwelling on 16 April 1995, and (c) anomalously intense upwelling on 9 April 1997.

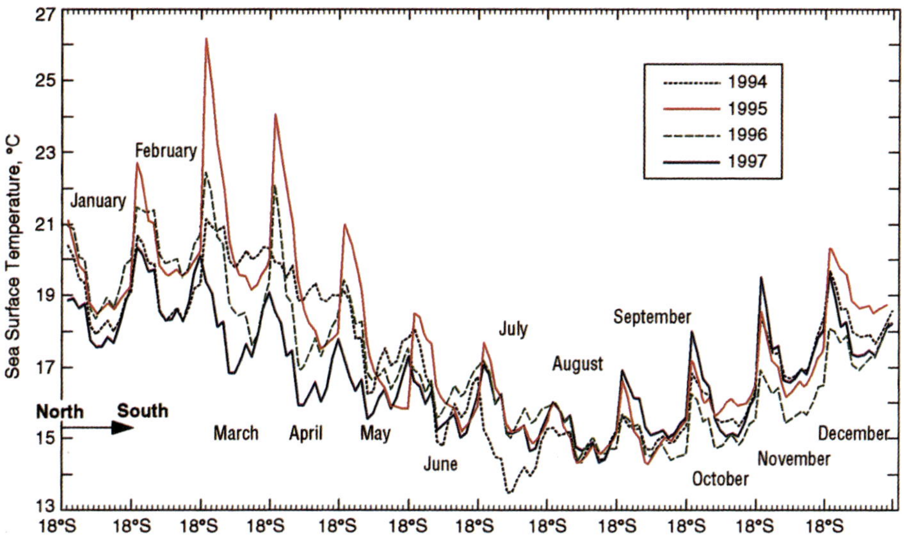

Figure 4. Monthly SST (°C) at 200-m isobath at 1°-latitude intervals (small tick mark on abscissa) along the Namibian coast from 18°S (large tick mark on abscissa) to 30°S during 1994–97.

4. Discussion

Although satellite-derived ocean-color data have been applied to fisheries research, the Coastal Zone Color Scanner (CZCS) was only operational from 1978 to 1986. The August 1997 launch of the Sea-viewing Wide Field-of-view Sensor (SeaWiFS), designed to be the successor to CZCS, opens up the possibility that remote sensing of ocean color will play a major role in the future of fisheries research. Nevertheless, applications of satellite remote sensing reviewed in this chapter showed that, in the last decade, the satellite data used by fisheries is mostly SST. In spite of the fact that the distribution of pelagic fish is closely related to the distribution of thermal features, the scarcity of scientific work on that relationship using satellite data is surprising.

Environmental conditions affect fisheries in different ways and degrees. For time scales less than a year, fish-catch variability may be due to fish migratory patterns and/or fish availability. Although no definitive conclusions can be drawn at this stage, we definitely know that hake populations respond, directly or indirectly, to environmental seasonal cycles. Direct response should be possible, because there are significant changes in the physical structures of the water masses inhabited by hake. In the northwest Atlantic, Perry and Smith (1994) found that silver hake (*Merluccius bilinearis*) is a temperature-keeper, following similar water temperatures in winter and summer by changing its seasonal depth distributions. Localization of the shelf-break current, which characterizes the coast of British Columbia, definitely affects hake (*M. productus*) distribution (Mackas et al. 1997), which were dispersed in summer when shelf-break upwelling is reduced.

We found a clear relationship between the SST cycle and hake availability. The availability of infrared satellite-derived SST has revealed, as in other areas worldwide, the mesoscale variability that characterizes the Benguela upwelling system (Lutjeharms and Stockton 1987; Lutjeharms et al. 1995). Seasonal and mesoscale variability of the current patterns of Namibian waters is related to the variability of upwelling intensity. Upwelling quiescent periods are characterized by an onshore movement of the oceanic front. It seems likely that changes in flow dynamics affect hake vertical distribution, as was previously hypothesized by Macpherson et al. (1991) or hake aggregation size (Dorn 1997).

Seasonal success of Namibian hake fisheries is not a simple function of stock size, but results from the complex interactions of hydrodynamics and fish behavior. Further research on different time and spatial fisheries/environment interactions is warranted.

Acknowledgments. We gratefully acknowledge the help and cooperation of Gorka Bidegain, Nuria Raventos, L. Burmesiter, and C. Bartholomae. This work was sponsored by the Agencia Internacional de Cooperación Española.

References

Alberson, D. L., Ocean temperature and its relation to albacore tuna (*Thunnus germo*) distribution in waters off the coast of Oregon, Washington and British Columbia, *J. Fish. Res. Board Canada, 18*, 1145–1152, 1961.

Barbieri, M. A., E. Yanez, and M. Fabrias, Remote sensing and the Chilean small-scale albacore and swordfish fishery: An example of technology transfer, *Colloq. Semin. Inst. Fr. Rech. Sci. Dev. Coop.*, *2*, 1991.

Bailey, K. M., R. C. Francis, and P. R. Stevens, The life history and fishery of Pacific whiting, *Merluccius productus, CalCOFI Rep.* XXIII, 81–98, 1982.

Bochenek, E. A, Virginia's pelagic recreational fishery: Biological, socioeconomic and fishery components, *Diss. Abst. Int. Pt. B Sci. Eng.*, *51*, 290, 1990.

Boyd, A. J., and J .J. Agenbag, Seasonal trends in the long shore distribution of surface temperatures off southwestern Africa 18–34°S, and their relation to subsurface conditions and currents in the area 21–24°S, In *International Symposium on the Most Important Upwelling Areas off Western Africa*, edited by C. Bas, R. Margalef, and P. Rubiés, 119–148, 1985.

Boyd, A., J. Salat, and M. Masó, The seasonal intrusion of relatively saline water on the shelf off northern and central Namibia, In *The Benguela and Comparable Ecosystems*, edited by J. I. L. Payne, J. A. Gulland, and K. H. Brink, *S. Afr. J. Mar. Sci.*, *5*, 107–120, 1987.

Bull, H. O., An evaluation of our knowledge of fish behavior in relation to hydrography. *Rapp. ICES*, *131*, 8–23, 1952.

Cushing, D. H., *Climate and Fisheries*, Academic Press, New York, 373 pp, 1982.

Dorn, M. W., Mesoscale fishing patterns of factory trawlers in the Pacific hake (*Merluccius productus*) fishery, *CalCOFI Rep.*, *38*, 77–89, 1997.

Espino Barr, E., R. Macias Zamora, M. Cruz Romero, and A. Garcia-Boa, Catch per unit effort trends in the coastal fishery of Manzanillo, Colima, Mexico, *Fish. Manage. Ecol.*, *4*, 255–261, 1997.

Fiedler, P. C., G. B. Smith, and R. M. Laurs, Fisheries applications of satellite data in the eastern North Pacific, *Mar. Fish. Rev.*, *46*, 1–13, 1985.

Fiedler, P. C., and H. J. Bernard, Tuna aggregation and feeding near fronts observed in satellite imagery, *Cont. Shelf Res.*, *7*, 871–881, 1987.

Fiúza, A. F. G., Application of satellite remote sensing to fisheries, In *Operations Research and Management in Fishing*, edited by A. G. Rodrigues, Kluwer Academic Publishers, Dordrecht, The Netherlands, 257–279, 1990.

Frechet, A., Catchability variations of cod in the marginal ice zone, *Can. J. Fish. Aquat. Sci.*, *47*, 1678–1683, 1990.

Gargett, A. E., Physis to fish: Interactions between physics and biology on a variety of scales, *J. Oceanogr.*, *10*, *3*, 128–131, 1997.

Gonzalez, A. F., P. N. Trathan, C. Yau, and P. G. Rodhouse, Interactions between oceanography, ecology and fishery biology of the ommastrephid squid *Martialia hyadesi* in the South Atlantic, *Mar. Ecol. Prog. Ser.*, *152*, 1–3, 205–215, 1997.

Gordoa, A., E. Macpherson, and M. P. Olivar, Biology and fisheries of Namibian hakes (*M. paradoxus* and *M. capensis)*, In *Hake Fisheries, Ecology and Markets*, edited by J. Alheit and T. Pitcher, Chapman and Hall, London, 49–79, 1995.

Gordoa, A., M. Masó, and L. Voges, Monthly variability in the catchability of Namibian hake and its relationship with environmental seasonality, *Fish. Res.*, in press, 2000.

Gower, J. F. R., General overview of the nature and use of satellite remote sensing data for fisheries application, *NAFO Science Council Studies*, *4*, 7–19, 1982.

Hamukuaya, H., M. J. O'Toole, and P. M. J. Woodhead, Observations of severe hypoxia and offshore displacement of Cape hake over the Namibian shelf in 1994, *S. Afr. J. Mar. Sci.*, *19*, 57–59, 1998.

Hela, L., and T. Laevastu, *Fisheries Hydrography*, Fishing New Books, London, 137 pp, 1963.

Klimley, A. P., and S. B. Butler, Immigration and emigration of a pelagic fish assemblage to seamounts in the Gulf of California related to water mass movements using satellite imagery, *Mar. Ecol. Prog. Ser.*, *49*, 11–20, 1988.

Kumari, B., M. Raman, A. Narain, and T. E. Sivaprakasam, Location of tuna resources in Indian waters using NOAA AVHRR data, *Int. J. Remote Sensing*, *14*, 3305–3309, 1993.

Laevastu, T., and Hayes, *Fisheries Oceanography*, Fishing New Books, Norwich, England, 199 pp, 1981.

Laurs, R. M., P. C. Fiedler, and D. R. Montgomery, Albacore tuna catch distributions relative to environmental features observed from satellites, *Deep-Sea Res.*, *31*, 1085–1099, 1984.

Laurs, R. M., and J. T. Brucks, Living marine resources applications, *Adv. in Geophys.*, *27*, 419–452, 1985.

Lehodey, P., M. Bertignac, J. Hampton, A. Lewis, and J. Picaut, El Niño Southern Oscillation and tuna in the western Pacific, *Nature*, *369*, 715–718, 1997.

Leming, T. D., and W. E. Stuntz, Zones of coastal hypoxia revealed by satellite scanning have implications for strategic fishing, *Nature*, *318*, 136-138, 1984.

Lutjeharms, J. R. E., and P. Stockton, Kinematics of the upwelling front off southern Africa, In *The Benguela and Comparable Ecosystems*, edited by A. I. L. Payne, J. A. Gulland, and K. H. Brink, *S. Afr. J. Mar. Sci*, *5*, 35–49, 1987.

Lutjeharms, J. R. E., D. J. Webb, B. A. de Cuevas, and S. R. Thompson, Large-scale modeling of the southeast Atlantic upwelling system, *S. Afr. J. Mar. Sci.*, *16*, 205–225, 1995.

Mackas, D. L., R. Kiesser, M. Saunders, D. R. Yelland, R. M. Brown, and D. F. Moore, Aggregation of euphausiids and Pacific hake (*Merluccius productus*) along the outer continental shelf off Vancouver Island, *Can. J. Fish. Aquat. Sci.*, *54*, 2080–2096, 1997.

Macpherson, E., M. Masó, M. Barange, and A. Gordoa, Relationship between measurements of hake biomass and sea surface temperature off southern Namibia, *S. Afr. J. Mar. Sci.*, *10*, 213–217, 1991.

Maul, G. A., F. A. Williams, M. A. Roffer, and F. M. Sousa, Remotely sensed oceanographic patterns and variability of bluefin tuna catch in the Gulf of Mexico, *Oceanol. Acta.*, *7*, 469–479, 1984.

Montgomery, D. R., Commercial applications of satellite oceanography, *Oceanus*, *24*, 56–65, 1981.

Marsac, F., Oceanographic research in relation with tuna fisheries assessment: The regional tuna project of the "Commission de l'Ocean Indien," *Andhra. Univ. Oceanogr. Mem.*, *3*, 158–175, 1996.

Narain, A., R. M. Dwivedi, B. Kumari, N. Chaturvedi, H. U. Solanki, P. C. Mankodi, and D. Sudarsan, Relationship between sea-surface temperature (SST) and fish catch data: A feasibility study, In Proc. Nat. Workshop on Fish. Res. Data and the Fishing Ind., Visakhapatnam, 19–261, 1991.

Nelson, G., and L. Hutchings, The Benguela upwelling area, *Prog. Oceanogr.*, *12*, 333–356, 1983.

Perry R. I., and S. J. Smith, Identifying habitat associations of marine fishes using survey data: An application to the northwest Atlantic, *Can. J. Fish. Aquat. Sci.*, *51*, 589–602, 1994.

Pitcher, T. J., and J. Alheit, What makes a hake? A review of the critical biological features that sustain global hake fisheries, In *Hake Fisheries, Ecology and Markets*, edited by J. Alheit and T. Pitcher, Chapman and Hall, London, 1–14, 1995.

Power, J. H., and L. N. May, Jr., Satellite observed sea-surface temperatures and yellowfin tuna catch and effort in the Gulf of Mexico, *Fish. Bull.*, *89*, 429–439, 1991.

Ramos, A., A. Delgado de Molina, J. Ariz, J. C. Santana, L. Garcia Weill, and M. Canton, Aggregations of yellowfin tuna (*Thunnus albacares*; Bonaterre 1788) and bigeye (*Thunnus obesus*; Lowe 1839) in oceanographic sub-mesoscale events in the Canary Islands area observed through infrared teledetection (SCRS/95/71), *Collect. Vol. Sci. Pap. Iccat. Recl. Doc. Sci. Cicta. Colecc. Doc. Cient. Cicaa*, *45*, *3*, 175–181, 1996.

Reid, D., D. Williams, A. Gambang, and J. Simmonds, Distribution of North Sea herring and their relationship to the environment, *ICES Council Meeting Papers*, 12 pp, 1993 (ICES-CM-1993/H:23).

Reddy, R., V. Lyne, R. Gray, A. Easton, and S. Clarke, An application of satellite-derived sea surface temperatures to southern bluefin tuna and albacore off Tasmania, *Sci. Mar. Barc.*, *59*, 3–4, 445–454, 1995.

Shannon, L. V., The Benguela ecosystem. 1. Evolution of the Benguela, physical features and processes, In *Oceanography and Marine Biology, An Annual Review*, *23*, edited by M. Barnes, University Press, Aberdeen, 105–182, 1985.

Shannon, L. V., J. J. Agenbag, and M. E. L. Buys, Large and mesoscale features of the Angola-Benguela front, In *The Benguela and Comparable Frontal Systems*, edited by A. I. L. Payne, J. A. Gulland, and K. H. Brink, *S. Afr. J. Mar. Sci.*, *5*, 11–34, 1987.

Shannon, L. V., R. J. M. Crawford, G. B. Brundrit, and L. G. Underhill, Responses of fish populations in the Benguela ecosystem to environmental change, *J. Cons. Int. Explor. Mer.*, *45*, 5–12, 1988.

Scheibling, R. E., and A. W. Hennigar, Recurrent outbreaks of disease in sea urchins *Strongylocentrotus droebachiensis* in Nova Scotia: Evidence for a link with large-scale meteorologic and oceanographic events, *Mar. Ecol. Prog. Ser.*, *152*, 1–3, 155–165, 1997.

Stretta, J. M., and M. Petit, Thonides tropicaux: La synthese ecologique. *Collect. Vol. Sci. Pap. Iccat Recl. Doc. Sci. Cicta Collecc. Doc. Cient. Cicaa.*, *39*, *1*, 307–321, 1992.

Tameishi, H., Y. Naramura, and H. Shinomiya, Role of warm streamers in the northward migration of Japanese sardine off Sanriku, *Nippon Suisan Gakkaishi Bull. Jap. Soc. Sci. Fish.*, *60*, *1*, 45–50, 1994.

Trinanes, J. A., J. Arias, J. M. Cotos, and J. Torres, Monitoring and detection system for operational use in tuna fisheries, *ICES Council Meeting Papers*, 7 pp, 1993 (ICES-CM-1993/H:13).

Uda, M., Water mass boundaries 'Siome': Frontal theory in oceanography, *Fish. Res. Board Canada*, *51*, 10–20, 1959.

Yamanaka, I., Application of satellite remote sensing to fishery studies in Japan, *NAFO Sci. Council Studies*, *4*, 41–50, 1982.

Yanez, E., V. Catasti, M. Barbieri, and G. Bohm, Relationships between the small pelagic resources distribution and the sea surface temperatures recorded by NOAA satellites from Chile central zone, *Invest. Mar.*, *24*, 107–122, 1996a.

Yanez, E., C. Silva, M. Barbieri, and K. Nieto, Artisanal swordfish fishery and sea surface temperatures from NOAA satellites in central Chile, *Invest. Mar.*, *24*, 131–144, 1996b.

Zusser, S. G. A., A contribution to the study of fish behavior, paper presented to the IPFC Symp. on Fish Behavior, 1958.

Ana Gordoa, Centro de Estudios Avanzados de Blanes (CEAB-CSIC), Camí de Sta. Barbara s/n, 17300 Blanes (Girona), Spain. (email, gordoa@ceab.csic.es; fax, +34-972-337-806)

Satellites, Oceanography and Society
edited by David Halpern

Chapter 11

Ocean-color satellites and the phytoplankton-dust connection

P. M. Stegmann
Graduate School of Oceanography, University of Rhode Island, Narragansett

Abstract. Results of a time series of satellite measurements of aerosol radiance made with two ocean-color sensors are presented. Data from the Coastal Zone Color Scanner (CZCS) were collected from 1978 to 1986. The follow-on sensor, the Sea-viewing Wide Field-of-view Sensor (SeaWiFS), has been transmitting data since September 1997. Both CZCS and SeaWiFS images successfully depicted regions of well-known, large-scale mineral aerosol plumes, the seasonality of which corresponds to that found by other satellite and land-based platforms. Aerosol radiance extractions were made for two subregions in the North Atlantic, both of which are recipients of regular mineral aerosol deposits originating from northwest Africa. In the almost eight-year time series obtained with CZCS, the annual cycle in both subregions follows a similar pattern each year and agrees well with results from the published literature. However, there is interannual variability and the observed fluctuations may be linked to climatic shifts associated with the North Atlantic Oscillation. The SeaWiFS annual cycle of aerosol radiance in both subregions closely followed that found in the CZCS climatology; SeaWiFS-measured aerosol optical thickness mirrors aerosol radiance to a high degree. The higher temporal resolution offered by the SeaWiFS data demonstrates the sporadic nature of dust events throughout the entire year and not only during the high dust season.

1. Phytoplankton Regulation

An important control of phytoplankton growth is nutrient supply. Nitrate, phosphate, and silicate are considered the macronutrients most vital for phytoplankton carbon assimilation in ocean ecosystems. In addition to these three major nutrients, however, there are micronutrients that have been shown to enhance phytoplankton growth, and in some regions, even be the controlling nutrient. The work done by the late John Martin and coworkers more than a decade ago formulated the hypothesis that iron was the limiting micronutrient in so-called high-nutrient, low-chlorophyll (HNLC) open-ocean regions (Martin and Gordon 1988; Martin 1991, 1992), which are identified as the equatorial

Pacific, the Southern Ocean, and the subarctic Pacific (Hutchins 1995). With the success of the iron fertilization experiments (IronEx I and II), Martin's hypothesis was confirmed (Martin et al. 1994; Kolber et al. 1994; Coale et al. 1996), leaving little doubt that phytoplankton growth in some open-ocean regions of the world's oceans is regulated by iron availability. However, open-ocean HNLC regions are not the only areas that can be iron-limited. The recent study by Hutchins and Bruland (1998) found iron-limited growth in a productive coastal upwelling regime, indicating that iron limitation may be more widespread than previously thought. And most recently, Behrenfeld and Kolber (1999) found a large expanse of the South Pacific gyre, a non-HNLC region, to be iron-limited as well. Thus, the study of, and interest in, the interaction between micronutrient input and phytoplankton ecosystem dynamics will certainly continue.

The supply of new iron to the photic zone of open-ocean regions has been shown to be predominantly derived from the atmosphere (Duce 1986; Duce and Tindale 1991), although upwelling and vertical advection have been found to supply iron to surface waters as well and, in fact, contribute substantially more iron than that estimated via aeolian deposits in two HNLC ecosystems (De Baar et al. 1995; Gordon et al. 1997). Atmospherically transported iron is found in mineral dust particles that originate in deserts or arid/semi-arid regions and can be carried long distances with predominant winds before being deposited on the sea surface (Duce et al. 1991). Mineral dust is an important component of atmospheric aerosols, which in turn can be an important forcing mechanism on global climate (Andreae and Crutzen 1997). In fact, after sea salt, mineral dust is the second-largest source of global atmospheric aerosols (Andreae 1995). Given its profound influence on the Earth's radiation budget, it becomes clear why it is crucial to monitor dust concentration and transport on a global scale. Furthermore, as surface-residing phytoplankton remove iron-laden aerosol particles, their rate of photosynthesis may increase, which can result in increased removal of carbon dioxide. Thus, changes in phytoplankton biomass as a result of aerosol input may have a direct feedback loop to changes in climate (Denman et al. 1996; Falkowski et al. 1998).

One of the major difficulties faced in determining such effects and linkages, however, has been the lack of long-term records of atmospheric aerosols on a global scale. Even though numerous ground-based stations do exist, a network of stations that spans the globe and provides adequate spatial and temporal coverage on a worldwide basis is not available and does not seem practical. The remedy to this problem may be satellite-based measurements. Earth-observing satellites can provide a platform for sensors to acquire synoptic as well as long-term records of atmospheric aerosols. Thus, satellite measurements of global aerosol loading can help us understand the role of the oceans in global climate patterns. The first step necessary in trying to study the relationship between phytoplankton growth and mineral aerosols via satellite is to find an Earth-observing platform with the ability to measure both parameters at relevant scales on a global basis.

This paper discusses the role of satellites, particularly ocean-color sensors, in studying the phytoplankton-dust-climate connection over the last two decades. Beginning with a brief overview of two of the most common platforms used to monitor the spatial and temporal variability of aerosols, some major results concerning global aerosol patterns and transport are presented, with the focus on mineral aerosols in the Atlantic and Pacific Oceans. This does not relegate the many other substances comprising atmospheric aerosols as unimportant; they are simply beyond the scope of this paper. For detailed overviews of aerosols and climate, the reader is referred to some excellent books on the subject (e.g., Charlson and Heintzenberg 1995; Hobbs 1993). Furthermore, this paper does not intend to be a comprehensive review of all the available literature on remote sensing of aerosols. Rather, the provided synopsis will be succinct, and is intended, first, to supply some basic background information on the seasonal cycle of mineral aerosol distributions and, second, to present those representative results that will be used when discussing mineral aerosol patterns obtained from ocean-color sensors. It is hoped that this type of format will highlight some of the major advances made in the tools used to study phytoplankton and aeolian input and their connection to climate change.

2. Measuring Aerosols

Several methods have been developed to measure atmospheric aerosol load and composition. The traditional mode collects samples at land- or ship-based towers that are then analyzed to determine aerosol concentration levels or elemental composition. Another, newer method utilizes data recorded with sensors located onboard satellites. Both are briefly presented here for the two most-studied oceans, the Atlantic and Pacific, with the aim of deriving the general picture of dust modes in these two ocean basins.

2.1 Ground-based platforms

Measurements of dust concentration, composition, and transport to the Atlantic and Pacific Oceans have been primarily conducted at island stations and as part of large-scale, international programs (see review by Duce 1995). One of the longest records in the Atlantic is from Barbados, where dust concentrations have been sampled continuously for over 30 years (Prospero and Nees 1986; Prospero 1996). Barbados is downwind from the desert and arid regions of northern Africa, the major source of mineral aerosols deposited at this island. Although dust outbreaks from sources in North Africa are episodic and can occur throughout the year, the highest (lowest) dust concentrations measured at Barbados are in summer (winter), with a similar seasonal pattern also observed at Bermuda and on the Canary Islands (Prospero 1996), indicating how far these dust events can be transported.

Sampling at Pacific Ocean sites has a shorter timeline, but their placement on a series of island stations spanning roughly 50°N to 30°S allows for a relatively detailed picture of dust activity and transport (Prospero 1996). From this time series, it was found that the seasonal cycle of dust deposition was highest in spring and corresponded well to the period of major dust storm activity emanating from the Asian continent (Prospero 1996). It was also found that dust transport to stations located in the South Pacific was much lower than to those in the North Pacific.

This network of sampling stations has undoubtedly provided invaluable information on the seasonal cycle of mineral aerosols (and other aerosol components), including elemental composition and the interannual variability of aerosol deposition patterns. However, a network of ground-based measurements is spatially restrictive and does not allow accurate synoptic coverage. This is where Earth-observing satellites offer a unique platform from which to obtain both global coverage and the possibility of establishing a long-term time series of the global atmospheric aerosol burden. A recently established ground-based aerosol monitoring network, Aerosol Robotic Network (AERONET) (Holben et al. 1998), began in 1993 and measures aerosol optical properties with spectral radiometers installed at over 60 locations across the globe, including several island stations. AERONET measurements are automated and data are transmitted via satellite to a global database. Easily accessible on the Internet, this database provides near real-time aerosol information that can be used in conjunction with satellite data and other aerosol measurement capabilities.

2.2 Satellite platforms

Kaufman (1995) presents a comprehensive review of satellite sensors used for aerosol applications. He also points out that no satellite sensors were explicitly built to study tropospheric aerosols; this was not the case for stratospheric aerosols, which used the Stratospheric Aerosol and Gas Experiment (SAGE) sensors. Nonetheless, several important aerosol properties have been derived with satellite data; three of the longer time series datasets will be summarized here. Observations of dust outbreaks and their movement across the oceans can also be tracked in visible images (snapshots) from the Geostationary Operational Environmental Satellite (GOES) and the geostationary Meteorological Satellite (METEOSAT), but these will not be detailed here.

Routine monitoring of global aerosol optical thickness (AOT) fields over the ocean derived from the Advanced Very-High Resolution Radiometer (AVHRR) sensor flown onboard National Oceanic and Atmospheric Administration (NOAA) satellites has been carried out since 1987 (Rao et al. 1989), with the retrieval algorithm recently undergoing its phase 2 revision (Stowe et al. 1997). This operational product gives an estimate of the amount of solar radiation backscattered over the oceans by aerosols, thereby providing a synoptic estimate of aerosol source regions and distribution patterns. Some of the most striking and expansive features visible in these global maps are large-scale dust plumes, especially those found in the tropical North Atlantic, the Arabian Sea, and the northwest

Pacific (Husar et al. 1997). The seasonal cycle of AOT fields is found to be consistent with aerosol concentrations measured in situ at island sites, with highest levels generally in spring/summer and lowest in winter (Husar et al. 1997).

Another long, global time series of remotely sensed aerosols was obtained with the Nimbus-7 Total Ozone Mapping Spectrometer (TOMS) for the years 1979–1993 (Herman et al. 1997). Although primarily designed to detect ozone, Herman et al. (1997) were able to determine the distribution patterns of ultraviolet-absorbing aerosols using TOMS data. Again, highest levels were predominantly observed during the spring/summer period, agreeing well with results obtained from other sampling platforms.

Numerous other studies using satellite-derived estimates of aerosol burden have also been carried out on regional or basin-scale levels. One example is an 11-year study (Moulin et al. 1997a, 1997b), which used METEOSAT data to estimate Saharan dust transport to the western Mediterranean. While the authors found the highest dust intensity to occur in spring/summer, consistent with previous datasets, they also found a high degree of variability in seasonal and interannual time scales, indicative of the sporadic nature of these intense dust episodes. They concluded that seasonal variability was forced by meteorological conditions, while interannual variability was strongly related to changes in the North Atlantic Oscillation (NAO) (Hurrell 1995). Furthermore, they found that changes in aerosol optical depth (derived from METEOSAT) were mirrored in the NAO index, and that increases in the NAO index corresponded to increases in mineral aerosol concentrations deposited at Barbados. Obviously, there is very close linkage between fluctuations in mineral dust transport out of the northwest African continent and climatic oscillations in the North Atlantic.

3. Ocean-Color Sensors

3.1 Coastal Zone Color Scanner

The first Earth-observing sensor whose primary objective was to capture changes in ocean color, i.e. changes in phytoplankton pigment concentration, was the Coastal Zone Color Scanner (CZCS) launched by the National Aeronautics and Space Administration (NASA) in 1978. Originally intended as a one-year proof-of-concept mission, CZCS remained functional until June 1986, outliving its planned lifetime by almost seven years. The many thousands of CZCS images revolutionized our view of the oceans as a domain far more dynamically active than previously thought. CZCS images continue to be analyzed even today, over two decades after its launch, and are now, for example, compared to new phytoplankton biomass estimates obtained from recent ocean-color sensors (e.g., Sea-viewing Wide Field-of-view Sensor (SeaWiFS) and Ocean Color and Temperature Scanner (OCTS)).

Besides providing estimates on the distribution and variability of phytoplankton pigment content, the CZCS dataset also included aerosol radiance intensities. One CZCS channel centered at 670-nm wavelength was exclusively used in the atmospheric correction process and then cast aside (Gordon and Castaño 1987). While a few studies had shown that aerosols over the ocean could be retrieved by CZCS at this wavelength (Stegmann and Tindale 1999), a global study spanning the duration of the CZCS mission was only recently completed (Stegmann and Tindale 1999). Global climatological maps of aerosol radiance showed the occurrence of large-scale aerosol plumes in all major ocean basins (Stegmann and Tindale 1999). Mineral aerosol particles are likely the major light scatterers in these global maps, given both the wavelength of the channel used as well as their sheer abundance (Seinfeld and Pandis 1998). What this means is that the CZCS aerosol signal may be more sensitive towards mineral dust and thus be more indicative of the dust load in the atmosphere. Indeed, the seasonal distribution pattern of aerosol radiance derived from CZCS (Figure 1) shows a remarkable correspondence to the seasonality of global AOT fields derived from AVHRR and which are attributed primarily to dust plumes (Husar et al. 1997). This good correspondence occurs despite the fact that these are two very different sensors (CZCS versus AVHRR); there is no temporal overlap between the two datasets (CZCS is 1978–1986, AVHRR is 1989–1991), and the compared aerosol products are not the same (one is AOT (AVHRR), while the other is aerosol radiance (CZCS)).

It is pretty remarkable that CZCS and AVHRR results are in such good accord, especially given all the caveats associated with this comparison. But equally impressive is that the spatial distribution and temporal development of observed mineral aerosols matched that found from land-based sampling sites (cf. Section 2). A time series of mineral aerosol load at two sites in the North Atlantic obtained from CZCS exemplifies this correspondence. Aerosol radiance (at 670 nm) was extracted from a region off the southeastern U.S. coast (SEC; approximately 25–34°N, 68–76°W) and Bermuda (BER; approximately 28–32°N, 59–65°W). Monthly mean radiance intensities in each of these regions for the 7.5-year time period are shown in Figure 2. What can be clearly seen is the repeated seasonality with which dust deposition patterns occurs at both sites; this is consistent with the long-term ground-based dust records (cf. 2.1) as well as the AOT fields measured from AVHRR (cf. 2.2). In fact, the mineral dust pattern observed with CZCS at the Bermuda site even coincides with dust flux measurements to a deep-ocean sediment trap in the Bermuda region during the same period (Jickells et al. 1998). This result is surprising considering that temporal and spatial sampling scales of CZCS and subsurface sediment traps are substantially different from each other.

Another interesting feature in Figure 2 is the observed increase in radiance intensities in the SEC region during the summer months from 1982 to 1983, and then again in 1985. This observation coincides with elevated concentrations of mineral dust measured at Barbados during this period (Prospero 1996). A comparable summertime increase in the

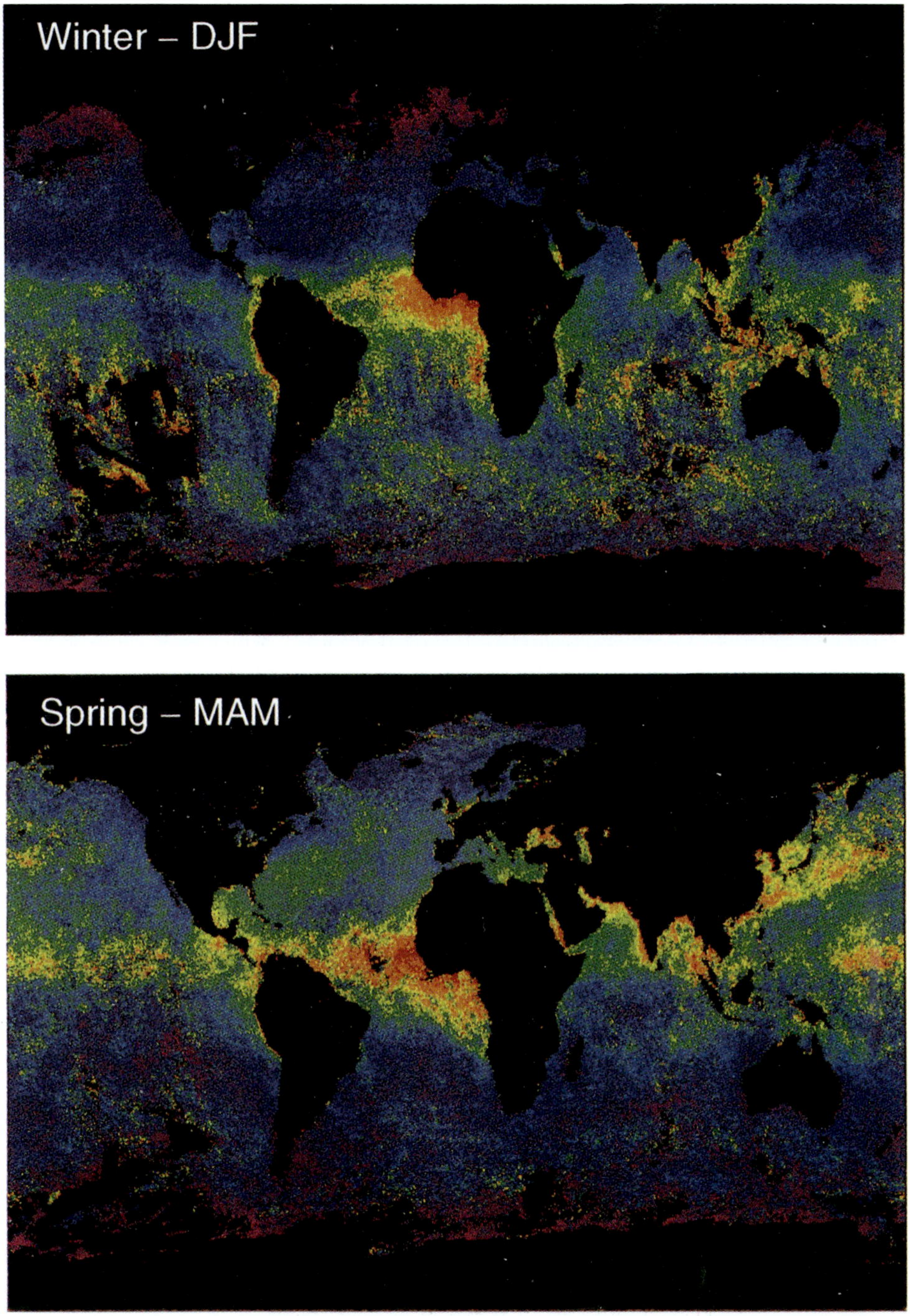

Figure 1a. Mean 1978–1986 CZCS aerosol radiances for Northern Hemisphere (top) winter and (bottom) spring. High levels are yellow-orange and low levels are blue-violet. Black ocean areas contain no data.

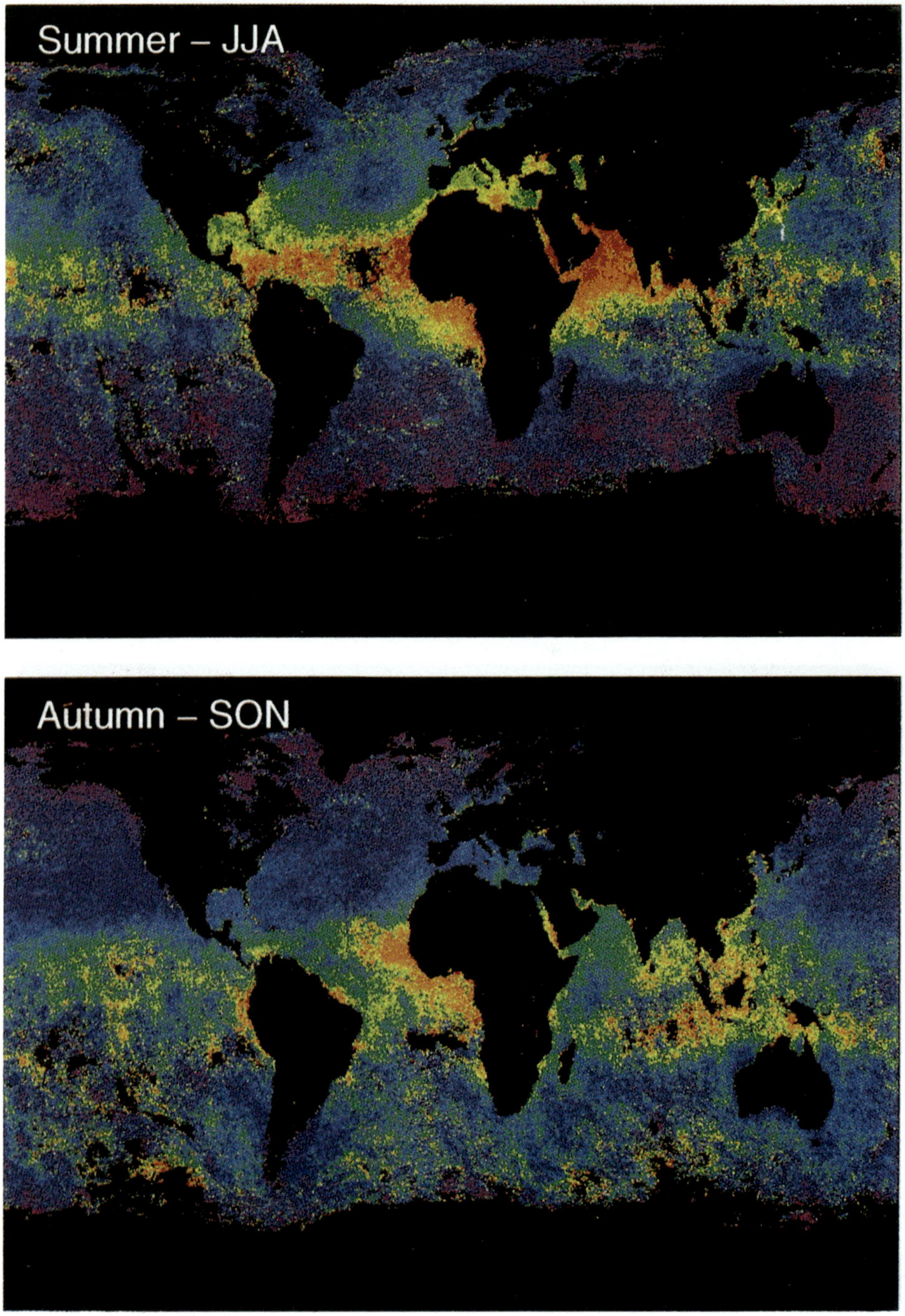

Figure 1b. Same as for Figure 1a, except for Northern Hemisphere (top) summer and (bottom) autumn.

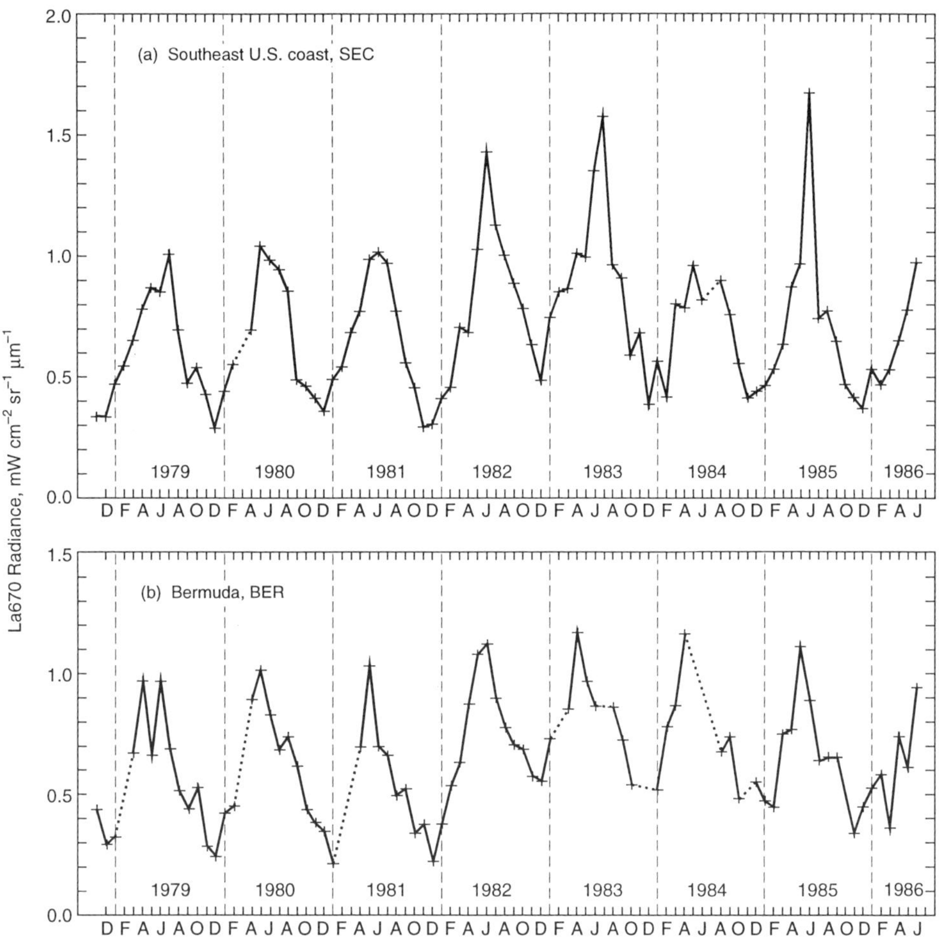

Figure 2. CZCS time series of monthly mean aerosol (670 nm) radiance for 1978–1986, extracted for the (a) SEC and (b) BER regions. Dotted lines indicate where 30% or less of the pixels were invalid.

BER region is not as pronounced, possibly suggesting that this increase may only have been a localized feature. However, an increased dust load does occur during winter in the BER region: November–February radiance levels from 1982 onward are almost twice as high as baseline winter intensities of the previous four years. This also is apparent at the SEC site, but not as distinct. Not only was the dust burden intensified during the summer months, when most dust outbreaks occur in northwest Africa, but the dust load did not return to low wintertime levels as had occurred in prior years in the western Atlantic. A similar observation was also measured at Barbados (Prospero 1996). Thus, it appears that a large-scale phenomenon was occurring that had a pronounced effect on mineral dust deposition patterns—this was successfully captured by CZCS. Moulin et al. (1997b) (cf. 2.2) found that the driving force responsible for changes in dust deposition patterns was fluctuations in the NAO. Indeed, a comparison between annual mean dust radiance intensities obtained with CZCS at both sites and the NAO index (Figure 3) shows that the two generally follow a similar trend; this supports the conclusion reached by Moulin et al. (1997b). CZCS was not only able to correctly estimate the seasonality and distribution patterns of dominant mineral aerosol plumes on a global scale, but it was also successful in directly capturing one aspect of what was, in all likelihood, a result of climatic variability attributed to the NAO: changes in aerosol radiance patterns that are related to dust activity in northwest Africa.

While Stegmann and Tindale (1999) have shown that an ocean-color sensor such as CZCS can produce global maps of mineral aerosol distributions and that these maps reproduce well-known aerosol distribution patterns evidenced via other platforms, they took their study one step further: They examined aerosol patterns vis-à-vis phytoplankton pigment distributions, also obtained from CZCS, to determine if a linkage between aerosol (i.e., dust) input and phytoplankton growth could be established from satellite. They found that the seasonal cycle of aerosol radiance in mid-latitudes followed that of phytoplankton biomass. In some regions of the Indian Ocean and subpolar zones, elevated phytoplankton pigment concentrations were observed one month after the aerosol load had increased in that region. This may indicate a connection between aerosol load and phytoplankton growth. Stegmann and Tindale (1999) hasten to point out that one cannot conclude that

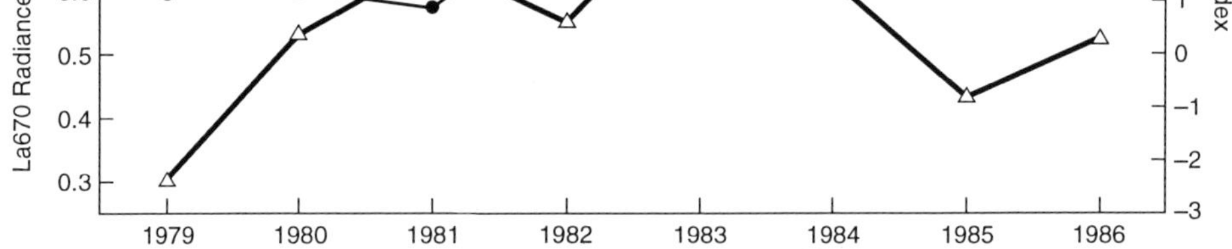

Figure 3. Mean annual CZCS aerosol radiance (670 nm) extracted from SEC and BER regions and the NAO index for 1979–1986. NAO index provided by J. W. Hurrell (personal communication 1999). Aerosol radiance average for 1986 is based on 6 months of available data.

the observed increase in pigment level was a direct result of mineral dust input, since the distribution of measured aerosol radiance intensities does not directly translate into dust deposition to the ocean surface. Also, the temporal resolution of the CZCS data was one-month binned images. Since dust outbreaks are known to be episodic and have a relatively short atmospheric residence time (larger particles fall out faster than smaller ones), a higher temporal resolution will be necessary to assess the response of phytoplankton uptake to aeolian input. Finally, caution should be exercised when interpreting CZCS-derived chlorophyll levels in the North Indian Ocean, especially the Arabian Sea, which is influenced by well-known dust outbreaks originating in the surrounding deserts. The presence of elevated levels of mineral aerosols in the atmosphere may reduce the reliability of the atmospheric correction algorithm used to obtain estimates of chlorophyll. As a result, the absolute chlorophyll concentrations may be overestimated in this basin, although the seasonality and observed patterns are not (Banse and English 1993).

3.2 Sea-viewing Wide Field-of-view Sensor

SeaWiFS, like its predecessor, CZCS, was designed and engineered to primarily be an ocean-color sensor and, as such, has some of the same characteristics and spectral channels had by CZCS. And like CZCS, it contains channels in the near-infrared that are destined for use in the atmospheric correction scheme of all ocean-color images. Unlike CZCS, SeaWiFS has an additional channel at 865-nm wavelength, thus extending the spectrum of available bands which can be used to study mineral aerosols. Furthermore, SeaWiFS provides global coverage every two days—much more complete coverage than CZCS, which was often turned off during each orbit, resulting in large areas of the global ocean receiving no coverage at all during its 7.5-year lifetime (cf. Figure 1).

Figure 4 shows eight-day variations during 1998 of two aerosol properties derived from SeaWiFS for the SEC and BER regions. As expected, the same general trend found by CZCS (and the other sampling platforms) for these regions is evident here. The AOT distribution patterns derived from SeaWiFS are in accord with concurrent AOT maps derived from AVHRR (Stegmann and Tindale 1998). Using eight-day binned data shows the episodic nature of North African dust events; this would not have been as evident using monthly binned data, which would have smoothed out such event-scale occurrences (Stegmann and Tindale 1998).

Aerosol radiance and AOT cycles run pretty much in parallel, although there are some periods when they diverge slightly; for example, in early May, June, and July at the SEC site. This is not surprising, given that they are different aerosol properties. Furthermore, although both are in the near-infrared, they do not share the same wavelength (670 vs. 865 nm). However, since both wavelengths are probably more sensitive to mineral aerosols than other aerosol species, the apparent discrepancy between 670-nm radiance intensity and AOT at 865 nm may indicate that there are other sources affecting the measurements.

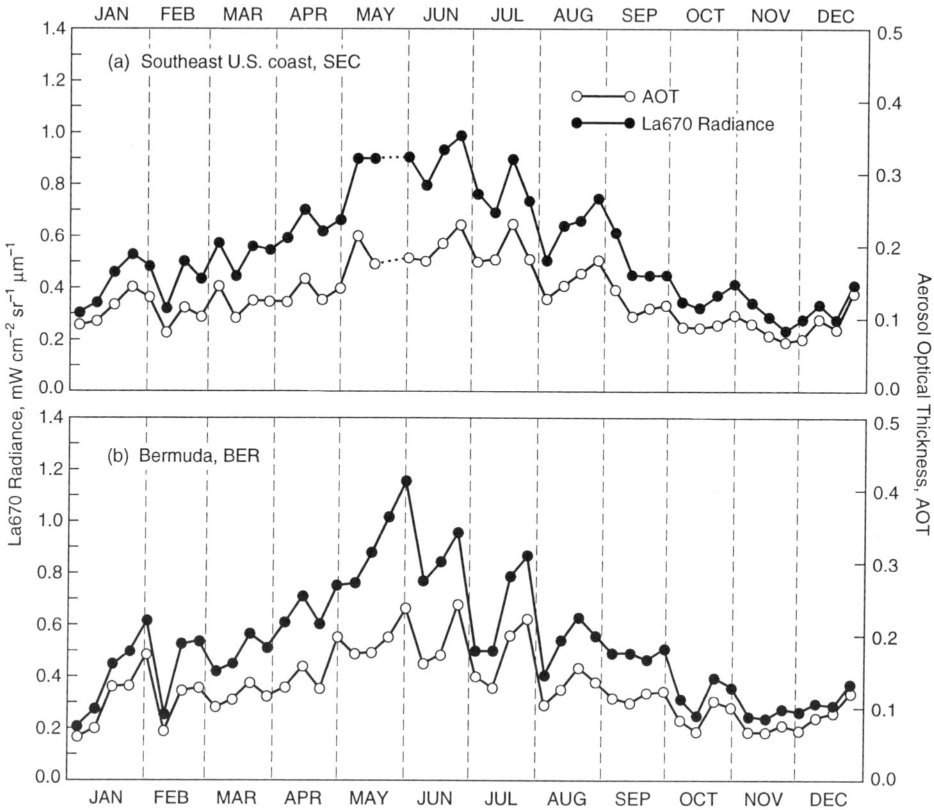

Figure 4. SeaWiFS aerosol radiance (670 nm) and aerosol optical thickness (τ_{865}) extracted for (a) SEC and (b) BER regions during 1998. Dotted lines indicate where 30% or less of the pixels were invalid. SeaWiFS Level-3 global area coverage (GAC), 8-day binned images, were used.

Aerosols could have originated in the continental United States (U.S.), just adjacent to the SEC. A recurring phenomenon along the U.S. East Coast is summertime haze, which is at its most turbid during the spring and summer months and contains a variety of anthropogenic pollutants (e.g., Husar et al. 1981). As the SEC is often downwind from the continental sources, it seems possible that aerosols with a different optical signature were transported to this region and that these substances caused the observed deviation.

However, it may also be possible that pollutants originating in Europe were transported to northern Africa and from there were carried with the dust plume across the Atlantic. Such a scenario has recently been suggested by Li-Jones and Prospero (1998). For now, it is not possible to accept one explanation over the other, nor can it be ruled out that this may just be 'noise' in the data. Further observations will be made from SeaWiFS during the summer months to determine if this deviation is a recurring feature.

As with CZCS, SeaWiFS is capable of mapping large-scale mineral aerosol fields on a synoptic scale. The spatial and temporal evolution of these distributions are in agreement with those obtained via other platforms normally used for aerosol detection. Furthermore, SeaWiFS offers two aerosol products (670 nm radiance and τ_{865}), which its predecessor did not. While the main goal of the Stegmann and Tindale (1999) study was to map mineral aerosol distribution patterns and their seasonality, the next step is quantification of the aerosol signal and estimating aeolian flux to the ocean surface. Periods where elevated aerosol radiance signals occurred due to large-scale dust outbreaks and which were followed by substantial growth in phytoplankton biomass (i.e., chlorophyll) have been recorded with SeaWiFS and are being examined in conjunction with independent datasets to address these questions.

3.3 Future ocean-color sensors

Several ocean-color sensors are scheduled for launch within the next few years, beginning with the Moderate-Resolution Imaging Spectroradiometer (MODIS), which is planned for launch in December 1999 by NASA. Towards the end of 2001, the Medium-Resolution Imaging Spectrometer (MERIS) and Global Imager (GLI) will be put into orbit by the European Space Agency and the National Space Development Agency of Japan, respectively. Numerous others are planned thereafter. In addition to the ocean-color channels necessary to measure phytoplankton chlorophyll, these sensors will have more channels in the near-infrared and infrared frequencies than are currently available on SeaWiFS. These bands are intended for aerosol studies, so that differentiation of aerosol species and size distributions may soon be possible. Furthermore, these missions have detailed aerosol science programs, in addition to those for ocean color. Thus, a suite of Earth-observing platforms with much-improved measuring capabilities with which to address the phytoplankton-dust-climate question will be available in the next few years.

4. Summary and Outlook

This paper has shown that SeaWiFS can successfully capture large-scale mineral aerosol plumes, as had its predecessor, CZCS. This study has also shown that ocean-color sensors can be used as a monitoring platform to establish a long-term time series of global atmospheric aerosol burden in parallel with phytoplankton chlorophyll concentrations. The advantage of using an ocean-color sensor to study aerosol patterns over the ocean is twofold: first, aerosols and chlorophyll are bundled in the same dataset, facilitating processing; second, since the orbital characteristics are identical, the spatial and temporal coverage is the same, as is the resolution, so that a direct comparison between aerosol and chlorophyll can be easily accomplished. Thus, SeaWiFS satisfies the requirement for studies of the relationship between phytoplankton and mineral aerosols via satellite.

While the emphasis has been on how to synoptically measure mineral dust distributions from satellite, there are several other reasons, beside the phytoplankton-dust-climate problem, why it is important to monitor global dust outbreaks and transport. Mineral aerosols can be vehicles, which transport pollutants, carcinogens, agricultural pests, or even pathogen-bearing particles long distances over the ocean before being deposited on land (e.g., Duce et al. 1991; Andreae 1995). The recent massive dust plume that originated in China and moved eastward across the Pacific Ocean attests to how quickly this transport process occurs—it took less than a week for the dust cloud to arrive in North America (e.g., Westphal 1998). There have even been signs of an increase in the occurrence of some diseases, as well as reappearance of others that were thought to have been eliminated (e.g., Epstein et al. 1998; Epstein 1999). As dust concentrations are on the increase (Andreae 1996), so may the transport of a suite of anthropogenic substances as they are rapidly mobilized from one part of the world to another.

Acknowledgments. The support provided by the National Aeronautics and Space Administration is gratefully acknowledged. Data used in this study were produced by the SeaWiFS Project at Goddard Space Flight Center and obtained from the Goddard Distributed Active Archive Center. Use of this data is in accord with the SeaWiFS Research Data Use Terms and Conditions Agreement. Two anonymous reviewers provided helpful comments and their input is greatly appreciated.

References

Andreae, M. O., Climatic effects of changing atmospheric aerosol levels, In *Future Climates of the World, Vol. 16, World Survey of Climatology*, edited by A. Henderson-Sellers, Elsevier, 341–392, 1995.

Andreae, M. O., Raising dust in the greenhouse, *Nature*, *380*, 389–390, 1996.

Andreae, M. O., and P. J. Crutzen, Atmospheric aerosols: Biogeochemical sources and role in atmospheric chemistry, *Science*, *276*, 1052–1058, 1997.

Banse, K., and D. C. English, Revision of satellite-based phytoplankton pigment data from the Arabian Sea during the northeast monsoon, *Mar. Res.*, *2*, 83–103, 1993.

Behrenfeld, J. J., and Z. S. Kolber, Widespread iron limitation of phytoplankton in the South Pacific Ocean, *Science*, *283,* 840–843, 1999.

Charlson, R. J., and J. Heintzenberg, editors, *Aerosol Forcing of Climate*, John Wiley, New York, 416 pp, 1995.

Coale, K. H., K. S. Johnson, S. E. Fitzwater, R. M. Gordon, S. Tanner, F. P. Chavez, L. Ferioli, C. Sakamoto, P. Rogers, F. Millero, P. Steinberg, P. Nightingale, D. Cooper, W. P. Cochlan, M. R. Landry, J. Constantinou, G. Rollwagen, A. Trasvina, and R. Kudela, A massive phytoplankton bloom induced by an ecosystem-scale iron fertilization experiment in the equatorial Pacific Ocean, *Nature*, *383*, 485–501, 1996.

DeBaar, H. J. W., J. T. M. DeJong, D. C. E. Bakker, B. M. Löscher, C. Veth, U. Bathmann, and V. Smetacek, Importance of iron for phytoplankton blooms and carbon dioxide drawdown in the Southern Ocean, *Nature*, *373*, 412–415, 1995.

Denman, K., E. Hoffman, and H. Marchant, Marine biotic responses to environmental change and feedbacks to climate, In *Climate Change 1995: The Science of Climate Change*, edited by J. T. Houghten, L. G. Meira Filho, B. A. Callander, N. Harris, A. Kattenberg, and K. Maskell, Cambridge Univ. Press, Cambridge, England, 483–516, 1996.

Duce, R. A., The impact of atmospheric nitrogen, phosphorous, and iron species on marine biological productivity, In *The Role of Air-Sea Exchange in Geochemical Cycling*, edited by P. Buat-Menard, D. Reidel, New York, 497–529, 1986.

Duce, R. A., Sources, distributions, and fluxes of mineral aerosols and their relationship to climate, In *Aerosol Forcing of Climate*, edited by R. J. Charlson and J. Heintzenberg, John Wiley, New York, 43–72, 1995.

Duce, R. A., and N. W. Tindale, Atmospheric transport of iron and its deposition in the ocean, *Limnol. Oceanogr.*, *36*, 1715–1726, 1991.

Duce, R. A., P. S. Liss, J. T. Merrill, E. L. Atlas, P. Buat-Menard, B.B. Hicks, J. M. Miller, J. M. Prospero, R. Arimoto, T. M. Church, W. Ellis, J. N. Galloway, L. Hansen, T. D. Jickells, A. H. Knap, K. H. Reinhardt, B. Schneider, A. Soudine, J. J. Tokos, S. Tsunogai, R. Wollast, and M. Zhou, The atmospheric input of trace species to the world ocean, *Global Biogeochem. Cycles*, *5*, 193–259, 1991.

Epstein, P. R., Climate and health, *Science*, *285*, 347–348, 1999.

Epstein, P. R., H. F. Diaz, S. Elias, G. Grabherr, N. E. Graham, W. J. M. Martens, E. Mosley-Thompson, and J. Susskind, Biological and physical signs of climate change: Focus on mosquito-borne diseases, *Bull. Amer. Meteorol. Soc.*, *79*, 409–417, 1998.

Falkowski, P. G., R. T. Barber, and V. Smetacek, Biogeochemical controls and feedbacks on ocean primary production, *Science*, *281*, 200–206, 1998.

Gordon, H. R., and D. J. Castaño, Coastal zone color scanner atmospheric correction algorithm: Multiple scattering effects, *Appl. Opt.*, *26*, 2111–2122, 1987.

Gordon, R. M., K. H. Coale, K. S. Johnson, Iron distributions in the equatorial Pacific: Implications for new production, *Limnol. Oceanogr.*, *42*, 419–431, 1997.

Herman, J. R., P. K. Bhartia, O. Torres, C. Hsu, C. Seftor, and E. Celarier, Global distribution of UV-absorbing aerosols from Nimbus 7/TOMS data, *J. Geophys. Res.*, *102*, 16911–16922, 1997.

Hobbs, P. V., editor, *Aerosol-Cloud-Climate Interactions*, Academic Press, New York, 233 pp, 1993.

Holben, B. N., T. F. Eck, I. Slutsker, D. Tanre, J. P. Buis, A. Setzer, E. Vermote, J. A. Reagan, Y. J. Kaufman, T. Nakajima, F. Lavenu, I. Jankowiak, and A. Smirnov, AERONET—A federated instrument network and data archive for aerosol characterization, *Remote Sensing Environ.*, *66*, 1–16, 1998.

Hurrell, J. W., Decadel trends in the North Atlantic Oscillation: Regional temperatures and precipitation, *Science*, *269*, 676–679, 1995.

Husar, R. B., J. M. Holloway, D. E. Patterson, and W. E. Wilson, Spatial and temporal pattern of eastern U.S. haziness: A summary, *Atmos. Environ.*, *15*, 1919–1928, 1981.

Husar, R. B., J. M. Prospero, and L. L. Stowe, Characterization of tropospheric aerosols over the oceans with the NOAA advanced very high resolution radiometer optical thickness operational product, *J. Geophys. Res.*, *102*, 16889–16909, 1997.

Hutchins, D. A., Iron and the marine phytoplankton community, *Prog. Phycological Res.*, *11*, 1–49, 1995.

Hutchins, D. A., and K. W. Bruland, Iron-limited diatom growth and Si:N uptake ratios in a coastal upwelling regime, *Nature, 393*, 561–564, 1998.

Jickells, T. D., S. Dorling, W. G. Deuser, T. M. Church, R. Arimoto, and J. M. Prospero, Airborne dust fluxes to a deep water sediment trap in the Sargasso Sea, *Global Biogeochem. Cycles, 12*, 311–320, 1998.

Kaufman, Y. J., Remote sensing of direct and indirect aerosol forcing, In *Aerosol Forcing of Climate*, edited by R. J. Charlson and J. Heintzenberg, John Wiley, New York, 297–332, 1995.

Kolber, Z. S., R. T. Barber, K. H. Coale, S. E. Fitzwater, R. M. Greene, K. S. Johnson, S. Lindley, and P. G. Falkowski, Iron limitation of phytoplankton photosynthesis in the equatorial Pacific Ocean, *Nature, 371*, 145–149, 1994.

Li-Jones, X., and J. M. Prospero, Variations in the size distribution of non-sea-salt sulfate aerosol in the marine boundary layer at Barbados: Impact of African dust, *J. Geophys. Res., 103,* 16073–16084, 1998.

Martin, J. H., Iron, Leibig's Law, and the greenhouse, *J. Oceanogr., 4*, 52–55, 1991.

Martin, J. H., Iron as a limiting factor in oceanic productivity, In *Primary Productivity and Biogeochemical Cycles in the Sea*, edited by P. G. Falkowski and A. D. Woodhead, Plenum, New York, 123–137, 1992.

Martin, J. H., and R. M. Gordon, Northeast Pacific iron distributions in relation to phytoplankton productivity, *Deep-Sea Res., 35*, 177–196, 1988.

Martin, J. H., K. H. Coale, K. S. Johnson, S. E. Fitzwater, R. M. Gordon, S. J. Tanner, C. N. Hunter, V. A. Elrod, J. L. Nowicki, T. L. Coley, R. T. Barber, S. Lindley, A. J. Watson, K. Van Scoy, C. S. Law, M. I. Liddicoat, R. Ling, T. Stanton, J. Stockel, C. Collins, A. Anderson, R. Bidigare, M. Ondrusek, M. Latasa, F. J. Millero, K. Lee, W. Yao, J. Z. Zhang, G. Friederich, C. Sakamoto, F. Chavez, K. Buck, Z. Kolber, R. Greene, P. Falkowski, S. W. Chisholm, F. Hoge, R. Swift, J. Yungel, S. Turner, P. Nightingale, A. Hatton, P. Liss, and N. W. Tindale, Testing the iron hypothesis in ecosystems of the equatorial Pacific Ocean, *Nature, 371*, 123–129, 1994.

Moulin, C., F. Guillard, F. Dulac, and C. E. Lambert, Long-term daily monitoring of Saharan dust load over ocean using Meteosat ISCCP-B2 data. 1. Methodology and preliminary results for 1983–1994 in the Mediterranean, *J. Geophys. Res., 102*, 16947–16958, 1997a.

Moulin, C., C. E. Lambert, F. Dulac, and U. Dayan, Control of atmospheric export of dust from North Africa by the North Atlantic Oscillation, *Nature, 387*, 691–694, 1997b.

Prospero, J. M., The atmospheric transport of particles to the ocean, In *Particle Flux in the Ocean*, edited by V. Ittekkot, P. Schäfer, S. Honjo, and P. J. Depetris, John Wiley, New York, 19–52, 1996.

Prospero, J. M., and R. T. Nees, Impact of the North African drought and El Niño on mineral dust in the Barbados trade winds, *Nature, 320,* 735–738, 1986.

Rao, C. R. N., L. L. Stowe, and E. P. McClain, Remote sensing of aerosols over the ocean using AVHRR data: Theory, practice and applications, *Int. J. Remote Sensing, 10*, 5743–5749, 1989.

Seinfeld, J. H., and S. N. Pandis, *Atmospheric Chemistry and Physics: From Air Pollution to Climate Change*, John Wiley, New York, 1326 pp, 1998.

Stegmann, P. M., and N. W. Tindale, Observations of aerosol events using the SeaWiFS platform, *Eos, Trans. Amer. Geophys. Un., 79*, F410, 1998.

Stegmann, P. M., and N. W. Tindale, Global distribution of aerosols over the open ocean as derived from the coastal zone color scanner, *Global Biogeochem. Cycles*, *13*, 383–397, 1999.

Stowe, L. L., A. M. Ignatov, and R. R. Singh, Development, validation, and potential enhancements to the second-generation operational aerosol product at the National Environmental Satellite, Data, and Information Service of the National Oceanic and Atmospheric Admininstration, *J. Geophys. Res.*, *102*, 16932–16934, 1997.

Westphal, D. L., Dynamical forcing of the Chinese dust storms of April 1998, *Eos, Trans. Amer. Geophys. Un.*, *79*, F100, 1998.

Petra M. Stegmann, Graduate School of Oceanography, University of Rhode Island, South Ferry Road, Narragansett, RI 02882-1197, U.S.A. (email, pstegmann@gso.uri.edu; fax, +1-401-874-6728)

Satellites, Oceanography and Society
edited by David Halpern

Chapter 12

An overview of temporal and spatial patterns in satellite-derived chlorophyll-*a* imagery and their relation to ocean processes

James A. Yoder
Graduate School of Oceanography, University of Rhode Island, Narragansett

Abstract. Satellite measurements of water-leaving radiance, from which estimates of phytoplankton chlorophyll-*a* are derived, began with the launch of the Coastal Zone Color Scanner (CZCS) in 1978. Global CZCS data were widely distributed beginning in 1988, providing oceanographers a new tool for studying temporal and spatial variability in surface waters of the global ocean. CZCS imagery, and that from more recently launched sensors, has been extensively used for studies of variability at spatial scales from <10 km to the global ocean and for temporal scales ranging from days to years. Ocean margin waters have particularly strong signatures in chlorophyll-*a* images, and the imagery has been effectively used to study a wide range of environments, including coastal upwelling systems and river plumes. Chlorophyll-*a* imagery is the only tool for observing the mean- and time-varying components of a biological variable at ocean basin to global ocean scales. Basin-to-global scale composite images are used as input to calculate primary production and to describe and quantify seasonal cycles on a global scale. Such images are also one of the important data sources for understanding how physical processes affect biological distributions on ocean basin and global scales, e.g., the impact of major ocean perturbations such as El Niño. Future sensors, as well as new capabilities to merge data from different sensors, will provide better accuracy and better coverage, thereby extending scientific and other applications in coastal and open ocean waters.

1. Introduction

The Coastal Zone Color Scanner (CZCS) demonstrated that satellite sensors can routinely measure sunlight spectral radiance backscattered out of the upper few meters of the ocean (water-leaving radiance, L_W, or "ocean color") for open ocean and some coastal waters (Hovis et al. 1980; Hooker et al. 1993). This was a major technical accomplishment because top-of-the-atmosphere (TOA) radiance is dominated by molecular and aerosol scattering in the atmosphere, and the atmospheric signal has to be removed before one can accurately determine the comparatively weak L_W signal (Gordon and Morel 1981, 1983). Bio-optical algorithms relating L_W to in-water constituents such as phytoplankton chlorophyll-*a* were developed and refined. One of the common and relatively simple approaches to calculate (retrieve) chlorophyll-*a* from L_W is a power curve of the form $Y = AX^b$, where Y = chlorophyll-*a* (mg m^{-3}) and X = "blue-green" ratio, e.g., L_W443/L_W550. In open ocean waters, where optical properties are controlled by phytoplankton and their degradation products (often referred to as Morel Case 1 waters), this relation yields an estimate of chlorophyll-*a* with an accuracy of about ±30% (Gordon and Morel 1981, 1983; Hooker et al. 1993). Assuming accurate atmospheric correction, this simple algorithm is robust in open ocean (Case 1) waters since L_W at 443 nm is dominated by phytoplankton chlorophyll-*a* absorption, whereas L_W at 550 nm is relatively insensitive to phytoplankton absorption (Gordon and Morel 1981). Calculations of chlorophyll-*a* concentration (mg m^{-3}) from satellite-determined L_W are often referred to as "CSAT" to distinguish satellite retrievals from in-situ measurements of chlorophyll-*a*.

Global CZCS data were widely distributed beginning in 1988 (Feldman et al. 1989), providing oceanographers a new tool for studying temporal and spatial variability in surface waters of the global ocean. Oceanographers for the first time had an observational tool for mapping phytoplankton chlorophyll-*a* distribution. CSAT images had a profound impact on biological oceanography, providing the first quasi-synoptic observations of a two-dimensional field directly related to biological distributions and processes (Longhurst 1998). CSAT images helped biological oceanographers understand the links between biological and physical processes (e.g., the latter inferred from satellite images of sea surface temperature) at regional and larger scales. Furthermore, chlorophyll-*a* (and CSAT) concentration is directly related to primary production, and CSAT images provide the single most important input to calculate primary production at regional-to-global spatial scales.

CZCS operated from 1978 to 1986 and was followed by a 10-year gap in global satellite L_W measurements and, hence, CSAT calculations. Satellite measurements of L_W began again in 1996 when the National Space Development Agency of Japan (NASDA) launched the Advanced Earth Observation Satellite (ADEOS), carrying both the Ocean Color and Temperature Scanner (OCTS) (Kawamura and OCTS Team 1998) and Polarization and Directionality of the Earth's Reflectance (POLDER), and continued with the launch of Orbital Science Corporation's Seastar satellite with the National Aeronautics and Space Administration (NASA) Sea-viewing Wide Field-of-view Sensor (SeaWiFS) in August 1997 (McClain et al. 1998). The European Space Agency (ESA), Centre National d'Etudes Spatiales (CNES), NASA, NASDA, and other space agencies will all launch new global sensors in the next few years with advanced capabilities for measuring L_W

from space (IOCCG 1998). The purpose of this chapter is to show how CSAT measurements are being used to quantify phytoplankton variability at regional-to-global scales and to better describe and understand its relation to ocean processes.

2. Frequency Distributions of In-Situ Chlorophyll-*a* and CSAT

Before discussing oceanographic applications, it is useful to consider a basic principle of CSAT imagery (as well as in-situ chlorophyll-*a*). From a statistical point of view, the ocean consists of many CSAT (and chlorophyll-*a*) populations separated by discrete boundaries. Thus, a single frequency distribution does not characterize the entire surface chlorophyll-*a* "population." For a given population, a log-normal frequency distribution is commonly observed for CSAT and in-situ chlorophyll-*a* data, particularly within ocean margin waters (Campbell 1995; Campbell et al. 1995). However, a power law distribution (i.e., probability of occurrence is inversely and linearly related to CSAT on a log-log scale) was shown to characterize CZCS imagery of the global ocean (Antoine and Morel 1995). The CSAT frequency distribution affects calculation of basic statistics, such as mean and variance, and needs to be considered when calculating spatial and temporal variability from satellite images or when building composite images from single scenes (Campbell et al. 1995). Furthermore, the CSAT frequency distribution is partly determined by the choice of bio-optical algorithm for calculating CSAT from L_W (O'Reilly et al. 1998). In other words, the choice of bio-optical algorithm (and to date, every mission has chosen a different algorithm) affects the CSAT frequency distribution, even when input L_W to the algorithms are identical. This makes it difficult to separate algorithm from environmental sources of variability when comparing CSAT data from different sensors as, for example, one has to do when comparing data from different sensors in different years.

3. CSAT Variability

3.1 Mesoscale (10 to ~100 km)

As illustrated by Figure 1, individual, cloud-free, regional-scale (1- to 4-km pixel resolution) imagery provides an extraordinary view of mesoscale variability. CZCS and other CSAT imagery revealed new and interesting and, in some cases, unanticipated features in many different ocean environments: equatorial Pacific (Feldman 1986; Feldman et al. 1984; Chavez et al. 1999); Caribbean Sea (Muller-Karger and Castro 1994); Adriatic Sea (Barale et al. 1986); eastern boundary currents (Abbott and Zion, 1985; Shannon et al. 1983); and western boundary currents, including frontal features (Ryan and Yoder 1996; Ryan et al. 1999) and rings (Garcia-Moliner and Yoder 1994; Saitoh et al. 1998). For example, CSAT and the National Oceanic and Atmospheric Administration (NOAA) Advanced Very-High Resolution Radiometer (AVHRR) sea surface temperature (SST)

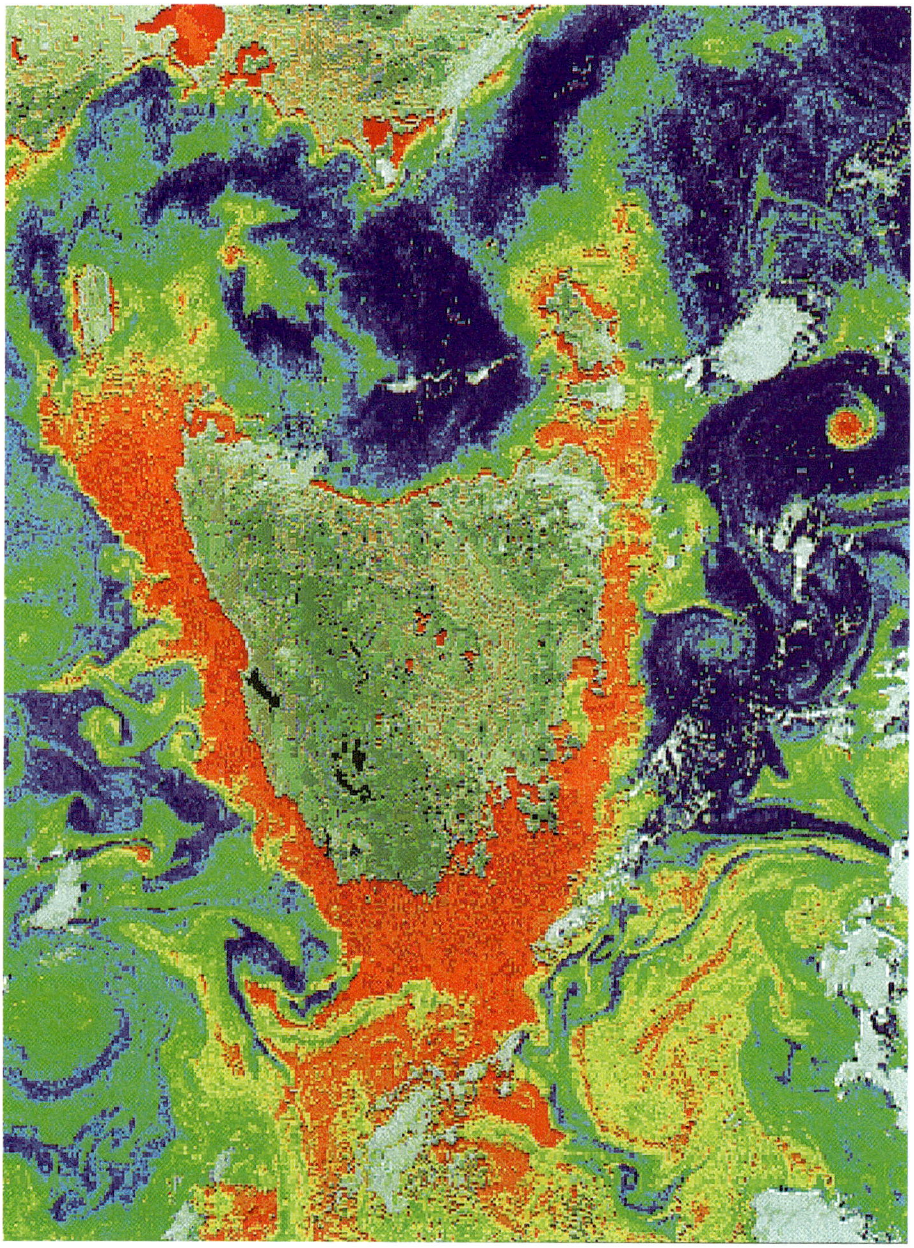

Figure 1. CZCS chlorophyll-*a* image on 27 November 1981, centered on Tasmania (off southeast Australia), illustrating complicated mesoscale patterns. Chlorophyll-*a* concentrations increase as color changes from blue-to-green-to-yellow-to-orange-to-red. (Image courtesy of G. Feldman, NASA Goddard Space Flight Center.)

imagery showed for the first time that filaments of pigment-rich, cold upwelled waters extend hundreds of kilometers offshore of the California and Oregon coasts (Abbott and Zion 1987). River plumes and the effects of river-borne nutrients also have a strong signal in CSAT imagery (Holligan et al. 1989), although the bio-optical complexity of these waters makes it very difficult to separate the phytoplankton optical signal from other constituents, such as suspended sediment, marine and terrestrial detritus, and colored dissolved organic matter (CDOM) (Doerffer and Fischer 1994; Hochman et al. 1994). Nevertheless, the imagery is very effective at showing the location of river plumes before they completely mix with ocean waters. A spectacular example is the Amazon River plume, which is visible in ocean-color imagery for hundreds of kilometers as it extends into the North Atlantic to be trapped along an equatorial wave (Muller-Karger et al. 1988). Another interesting and unusual feature in L_W patterns is large-scale coccolithophorid (a type of phytoplankton) blooms. Spatial structure of highly reflecting surface waters caused by very high densities of coccolithophorids or their by-products (coccoliths) were first observed with CZCS and AVHRR visible-band imagery. CZCS imagery revealed episodic surface blooms of the coccolithophorid, *Emiliania huxleyi*, in the eastern and western North Atlantic (Holligan et al. 1983; Brown and Yoder 1994) and in the North Pacific. Recent SeaWiFS imagery captured the spatial extent of an unusual coccolithophorid bloom in the Bering Sea that may be an early indicator of significant change to the Bering Sea ecosystem (Vance et al. 1998).

In addition to a synoptic view of mesoscale features, ocean-color imagery (and in combination with SST imagery) provides quantitative information on the important spatial scales in the mesoscale range (McClain et al. 1983; Yoder et al. 1987; Smith et al. 1988). For example, Figure 2 illustrates how CSAT imagery opened up a new spectral range for direct observations, and how the variability at the longer spatial scales observed in CSAT imagery compares with results from shipboard observations. In general, these studies showed close relations between CSAT and circulation features and processes, and thus contributed to the growing awareness of the importance of physical-biological interactions in marine ecology. CSAT imagery has also been used to infer frontal processes (Yoder et al. 1987; McClain et al. 1988; Garcia-Moliner and Yoder 1994; Ryan and Yoder 1996) and coastal processes, such as the effects of shoreline and bottom topography (McClain et al. 1988; Yentsch et al. 1994).

Regional-scale ocean-color imagery are particularly useful for space-time studies of ocean margins because of the importance and dominance of mesoscale features (e.g., upwelling plumes, eddies, filaments, and river plumes) in continental shelf and slope waters. Among the first quantitative studies were those of Denman and Abbott (1988, 1994), who used two-dimensional auto-spectrum and cross-spectrum analyses on CZCS and AVHRR-SST image sequences to calculate the rate at which CSAT (and SST) features decorrelated (in a Eulerian sense) in ocean margin waters off the west coast of North

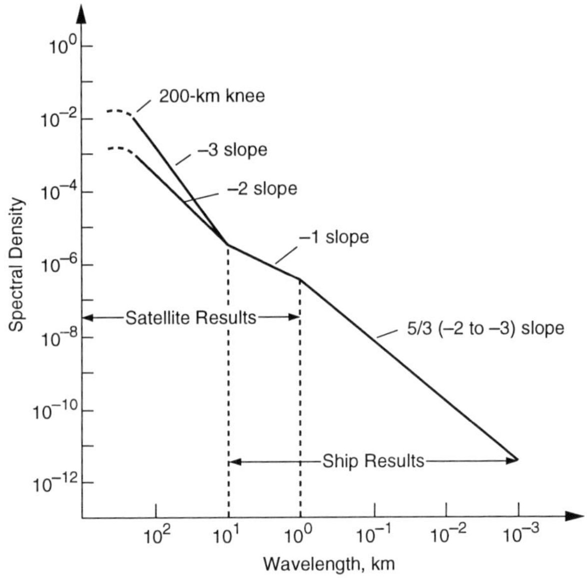

Figure 2. Schematic power spectrum for horizontal variability of chlorophyll-*a*, where satellite results were based on spectral analysis of 48 CZCS images from the California Current (from Smith et al. 1988).

America. The studies provided an excellent example of mesoscale physical and biological coupling and showed quantitatively the linkage between mesoscale temporal and spatial scales. In brief, CSAT features having wavelengths of 50–150 km lost coherence after 7–10 days, features having wavelengths of 25–50 km lost coherence after 5–7 days; only in the offshore region did features of less than 25 km wavelength maintain coherence for 1 day or longer.

Another approach for assessing spatial and temporal variability is to calculate empirical orthogonal functions (EOF) on image time series. Depending upon data pretreatment, one can focus on temporal or spatial variability or both. EOFs are generally calculated from continuous time series with no or few missing values. Applications to CSAT imagery are thus restricted to a composite image (week to month) and generally with a high degree of spatial averaging or subsampling to reduce the size of the arrays. The results have been used to quantify seasonal and interannual CSAT variability off the U.S. West (Thomas and Strub 1990) and East Coasts (Yoder et al. 2000), temporal change in spatial CSAT patterns in a Gulf Stream ring (Garcia-Moliner and Yoder 1994), along- and across-shelf CSAT variability (Eslinger et al. 1989), mesoscale variability in CSAT in relation to wind stress curl (Abbott and Barksdale 1991), and to show the seasonal significance of CSAT variability associated with the shelf-slope front off the U.S. East Coast

(Ryan et al. 1999). EOFs were also used to quantify spatial-temporal variability in CSAT imagery from the Gulf of St. Lawrence, Canada, where river runoff and wind stress explained the first EOF mode, whereas other processes, including the spring bloom, were of less significance (Fuentes-Yaco et al. 1997).

3.2 Basin-to-global scale

CSAT imagery is the only tool for observing the mean- and time-varying components of a biological variable at basin-to-global scales. One of the important uses of basin-to-global scale composite CSAT imagery is in the calculation of primary production. Large-scale composite imagery has also been used to describe and quantify seasonal cycles on a global scale, as well as the impact of major ocean perturbations such as El Niño. El Niño has particularly dramatic effects in CSAT imagery of the equatorial Pacific. The 1997–1998 El Niño caused very striking changes along the equator in the central Pacific during the transition from El Niño conditions in December 1997 to La Niña by July 1998 (Chavez et al. 1999). Near-surface chlorophyll-*a* and CSAT concentrations during this El Niño and the subsequent La Niña were among the lowest and highest, respectively, ever recorded in the equatorial Pacific (Strutton et al. 1998; Chavez et al. 1998). CSAT imagery from OCTS, POLDER, and SeaWiFS are currently being used to clarify important details of the 1997–1998 El Niño and La Niña cycle. With international attention focused on the possibility of human-induced changes to the global environment, there is considerable interest in using long time series of CSAT imagery from multiple satellite sensors as one of the tools to observe coastal and open ocean ecosystems and thereby distinguish long-term change from periodic phenomena such as El Niño (Karl 1999).

Esaias et al. (1986) published the first ocean basin composite image showing the extent of high CSAT concentrations associated with the spring (May) bloom in the North Atlantic, which dramatically illustrated the potential for using composite CSAT images to observe and quantify seasonal cycles in the global ocean. Campbell and Aarup (1992) used composite CSAT imagery to estimate new production in the North Atlantic from the timing of the onset of oligotrophic conditions. Yoder et al. (1993) showed that the global CSAT patterns in temperate and subpolar waters were generally consistent with simple conceptual models linking seasonality with changes in the irradiance and nutrient environment (Figure 3). Banse and English (1994) indicated that seasonality of pigment is driven almost equally by the interaction between phytoplankton cell growth and losses from grazing. Deuser et al. (1990), in a remarkable study, showed close relations (with a 1.5-month lag) during a 7-year time series between CSAT concentration peaks and particulate carbon flux to the deep sea off Bermuda.

CSAT imagery is also proving useful for classifying the ocean into biogeochemical provinces, i.e., a methodology for dividing the global ocean into regions having similar biogeochemical characteristics. The availability of the global composite CSAT imagery

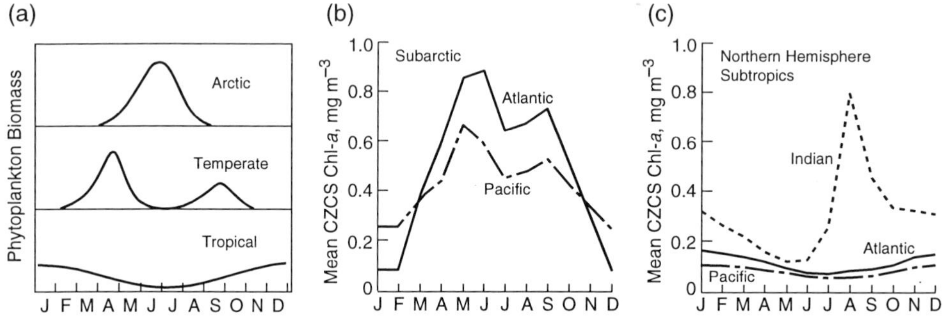

Figure 3. (a) Annual cycles of phytoplankton biomass at different latitudinal regions in the Northern Hemisphere, as predicted from one of the first mathematical models of plankton dynamics (Cushing 1959); (b) CZCS data from the subarctic region, which encompasses the arctic and temperate regions in (a), showing a seasonal pattern representing a combination of features from the latter two regions; (c) CZCS data from the subtropical North Atlantic has a winter maximum and summer minimum, as predicted by the Cushing (1959) model for the tropical region shown in (a). The subtropical North Indian Ocean has an anomalous pattern caused by monsoon wind-induced upwelling (from Yoder et al. 1993).

was the key source of information leading to the definition of such provinces (Longhurst 1995, 1998). The biogeochemical province concept led to new and improved estimates of global primary production and to a new conceptual model for understanding seasonal phytoplankton and nutrient cycles in the global ocean. Furthermore, the Longhurst (1998) biogeochemical provinces can be the basis for a global ocean synthesis of carbon and other biogeochemical fluxes, although it has not yet been widely used for such purposes.

CSAT imagery, data products from other satellite sensors, and in-situ observations are used to understand the physical mechanisms affecting biological distributions on basin-to-global scales. For example, CZCS imagery in combination with meteorological and oceanographic data sets were used to study wind mixing and upwelling as controls on CSAT distributions at monthly-to-seasonal time scales. McClain et al. (1990) and McClain and Firestone (1993) suggested that Ekman upwelling in the tropical Atlantic explains a relatively high fraction (about 20–60%, depending on season) of basin-wide phytoplankton biomass variability. With the launch of the ADEOS satellite, concurrent measurements of ocean color and ocean winds were made for the first time with the OCTS and the NASA Scatterometer (NSCAT). This unique suite of data is the first such dataset available to study biological and physical coupling at ocean basin scales. Future observations from similar satellite sensors will provide opportunities to further test and refine models of CSAT temporal and spatial variability at basin-to-global scales.

Figures 4 and 5 show the results of an EOF analysis of 10-day averaged POLDER CSAT and NSCAT ocean wind stress patterns in the Atlantic from September 1996 to June 1997 when the ADEOS mission ended prematurely. The EOF mode 1 represents the ocean basin spatial pattern showing high wind stress and high chlorophyll-*a* in high latitudes and low wind stress and low chlorophyll-*a* in low latitudes throughout the time

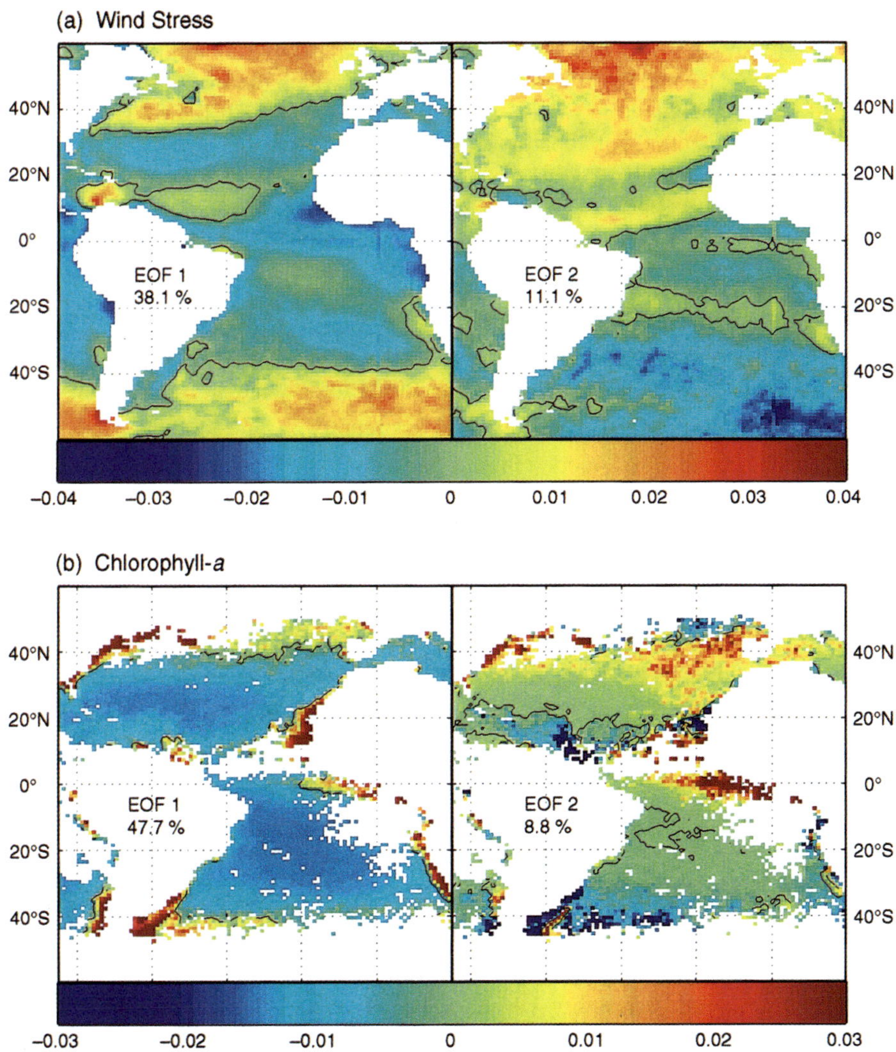

Figure 4. Normalized EOF modes 1 and 2 for (a) NSCAT wind stress and (b) POLDER CSAT.

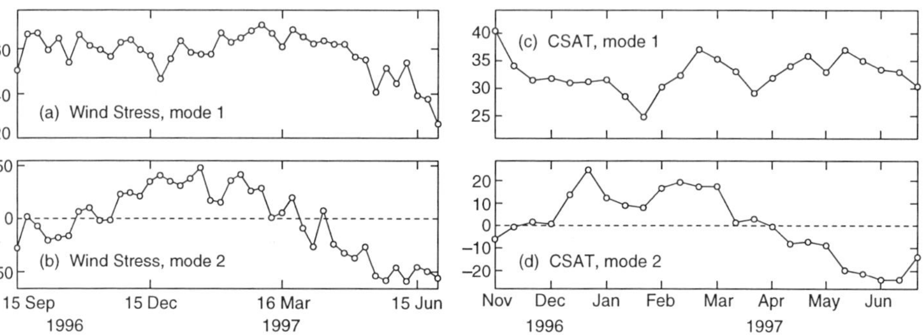

Figure 5. Amplitude time series for normalized EOF modes 1 and 2: (a) and (b) NSCAT wind stress, for which the spatial patterns are shown in Figure 4a; and (c) and (d) POLDER CSAT, for which the spatial patterns are shown in Figure 4b.

interval. High chlorophyll-*a* concentrations are generally found where wind mixing is strong to provide a source of deep-water nutrients to the euphotic zone to sustain seasonal plankton blooms. Mode 2 is more subtle and shows that when Northern Hemisphere mid-latitude wind stress, which is generally stronger during winter than during summer, is large, then Southern Hemisphere mid-latitude wind stress is small, i.e., the annual cycle. The effect of strong seasonal asymmetry in mid-latitude wind stress is reflected in the second CSAT mode. In mid-latitudes, CSAT mode 2 shows much higher chlorophyll-*a* concentrations in winter than in summer, particularly where winter wind stress is strongest, i.e., in the eastern North Atlantic and western South Atlantic.

4. Conclusions

CSAT imagery is now a key tool for understanding and quantifying mesoscale features in the ocean. Applications range from river plumes to boundary current phenomena (e.g., eddies) and the list is growing, especially for applications in ocean margin waters. The combination of CSAT image analysis with an in-situ measurement program is a particularly powerful combination for understanding mesoscale physical and biological coupling and is becoming a standard approach in modern oceanography. CSAT imagery has an equally, if not more, important role for understanding biological processes at ocean basin-to-global scales, as it is the only routine biological measurement at these large scales. CZCS, POLDER, OCTS, and SeaWiFS imagery provided biological oceanographers with new insights into large-scale processes. In addition, the global imagery provided a means for classifying the global ocean into biogeochemical provinces—an important step towards a new global synthesis of the role of ocean biota in carbon, nutrient, and related biogeochemical cycles.

CSAT imagery from new generation sensors will provide new technical capabilities and extend applications in new directions. The NASA Moderate-Resolution Imaging Spectroradiometer (MODIS) instrument was launched on the Terra satellite in December 1999. Several additional new sensors for measuring L_W from space will be launched in 2000 and 2001 (IOCCG 1998, 1999, and http://www.ioccg.org/). The Australian Resource Information and Environmental Satellite (ARIES) and the United States Naval Earth Map Observer (NEMO) will provide hyperspectral imaging of coastal waters at high spatial resolution. The ESA Medium Resolution Imaging Spectrometer (MERIS), and NASDA Global Imager (GLI) will measure sunlight-stimulated chlorophyll-*a* fluorescence. The increased number of spectral bands and higher spatial resolution on future satellite sensors will lead to better atmospheric and bio-optical algorithms, more accurate CSAT calculations, and improved assessment of time-space variability. In particular, MERIS will accurately measure L_W in complex coastal waters, a capability which is not routinely possible with SeaWiFS. Once we learn how to accurately merge CSAT and L_W data from new and old sensors, biological oceanographers will have an unprecedented opportunity to study ocean chlorophyll-*a* distributions at temporal scales ranging from days to decades and at spatial scales from 1 km to ocean basins.

References

Abbott, M. R., and P. M. Zion, Satellite observations of phytoplankton variability during an upwelling event, *Cont. Shelf Res.*, *4*, 661–680, 1985.

Abbott, M. R., and P. M. Zion, Spatial and temporal variability of phytoplankton pigment off northern California during coastal ocean dynamics experiment, *J. Geophys. Res.*, *92*, 1745–1755, 1987.

Abbott, M. R., and B. Barksdale, Phytoplankton pigment patterns and wind forcing off central California, *J. Geophys. Res.*, *96*, 14649–14667, 1991.

Antoine, D., J.-M. Andre, and A. Morel, Oceanic primary production: 2. Estimation at global scale from satellite (coastal zone color scanner) chlorophyll, *Global Biogeochem. Cycles*, *10*, 57–69, 1995.

Banse, K., and D. C. English, Seasonality of coastal zone color scanner phytoplankton pigment in the offshore oceans, *J. Geophys. Res.*, *99*, 7323–7345, 1994.

Barale, V., C. R. McClain, and P. Malanotte-Rizzoli, Space and time variability of the surface color field in the Northern Adriatic Sea, *J. Geophys. Res.*, *91*, 12957–12974, 1986.

Brown, C. W., and J. A. Yoder, Distribution patterns of coccolithophorid blooms in the Western North Atlantic, *Cont. Shelf Res.*, *14*, 175–197, 1994.

Brown, O. B., R. H. Evans, J. W. Brown, H. R. Gordon, R. C. Smith, and K. S. Baker, Phytoplankton blooming off the U.S. east coast: A satellite description, *Science*, *229*, 163–167, 1985.

Campbell, J. W., The lognormal distribution as a model for bio-optical variability in the sea, *J. Geophys. Res.*, *100*, 13237–13254, 1995.

Campbell, J. W., and T. Aarup, New production in the North Atlantic derived from seasonal patterns of surface chlorophyll, *Deep-Sea Res.*, *39*, 1669–1694, 1992.

Campbell, J. W., J. M. Blaisdell, and M. Darzi, Level-3 SeaWiFS data products: Spatial and temporal binning algorithms. SeaWiFS Tech. Rep. 32, NASA Tech. Mem. 104566, Goddard Space Flight Center, Greenbelt, Maryland, 73 pp plus 3 plates, 1995.

Chavez, F. P., P. G. Strutton, and M. J. McPhaden, Biological-physical coupling in the central equatorial Pacific during the onset of the 1997–98 El Niño, *Geophys. Res. Lett.*, *25*, 3543–3546, 1998.

Chavez, F. P., P. G. Strutton, G. E. Friederich, R. A. Feely, G. C. Feldman, D. G. Foley, and M. J. McPhaden, Biological and chemical response of the equatorial Pacific Ocean to the 1997–98 El Niño, *Science*, *286*, 2126–2131, 1999.

Cushing, D. H., The seasonal variation in oceanic production as a problem in phytoplankton dynamics, *J. Cons. Int. Explor. Mer.*, *24*, 455–464, 1959.

Denman, K. L., and M. R. Abbott, Time evolution of surface chlorophyll patterns from cross-spectrum analysis of satellite color, *J. Geophys. Res.*, *93*, 6789–6998, 1988.

Denman, K. L., and M. R. Abbott, Time scales of pattern evolution from cross-spectrum analysis of advanced very high resolution radiometer and coastal zone color scanner imagery, *J. Geophys. Res.*, *99*, 7433–7442, 1994.

Deuser, W. G., F. E. Muller-Karger, R. H. Evans, O. B. Brown, W. E. Esaias, and G. C. Feldman, Surface-ocean color and deep-ocean carbon flux: How close a connection?, *Deep-Sea Res.*, *37*, 1331–1343, 1990.

Doerffer, R., and J. Fischer, Concentrations of chlorophyll, suspended matter, and gelbstoff in case II waters derived from satellite coastal zone color scanner data with inverse modeling methods, *J. Geophys. Res.*, *99*, 7457–7466, 1994.

Esaias, W. E., G. C. Feldman, C. R. McClain, and J. A. Elrod, Monthly satellite-derived phytoplankton pigment distribution for the North Atlantic Ocean basin, *Eos, Trans. Amer. Geophys. Un.*, *67*, 835–37, 1986.

Eslinger, D. L., J. J. O'Brien, and R. L. Iverson, Empirical orthogonal function analysis of cloud-containing Coastal Zone Color Scanner images of northeastern North American coastal waters, *J. Geophys. Res.*, *94*, 10884–10890, 1989.

Feldman, G. C., Variability of the productive habitat in the eastern equatorial Pacific, *Eos, Trans. Amer. Geophys. Un.*, *67*, 106–108, 1986.

Feldman, G. C., D. Clark, and D. Halpern, Satellite color observations of the phytoplankton distribution in the eastern equatorial Pacific during the 1982–83 El Niño, *Science*, *226*, 1069–1071, 1984.

Feldman, G. C., N. Kuring, C. Ng, W. Esaias, C. McClain, J. Elrod, N. Maynard, D. Endres, R. Evans, J. Brown, S. Walsh, M. Carle, and G. Podesta, Ocean color: Availability of the global data set, *Eos, Trans. Amer. Geophys. Un.*, *70*, 634 and 640–641, 1989.

Fuentes-Yaco, C., A. F. Vezina, P. Larouche, Y. Gratton, and M. Gosselin, Phytoplankton pigment in the Gulf of St. Lawrence, Canada, as determined by the Coastal Zone Color Scanner. Part II: Multivariate analysis, *Cont. Shelf Res.*, *17*, 1441–1459, 1997.

Garcia-Moliner, G., and J. A. Yoder, Variability of pigment concentration in warm-core rings as determined by Coastal Zone Color Scanner satellite imagery from the Mid-Atlantic Bight, *J. Geophys. Res.*, *99*, 14277–14290, 1994.

Gordon, H. R., and A. Y. Morel, Water color measurements—An introduction, In *Oceanography from Space*, edited by J. F. R. Gower, Plenum Publishing, New York, 207–212, 1981.

Gordon, H. R., and A. Y. Morel, *Remote Assessment of Ocean Color for Interpretation of Satellite Visible Imagery: A Review*, Springer-Verlag, Berlin, 114 pp, 1983.

Hochman, H. T., F. E. Muller-Karger, and J. J. Walsh, Interpretation of the Coastal Zone Color Scanner signature of the Orinoco River plume, *J. Geophys. Res.*, *99*, 7443–7455, 1994.
Holligan, P. M., M. Viollier, C. Dupony, and J. Aiken, Satellite studies on the distributions of chlorophyll and dinoflagellate blooms in the western English Channel, *Cont. Shelf Res.*, *2*, 81–96, 1983.
Holligan, P. M., T. Aarup, and S. B. Groom, The North Sea satellite colour atlas, *Cont. Shelf Res.*, *9*, 665–765, 1989.
Hooker, S. B., C. R. McClain, and A. Holmes, Ocean color imaging: CZCS to SeaWiFS, *MTS Journal*, *27*, 3–15, 1993.
Hovis, W. A., D. K. Clark, E. Anderson, R. W. Austin, W. H. Wilson, E. T. Baker, D. Ball, H. R. Gordon, J. L. Mueller, S. Z. El-Saveo, B. Sturm, R. C. Wrigley, and C. S. Yentsch, Nimbus-7 Coastal Zone Color Scanner systems: Description and initial imagery, *Science*, *210*, 60–63, 1980.
IOCCG, Minimum requirements for an operational, ocean-colour sensor for the open ocean. Rep. 1, International Ocean Color Coordinating Group, Dept. Oceanography, Dalhousie University, Dartmouth, Nova Scotia, 46 pp, 1998.
IOCCG, Status and plans for satellite ocean-colour missions: Considerations for complementary missions, Rep. 2, International Ocean Color Coordinating Group, Dept. Oceanography, Dalhousie University, Dartmouth, Nova Scotia, 43 pp, 1999.
Karl, D. M., A sea of change: Biogeochemical variability in the North Pacific subtropical gyre, *Ecosystems*, *2*, 181–214, 1999.
Kawamura, H., and OCTS Team, OCTS mission overview, *J. Oceanogr.*, *54*, 383–399, 1998.
Longhurst, A., Seasonal cycles of pelagic production and consumption, *Prog. Oceanogr.*, *36*, 77–167, 1995.
Longhurst, A., *Ecological Geography of the Sea*, Academic Press, New York, 398 pp., 1998.
McClain, C. R., and J. Firestone, An investigation of Ekman upwelling in the North Atlantic, *J. Geophys. Res.*, *98*, 12327–12339, 1993.
McClain, C. R., L. J. Pietrafesa, and J. A. Yoder, Observations of Gulf Stream-induced and wind-driven upwelling in the Georgia Bight using ocean color and infrared imagery, *J. Geophys. Res.*, *89*, 3705–3723, 1984.
McClain, C. R., J. A. Yoder, L. P. Atkinson, J. O. Blanton, T. N. Lee, J. J. Singer, and F. Muller-Karger, Variability of surface pigment concentrations in the South Atlantic Bight. *J. Geophys. Res.*, *93*, 10675–10697, 1988.
McClain, C. R., W. E. Esaias, G. C. Feldman, J. Elrod, D. Endres, J. Firestone, M. Darzi, R. Evans, and J. Brown, Physical and biological processes in the North Atlantic during the first GARP global experiment, *J. Geophys. Res.*, *95*, 18027–18048, 1990.
McClain, C. R., M. L. Cleave, G. C. Feldman, W. W. Gregg, S. B. Hooker, and N. Kuring, Science quality SeaWiFS data for global biosphere research, *Sea Technology*, *39*, 10–16, 1998.
Muller-Karger, F., C. R. McClain, and P. L. Richardson, The dispersal of the Amazon's water, *Nature*, *333*, 56–59, 1988.
Muller-Karger, R. E., and R. A. Castro, Mesoscale processes affecting phytoplankton abundance in the southern Caribbean Sea, *Cont. Shelf Res.*, *14*, 199–221, 1994.
O'Reilly, J. E., S. Maritorena, B. G. Mitchell, D. A. Siegel, K. L. Carder, S. A. Garver, M. Kahru, and C. R. McClain, Ocean color chlorophyll algorithms for SeaWiFS, *J. Geophys. Res.*, *103*, 24937–24953, 1998.

Perry, M. J., Assessing marine primary production from space, *Biosci.*, *36*, 461–467, 1986.

Platt, T., and S. Sathyendranath, Oceanic primary production: Estimation by remote sensing at local and regional scales, *Science*, *241*, 1613–1620, 1988.

Ryan, J. P., and J. A. Yoder, Long-term mean and event-related pigment distributions during the unstratified period in South Atlantic Bight outer margin and middle shelf waters, *Cont. Shelf Res.*, *16*, 1165–1183, 1996.

Ryan, J. P., J. A. Yoder, and P. C. Cornillon, Enhanced chlorophyll at the shelfbreak of the Mid-Atlantic Bight and Georges Bank during the spring transition, *Limnol. Oceanogr.*, *44*, 1–11, 1999.

Saitoh, S.-I., D. Inagake, K. Sasaoka, J. Ishizaka, Y. Nakame, and T. Saino, Satellite and ship observations of Kuroshio warm-core ring 93A off Sanriku, northwestern North Pacific, in spring 1997, *J. Oceanogr.*, *54*, 495–508, 1998.

Shannon, L. V., S. A. Mostert, N. M. Walters, and F. P. Anderson, Chlorophyll concentration in the southern Benguela Current region as determined by satellite (Nimbus-7 Coastal Zone Color Scanner), *J. Plankton. Res.*, *5*, 565–583, 1983.

Smith, R. C., X. Zhang, and J. Michaelsen, Variability of pigment biomass in the California Current system as determined by satellite imagery: 1, Spatial variability, *J. Geophys. Res.*, *93*, 10863–10882, 1988.

Strutton, P. G., F. P. Chavez, and M. J. McPhaden, Biological-physical coupling in the central equatorial Pacific during the 1997–1998 El Niño, In Proc. Ocean Optics XIV, Office of Naval Research, U. S. Navy, Arlington, Virginia, 1998.

Thomas, A. C., and P. T. Strub, Seasonal and interannual variability of pigment concentrations across a California Current frontal zone, *J. Geophys. Res.*, *95*, 13023-13042, 1990.

Vance, T., C. T. Baier, R. D. Brodeur, K. O. Coyle, M. B. Decker, G. L. Hunt, Jr., J. M. Napp, J. D. Schumacher, P. J. Stabeno, D. Stockwell, C. T. Tynam, T. E. Whitledge, T. Wyllie-Echeverria, and S. Zeeman, Anomalies in the ecosystem of the Eastern Bering Sea: Including blooms, birds, and other biota, *Eos, Trans. Amer. Geophys. Un.*, *79*, 121 and 126, 1998.

Yentsch, C. S., D. A. Phinney, and J. W. Campbell, Color banding on Georges Bank as viewed by Coastal Zone Color Scanner, *J. Geophys. Res.*, *99*, 7401–7410, 1994.

Yoder, J. A., and M. A. Kennelly, Wind stress and satellite chlorophyll patterns in the Atlantic Ocean, In Abstract Volume, The Oceanography Society's Scientific Meeting, The Oceanography Society, Washington, D.C., *25*, 1999.

Yoder, J. A., C. R. McClain, J. O. Blanton, and L. Y. Oey, Spatial structure in CZCS-chlorophyll imagery of the southeastern U.S. continental shelf, *Limnol. Oceanogr.*, *32*, 929–941, 1987.

Yoder, J. A., C. R. McClain, G. C. Feldman, and W. E. Esaias, Annual cycles of phytoplankton chlorophyll concentrations in the global ocean: A satellite view, *Global Biogeochem. Cycles*, *7*, 181–194, 1993.

Yoder, J. A., J. E. O'Reilly, A. H. Barnard, T. S. Moore, and C. M. Ruhsam, Variability in Coastal Zone Color Scanner (CZCS) chlorophyll imagery in ocean margin waters off the U.S. East Coast, *Prog. Oceanogr.*, submitted, 2000.

James Yoder, Graduate School of Oceanography, University of Rhode Island, South Ferry Road, Narragansett, RI 02882, U.S.A. (email, yoder@emu.uri.gso.edu; fax, +1-401-874-6728)

Satellites, Oceanography and Society
edited by David Halpern

Chapter 13

Remote-sensing studies of the exceptional summer of 1997 in the Baltic Sea: The warmest August of the century, the Oder flood, and phytoplankton blooms

Herbert Siegel and Monika Gerth
Baltic Sea Research Institute, Warnemünde, Germany

Abstract. The summer of 1997 was exceptional in the Baltic Sea. Stable high pressure over Scandinavia in the first 10 days of August, associated with high solar radiation, resulted in the highest air and water temperatures of this century. These warm and calm conditions were favorable for a strong bloom of cyanobacteria in the southern Gotland Sea. In addition, two strong rainfall periods in the drainage area of the Oder River caused the century's most extensive regional flooding, with floodwaters reaching the Pomeranian Bight in August.

Satellite data from different sensors and spectral ranges were used to study these special events. The monthly mean summer sea surface temperature (SST) distribution for the years 1990–1998, as derived from the National Oceanic and Atmospheric Administration (NOAA) Advanced Very-High Resolution Radiometer (AVHRR) data, showed a large interannual variation. The mean SST in the Baltic Sea exceeded 22°C in August 1997, the warmest month of the 1990s. In addition to SST, visible data from NOAA AVHRR and Indian Remote-Sensing Satellites (IRS-P3 and IRS-1C) Multispectral Optical Sensor (MOS) and Wide Field Sensor (WiFS) were used to investigate the blooms of cyanobacteria and coccolithophores and the distribution of river discharge during the Oder flood.

1. Introduction

The Baltic Sea is a marginal sea with a positive water balance and limited water exchange with the North Sea through the Danish straits, Kattegat and Skagerrak. The Baltic has a permanent halocline. During the warming phase in spring a thermocline develops, yielding conditions for the phytoplankton spring bloom. Wind-induced mixing and upwelling processes transport nutrients from the cold intermediate waters into the surface layer until the nutrients are consumed by biological uptake. After a stagnation period in June, the cyanobacteria grow in July and August. During calm conditions the cyanobacteria accumulate at the water surface and are transported by circulation processes. Investigations of cyanobacteria blooms were the first focus for application of remote-sensing data in the Baltic Sea in the 1970s and were initially conducted with Landsat data. During the last 20 years, support for Baltic Sea research using remote sensing has increased continuously. The variety of satellite data products have also increased: 1978–1986, Coastal Zone Color Scanner (CZCS) data; 1981–present, AVHRR SST; 1972–present, Landsat-MSS; 1982–present, Landsat-TM; 1996–present, MOS and WiFS data; 1997–present, Sea-viewing Wide Field-of-view Sensor (SeaWiFS) data; and 1991–present, European Remote-sensing Satellite (ERS-1/2) Synthetic Aperture Radar (SAR).

Applications of remote-sensing data in Baltic Sea research cover a wide range of topics, such as the support of monitoring programs, studies of dynamic processes, coastal zone and open sea interactions in river discharge areas, phytoplankton blooms, ice coverage, oil films, wind field determination, and climatological phenomena.

Horstmann (1983) used AVHRR and CZCS data to discuss eddy developments. Gidhagen (1984) examined upwelling processes along the Swedish coast and Hansen et al. (1993) dealt with Danish waters. Lehmann (1992) compared results of numerical modeling and SST-maps. Dynamical features and especially upwelling processes in the southwestern Baltic Sea were studied by Siegel et al. (1994).

Investigation of water exchange between the coastal zone and the open sea started with the application of AVHRR, CZCS, and Landsat data in the southeastern Baltic Sea (Brosin et al. 1988; Horstmann et al. 1986; Horstmann 1988). Oder River discharge in the Pomeranian Bight of the southwestern Baltic Sea was studied by Siegel et al. (1994, 1996, 1998a, 1998b), who combined ship measurements, numerical ocean model simulation, and data from AVHRR, CZCS, Landsat, MOS, and WiFS. Siegel et al. (1996) derived distribution patterns of Oder River discharge in the Pomeranian Bight for different wind directions. Dominating westerly and easterly winds transported the river water in a coastal jet along the Polish coast into Gdansk Bay and along the German coast into the Arkona Sea, respectively, identified as the main accumulation areas for Oder River load.

Application of satellite data from the visible spectral range for the investigation of cyanobacteria blooms was begun by Ulbricht and Schmidt (1977) and continued by Horstmann et al. (1986). Kahru et al. (1994) made systematic studies of cyanobacteria in the Baltic using data from AVHRR visible channels. Using CZCS data, Dowell (1996) described phytoplankton features in the southeastern Baltic Sea. Siegel et al. (1999b) applied CZCS data to study temporal and spatial developments of the spring bloom and the summer cyanobacteria bloom in the Baltic, and showed the influence on the distribution of phytoplankton of different processes transporting nutrients into the surface layer.

Alpers (1995) and Gade and Alpers (1998) used SAR data for studies of oil films on the sea surface. SAR data was used by Melsheimer and Gade (1998) to identify rainfall over the sea, and by Horstmann (1998) to compute a wind field over the Szczecin Lagoon and Pomeranian Bight.

Satellite data have become increasingly important for monitoring Baltic Sea SST (Matthäus et al. 1998) and ice distribution (Strübing 1990, 1995) and, in particular, for studying exceptional events, such as cyanobacteria blooms (Kahru et al. 1994; Kahru 1997; Gade et al. 1998), extremely high SST, and the Oder flood (Siegel et al. 1998a, 1998b). Siegel et al. (1999c) applied SST data for climatological studies of seasonal and interannual variations in the entire Baltic during the 1990s and investigated the influence of extreme situations on the nutrient composition in the Szczecin Lagoon and Swine River (Pastuczak et al. 1999).

The present study focuses on the application of different satellite datasets to the investigation of various events in the exceptional summer of 1997 in the Baltic Sea and is organized as follows: Section 2 summarizes the satellite data and methods used; Section 3 presents results and discusses the peculiarities of 1997, focusing on the warm summer, Oder flood, and the cyanobacteria and coccolithophores bloom. Some brief concluding remarks are presented.

2. Satellite Data and Methods

The entire Baltic Sea was examined, focusing on the southern Gotland Sea, the Skagerrak as part of the transition area to the open ocean, and the Pomeranian Bight as the discharge area of the largest freshwater supply into the southwestern Baltic. Figure 1a shows the area of investigation, locations of weather stations in Warnemünde and Arkona, and of the biological station of the University of Rostock in Zingst, and locations of two Baltic Sea Research Institute (IOW) Oceanographic Data-Acquisition Systems, ODAS DS (Figure 1b) and ODAS OB (Figure 1c).

The AVHRR SST data (Siegel et al. 1994) were provided by the German Federal Maritime and Hydrographic Agency (BSH), Hamburg. The BSH operates a SeaSpace High-Resolution Picture Transmission receiving station. The applicability of SST data to process studies in the Baltic Sea was discussed by Siegel et al. (1996). For the investigation of seasonal and interannual variations of SST, monthly mean SST maps were calculated from all cloud-free pixels during 1990–1998. The number of usable pixels depends on the quantity of received data from satellites and is influenced by cloud coverage, misinterpretations, and ice coverage (Siegel et al. 1999c).

The NOAA AVHRR visible channels were used to describe the blooms of coccolithophores in May and June 1997 and of cyanobacteria in August 1997. To minimize atmospheric effects, a combination of Channels 1 and 2 was used. In addition to the AVHRR data, data from MOS and WiFS at IRS-P3 and IRS-1C, which had pixel resolutions of 500 m and 185 m, respectively, were applied to investigate algae blooms and the

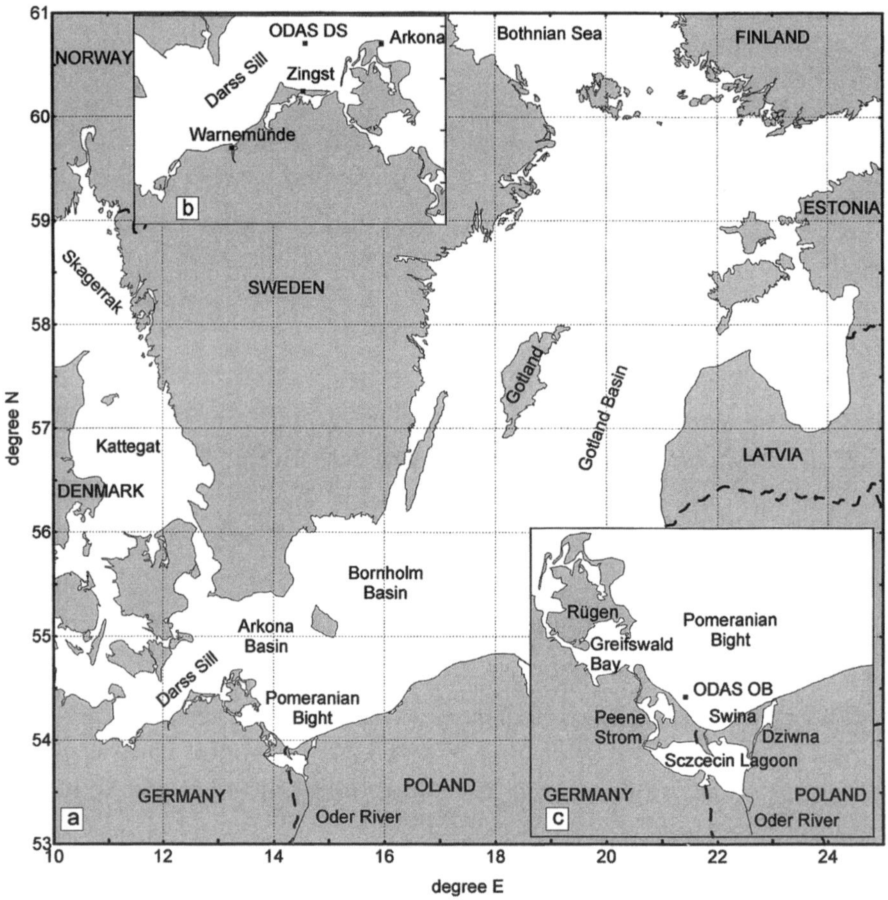

Figure 1. Maps of (a) Baltic Sea region; (b) western Baltic region with locations of ODAS DS at Darss Sill and of the weather stations at Warnemünde, Arkona, and Zingst; (c) Pomeranian Bight with location of ODAS OB.

distribution of suspended matter in the Pomeranian Bight. MOS and WiFS data were received from the German Aerospace Center (DLR). The optically active water constituents (chlorophyll, suspended matter, and yellow substances) were derived with algorithms developed for the Pomeranian Bight by Siegel et al. (1998a).

The water temperature is controlled by solar radiation and air temperature, wind that causes redistribution of heat in the water column due to vertical mixing, and ocean currents. The air temperatures measured at the Warnemünde weather station were used to calculate heat sums as the differences between daily mean values and a reference temperature of 16°C (Siegel et al. 1999c). Subsurface temperature measurements at Darss Sill

(21 m) were made at 10-minute intervals at 7-, 12-, 17-, and 19.5-m depths, with 0.01 K resolution (Matthäus et al. 1998).

3. Results

3.1 The hottest summer of the 1990s

Seasonal variations in water temperature in 1997 and interannual differences in July and August were studied using monthly mean SST maps (Figure 2). The annual cycle (Figure 2a) was characterized by temperature minimums in January and February and a

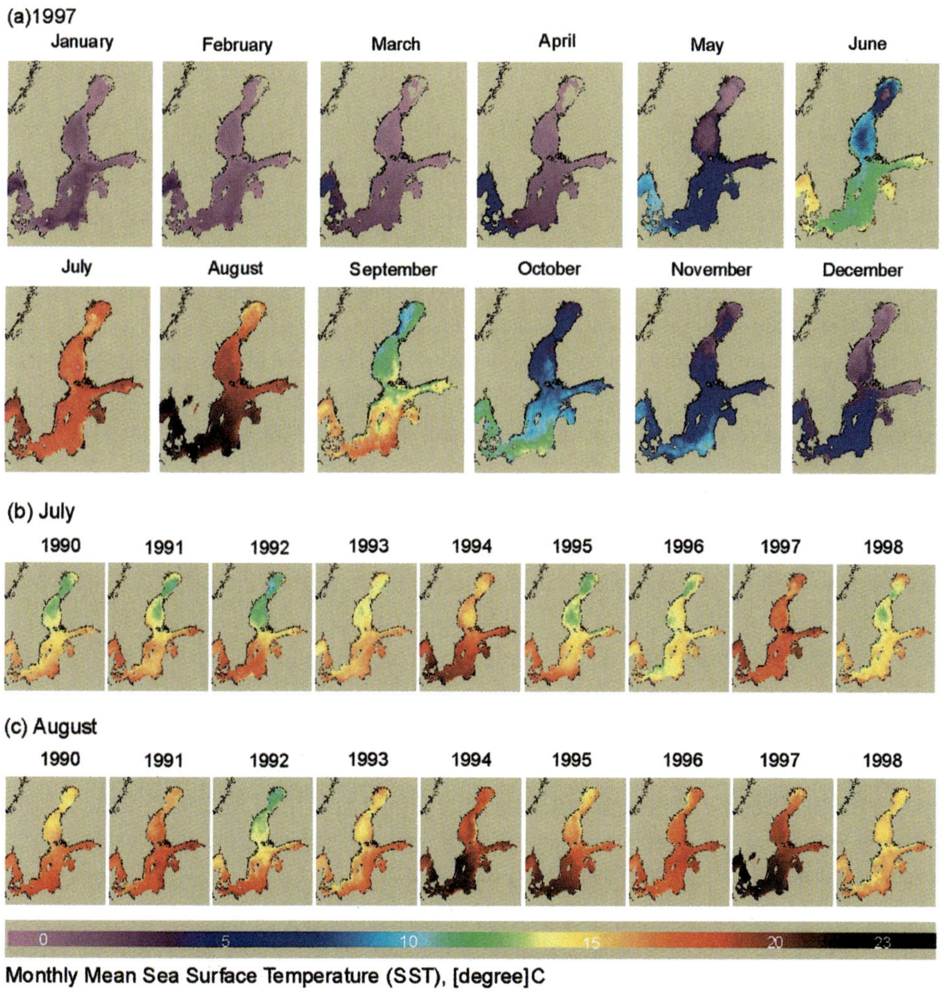

Figure 2. Monthly mean AVHRR SST in (a) 1997; (b) July 1990–1998; (c) August 1990–1998.

Table 1. Heat sum of air temperature in Warnemünde in summer (sum of temperature >16°C of daily mean temperatures)

Year	*Apr*	*May*	*Jun*	*Jul*	*Aug*	*Sep*	*Oct*	*Sum*
1990	0.0	5.9	20.9	30.2	69.8	1.4	0.6	128.8
1991	0.0	0.0	1.5	73.6	48.0	12.0	0.0	135.1
1992	0.0	16.2	**72.0**	103.6	80.9	2.6	0.0	275.3
1993	0.0	8.5	8.0	12.6	26.6	19.0	0.3	75.0
1994	0.0	0.1	19.4	**146.6**	78.0	0.8	0.0	244.9
1995	3.7	7.6	16.3	103.3	117.5	4.6	4.7	257.7
1996	11.0	2.3	34.1	22.2	78.3	0.2	0.0	148.1
1997	0.0	1.0	35.3	62.4	**173.0**	21.2	0.0	**292.9**
1998	1.8	7.0	23.0	29.0	34.5	9.9	0.0	105.2

maximum in August. The same pattern is seen in the upper-ocean temperature measurements from Darss Sill (Figure 3). The winter of 1996–1997 was marked by a cold December and January 1997, which led to low water temperatures in January. However, February 1997 was relatively warm and, consequently, spring 1997 water temperatures were similar to those in other years (Siegel et al. 1999c).

The third-warmest June in the 1990s was the first basis for higher summer temperatures in 1997. In July, the SST were 17–18°C in the entire Baltic Sea, beginning at Skagerrak and proceeding to the Gulf of Bothnia. In August, very intensive solar heating occurred (Table 1). As a result, SST reached more than 25°C in the western and southern Baltic Sea on some days, averaging more than 22°C for August. The warmest day was 21 August. As a result of upwelling processes, the southern and eastern coasts were an exception to the high temperatures. Strong storms in September led to fast cooling.

The summer of 1997 was not only the warmest of the 1990s, as expressed in the heat sum of air temperatures (Table 1), but also of this century, followed by 1992, 1995, 1994, and 1975 (Siegel et al. 1999). There seems to be no obvious climatological explanation for the sporadic occurrence of extreme summers during the past hundred years (1911, 1921, 1944, 1947, and 1975). The high frequency of hot temperatures in the 1990s could be due to the influence of the greenhouse effect.

Extreme monthly values of the heat sum of air temperatures were observed for June 1992, July 1994, and August 1997 (Table 1), and were partly due to solar radiation. The highest SSTs, exceeding 22°C, occurred in August 1994, 1995, and 1997 (Figure 2b). While SSTs were over 17°C in the Bothnian Sea in July 1994 and 1997, the July 1995 SST did not differ from other years. The highest SSTs in July were observed in 1994 and were related to the highest July 1994 air temperatures (Table 1). Similarly, the highest air

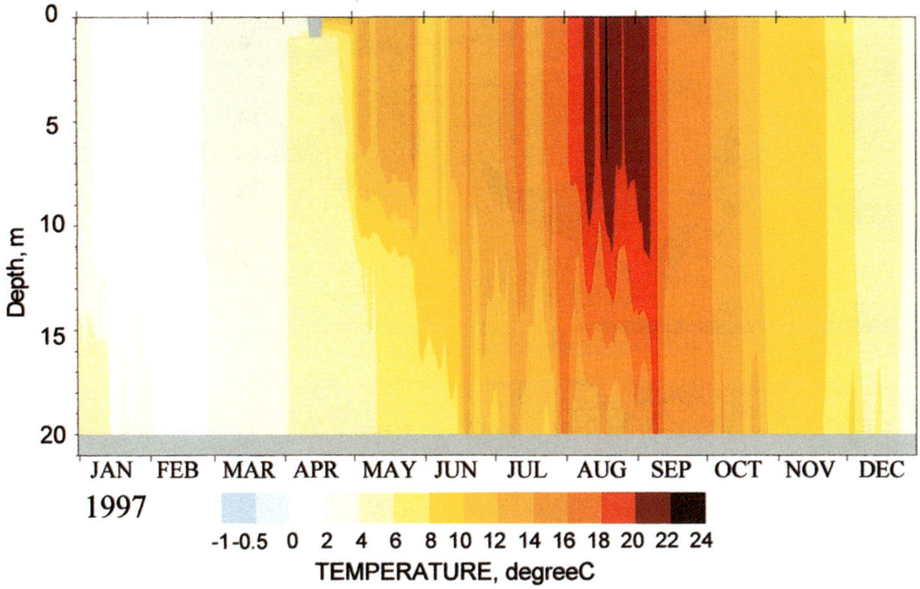

Figure 3. Time series of subsurface temperature in 1997 measured at ODAS DS (Lass personal communication).

temperatures in August 1997 caused the highest water temperatures. Subsurface temperatures at Darss Sill (Figure 3) indicated that the high temperature at the sea surface in August had extended to a depth of about 10 m.

The summer of 1992 was essentially different. The second-warmest summer after 1997 was not characterized by extreme temperatures in July and August, but by the warmest May and June. In this case, the highest SSTs, exceeding 15°C in the central Baltic Sea, occurred in June. July was also one of the warmest during this period. There was only little additional temperature increase in August because the heat sum and solar radiation were relatively low.

3.2 The Oder flood

The Oder River discharge (Figure 1c) is the fifth-largest water runoff into the Baltic Sea, with an annual flow of about 18 km^3 of water from a drainage area of 11.9×10^4 km^2.

In the summer of 1997, the drainage area of the Oder River was affected by an exceptional flood caused by two periods of heavy rain: 4–7 July and 18–21 July (Siegel et al. 1998b). The July precipitation rate was more than 5 times the long-term average in places. This caused the worst flooding of the Oder River in 50 years, with a maximum discharge at Eisenhüttenstadt, Germany, of 2500 m^3 s^{-1}, which was more than 6 times greater than normal. The water level increased by approximately 4 m by mid-July and

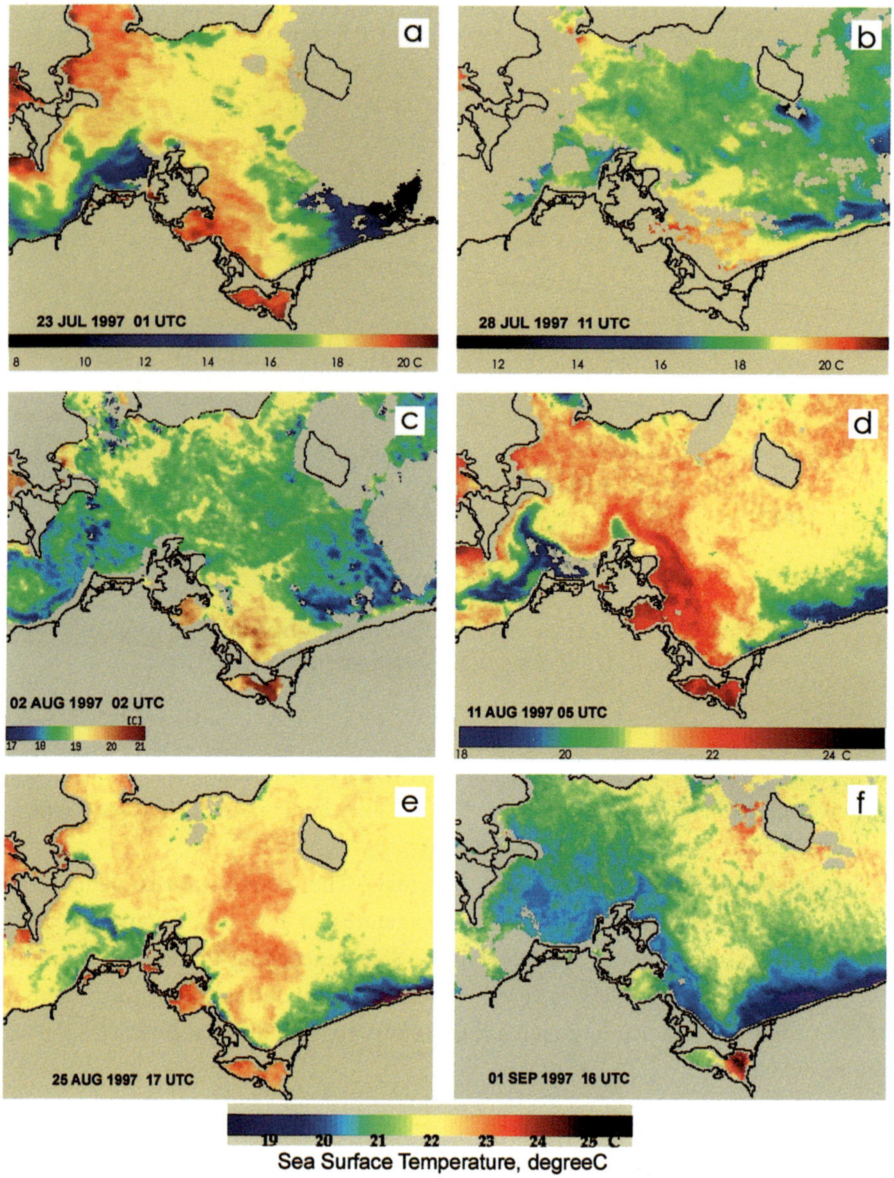

Figure 4. AVHRR SST of the Oder River discharge in the Pomeranian Bight, July–September 1997. Temperature scale is not the same in (a–f).

remained until 5 August. The flood reached Szczecin Lagoon on 20 July and the Pomeranian Bight on 28 July. As a result, about 9 km^3 of river water was transported into the Pomeranian Bight—6.5 km^3 more than the long-term mean value.

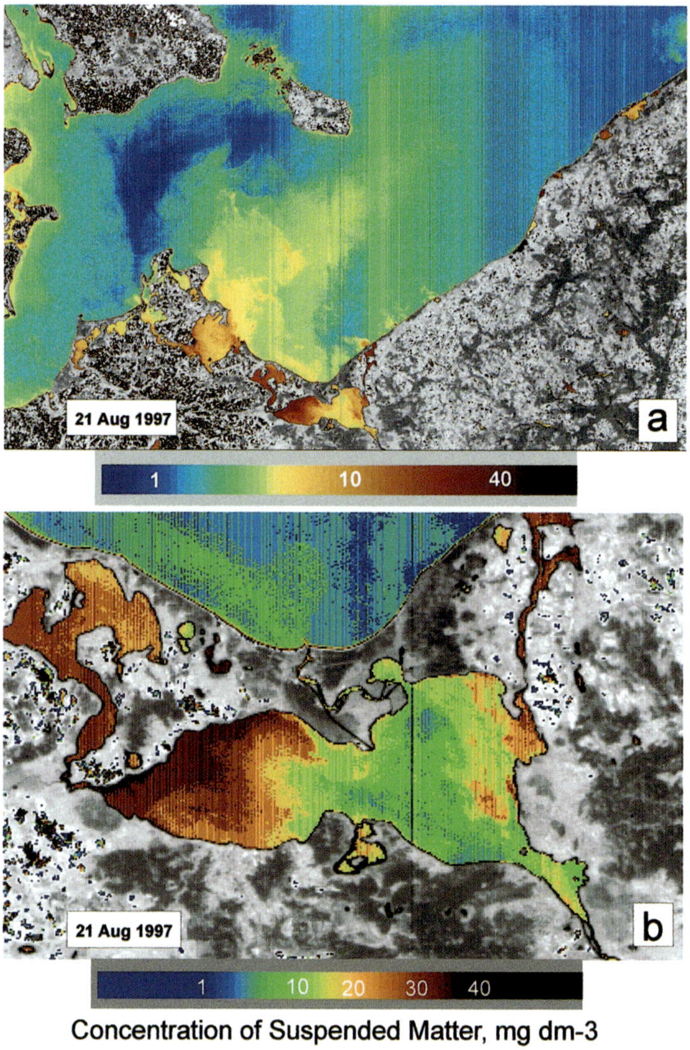

Figure 5. The 21 August 1997 distribution of WiFS-derived suspended matter in (a) Pomeranian Bight and (b) Szczecin Lagoon.

The spreading of the Oder River plume—with its higher temperatures, lower salinity, and larger quantity of constituents—in the Pomeranian Bight was studied using about 80 AVHRR SST images, shipborne measurements, and satellite ocean-color data. It was found that variations in the distribution patterns were caused by changes in wind direction and sea level differences between the lagoon and the southern Bight. The data revealed that, from 25 to 31 July, westerly winds generated an eastward transport along the Polish coast. During 4–28 August, a static high-pressure created easterly to southeasterly winds to transport the Oder plume along the German coast into the Arkona Sea. The SST (Figure 4), AVHRR visible channels, and WiFS and MOS images showed the spreading of the Oder discharge into the Arkona Sea (Siegel and Gerth 1998, 2000).

The Oder plume reached a thickness of 5-7 m (Siegel et al. 1998b). During the period of maximum extension on 21 August 1997, the discharge, with its high concentrations of suspended matter, was in the western Pomeranian Bight and the southern Arkona Sea, covering an area more than 3000 km^2 (Figure 5a). A further westward spreading was limited by upwelling off Hiddensee island.

The WiFS data of Szczecin Lagoon showed the Oder plume development in the lagoon (Figure 5). After the strong outflow on 29 July, differences occurred between the eastern and western parts of the lagoon. Specifically, the western part was mainly excluded from the water exchange (Figure 5b) (Siegel and Gerth 1998, 2000) and contained the "old" lagoon water with high concentrations of suspended constituents. The lower concentrations of suspended matter in the eastern lagoon were due to the diluting effect of the flood water, which reached the center of the western lagoon around 14 August. Complete mixing of the flood water in the lagoon did not take place until the end of August and the beginning of September.

The influence of the exceptional Oder River flood in the summer of 1997 was restricted to the Pomeranian Bight and the southern Arkona Sea. The large number of chemical and biological measurements demonstrated that the flood had no long-lasting effects or harmful consequences on the ecosystem of the Pomeranian Bight and the entire Baltic Sea (Humborg et al. 1998; Mohrholz et al. 1998; Siegel et al. 1998b).

3.3 Coccolithophore bloom in the Skagerrak

Coccolithophore blooms are a typical phenomenon in middle latitudes. The dominant species, *Emiliana huxleyi*, is a well-known producer of calcite and plays an important role in the carbon cycle (Ackleson et al. 1988). Growth conditions of this species are influenced by high-incident solar radiation, stable stratification, and iron content (Saydam and Polat 1998). Coccolithophore blooms in the Skagerrak and the North Sea were observed by Holligan et al. (1989) using CZCS and AVHRR data.

From 13 May to 12 June 1997, using AVHRR (Figure 6a-b) and MOS data, a strong bloom of coccolithophores was observed in the Skagerrak area. A relatively stable filament indicates low transport by the surface current and a low-wind period. The dominant species was *Emiliana huxleyi* (Dahl personal communication), which contain white calcite scales (Figure 6i). The scales cause high reflectivity and thus aid detection by ocean-color sensors and AVHRR broad-band visible channels. Concentrations measured during this period reached up to 3.2×10^6 cells/liter, reducing the Secchi disk depth to 3–4 m (Dahl personal communication). During its maximum extension the bloom covered up to 6×10^4 km^2 in the central Skagerrak and the region adjacent to Norway.

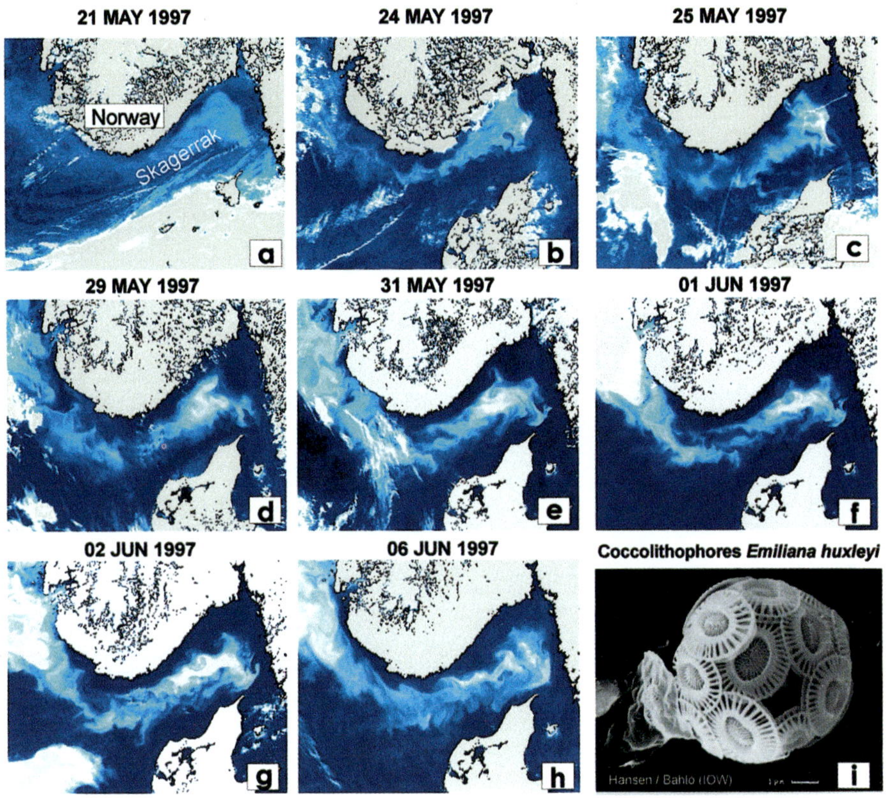

Figure 6. (a–h) Distribution of AVHRR-derived coccolithophores in May and June 1997 in Skagerrak; (i) electron microscope image (Hansen and Bahlo personal communication).

3.4 Cyanobacteria bloom in the southern Gotland Sea

The recording of cyanobacteria is usually complicated because of patchy distribution and sampling problems. However, satellite remote sensing in the visible spectral range helps overcome these problems. It is a powerful tool to study the spatial and temporal development of surface accumulations of cyanobacteria in the Baltic Sea.

The cyanobacteria, with the dominant species *Nodularia spumigena* and *Aphanizomenon* sp., bloom every summer in the Baltic during stable thermal stratification and nutrient depletion in the near-surface layer. Cyanobacteria are able to use dissolved molecular nitrogen to produce biomass. Water temperatures of about 15–17°C (Kahru et al. 1994) seem to be the favorable starting conditions for cyanobacteria growth, which normally takes place in July and August for about 4–6 weeks, depending on meteorological and hydrographic conditions. When winds are calm, cyanobacteria form large aggregates and

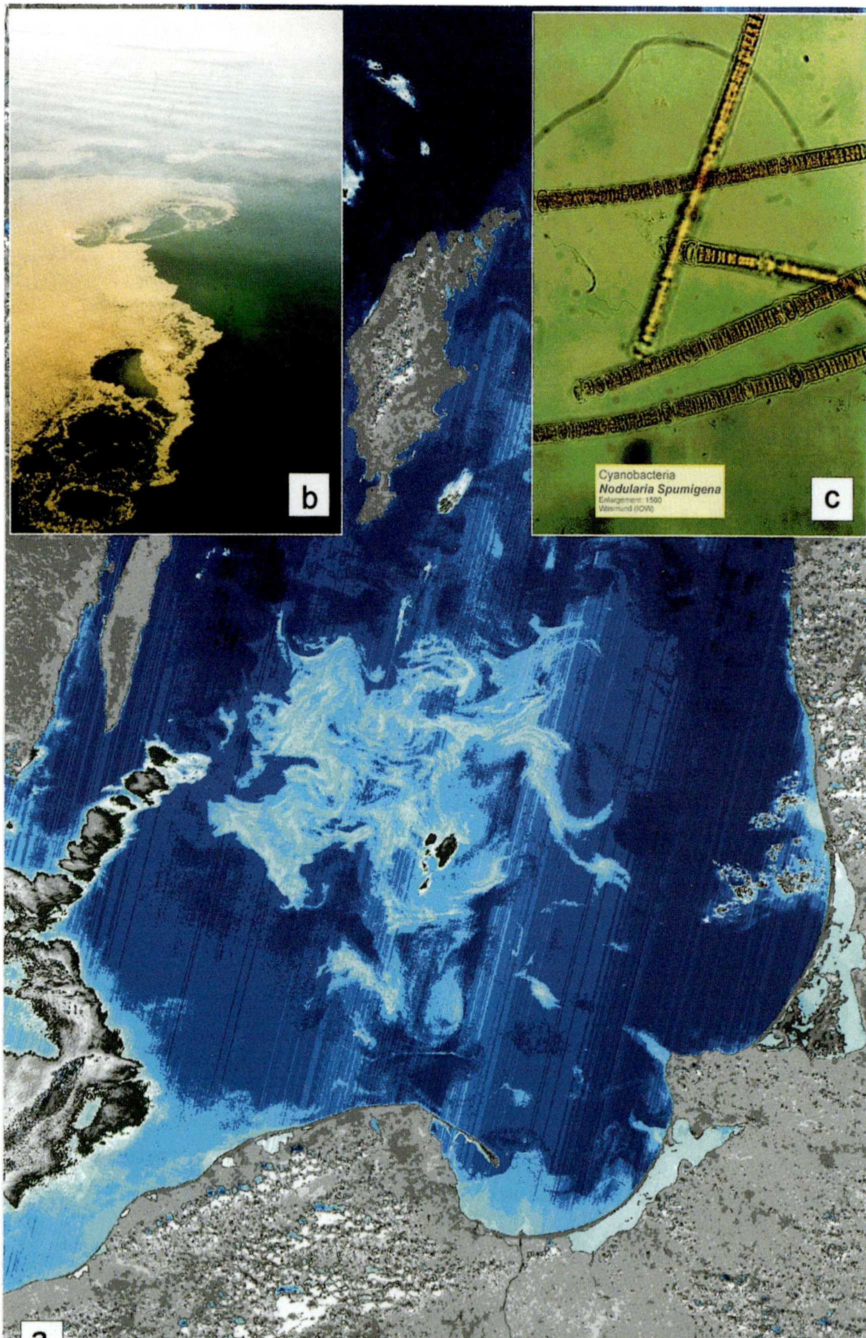

Figure 7. (a) WiFS-derived cyanobacteria pattern on 9 August 1997; (b) photograph by H. Siegel of cyanobacteria bloom; (c) microscopic photograph of *Nodularia spumigena* (provided by Wasmund).

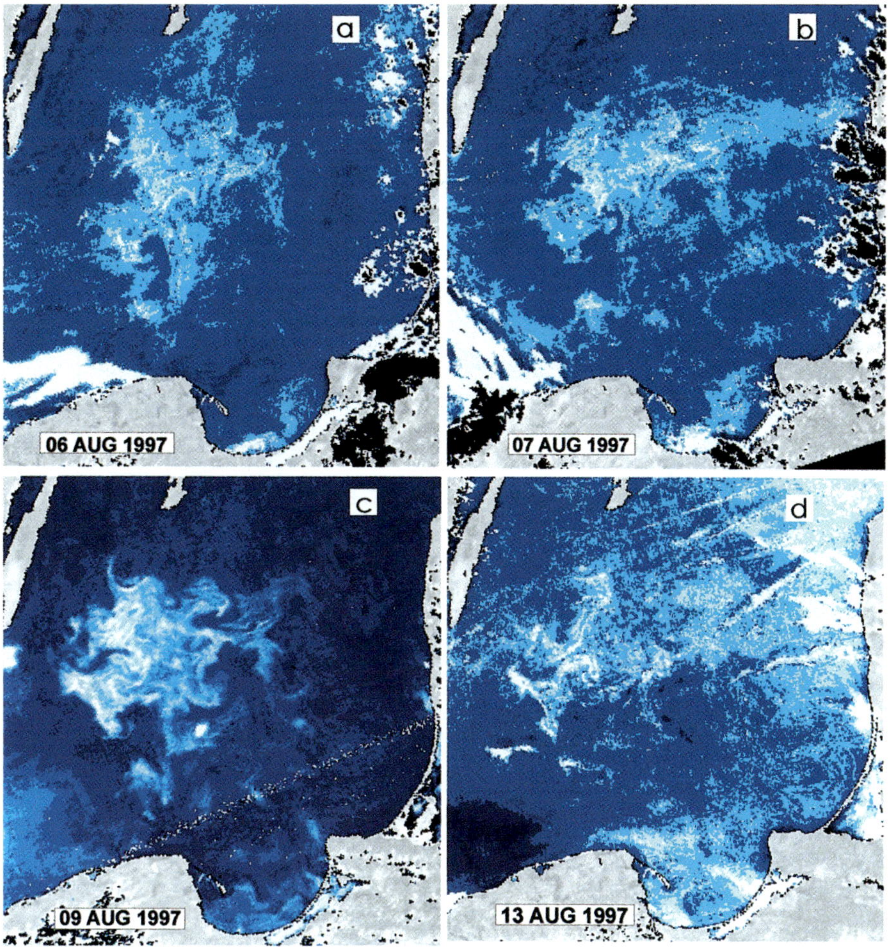

Figure 8. Development of an AHVRR-derived cyanobacteria bloom south of Gotland Island in August 1997.

float up to the surface due to their buoyancy. At the surface they are transported by currents and wind, forming stripes of high concentrations ending in eddy-like structures (Kononen 1992; Kononen and Leppänen 1997). Extreme gradients in the concentrations of cyanobacteria result in strong changes of the optical properties of water, especially the backscattering and, consequently, ocean color. Vertical mixing induced by strong wind and convective overturning may interrupt or end the bloom.

The cyanobacteria bloom in the summer of 1997 was one of the longest and, particularly in August, had one of the most intensive surface accumulations. The first surface cyanobacteria accumulation appeared in July, which was caused by a temperature

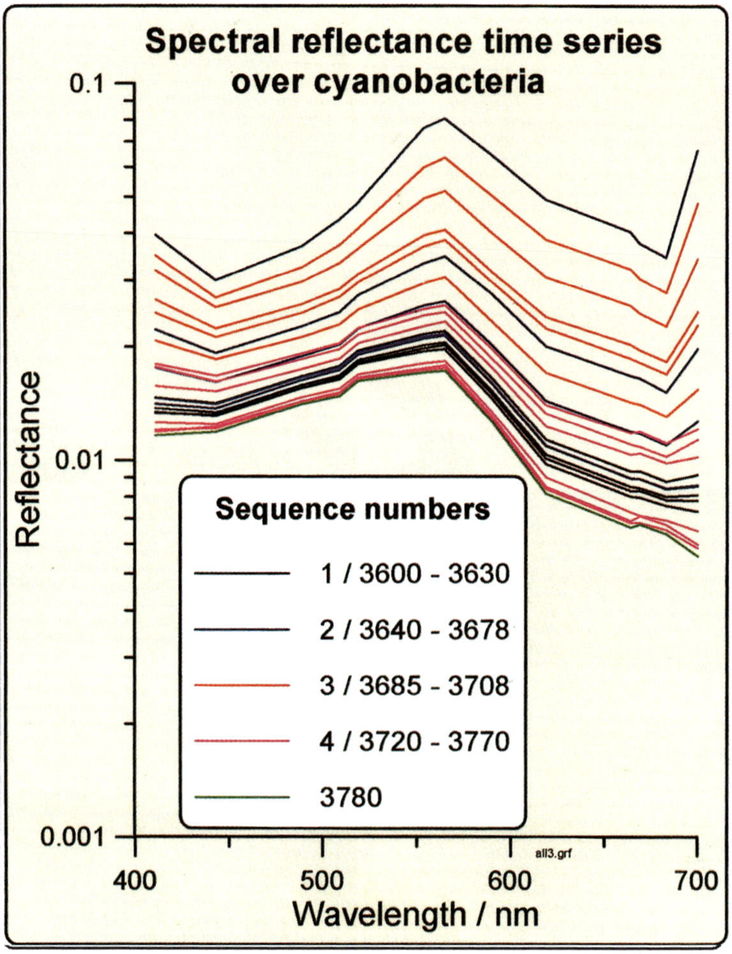

Figure 9. Time series of spectral reflectances measured over the drifting patch of cyanobacteria presented in Figure 8b, located at the west side of the covered area in Figure 8a, with a strong increase around 700 nm, which is similar to green plants without water coverage.

increase and calm conditions. During the low-wind period in August, intensive surface accumulations of the cyanobacteria *Nodularia spumigena* (Figure 7c) were observed south of Gotland from 6 to 13 August in AVHRR images (Figure 8).

The maximum areal extension exceeded 19×10^3 km^2, and the features were also noted in the WiFS image (Figure 7a). The bloom was also observed from the deck of a vessel (Figure 7b). The thickness of the compact surface accumulation was about 5 cm and the concentration was approximately 450 mgChl m^{-3}. The total reflectance measured above the sea surface (Figure 9) demonstrates the strong variation between the very clear water outside the patch and the cyanobacteria layer. The spectral distribution also changed. The high absorption of chlorophyll-*a* decreased reflectance near 440 nm and reflectance increased near 700 nm. This effect is known from land plants without water influence.

4. Summary and Conclusions

Owing to their synoptic character and high spatial and temporal resolutions, the application of remote-sensing data of different spectral ranges allowed advancements in Baltic Sea research concerning investigations of dynamical processes, river discharges, phytoplankton blooms, ice coverage, oil films, climatological studies, and event monitoring. The year 1997 provided an exemplary case with special events to demonstrate the usefulness of satellite-based remote sensing in Baltic Sea research. The exceptionally hot summer of 1997 was the warmest of this century. In August, the SST in the Baltic Sea exceeded 22°C in monthly mean values and 25°C in single AVHRR images.

During the Oder River flood in July 1997, the spreading of an additional volume of 6.5 km^3 water into the western Pomeranian Bight and southern Arkona Sea was monitored by satellite data. At the time of maximum extension the riverine water covered an area of more than 3×10^3 km^2. Due to the effects of dilution, the flood had no long-lasting or harmful consequences on the ecosystem.

A low-wind period in May and June induced a bloom of coccolithophores *Emiliana huxleyi* in Skagerrak for about 20 days, with concentrations up to 3.2×10^6 cells per liter and a maximum extention of approximately 6×10^4 km^2.

An exceptional bloom of the cyanobacteria *Nodularia spumigena* was observed south of Gotland Island for one week in August 1997. The bloom thickness was 5 cm, had a concentration of 450 mgChl m^{-3}, and a maximum extent of about 19×10^3 km^2.

Acknowledgments. The authors wish to thank Mrs. G. Tschersich of the BSH, Hamburg, for providing the NOAA AVHRR data, the Remote Serving Center Neustrelitz of the DLR, the DLR project "Oderflood," and the company EUROMAP for IRS-P3 WiFS scenes. We thank Dr. H. U. Lass for data from Darss Sill, Dr. Tiesel for the heat sum of air temperatures in Warnemünde, Dr. E. Dahl for information on coccolithophores in the Skagerrak, and R. Bahlo, R. Hansen, and N. Wasmund of the IOW for their support with the microscopic pictures of algae. Final thanks go to the reviewers for their helpful comments and to Prof. Dr. M. S. McLachlan for editorial support. The investigations were supported by the DLR (50 EE 92168).

References

Ackleson, S., W. M. Balch, and P. M. Holligan, White waters of the Gulf of Maine, *J. Oceanogr.*, *1/2*, 18–22, 1988.

Alpers, W., Measurement of mesoscale oceanic and atmospheric phenomena by ERS-1 SAR, *Radio Sci. Bull.*, *275*, 14–22, 1995.

Bernstein, R. L., Sea surface temperature estimation using the NOAA-6 satellite Advanced Very High Resolution Radiometer, *J. Geophys. Res.*, *87*, 9455–9465, 1982.

Brosin, H.-J., L. Gohs, T. Seifert, H. Siegel, I. A. Bychkova, S. V. Viktorov, M. D. Demina, V. J. Lobanov, V. N. Losinskij, and V. M. Smoljanickij, Mesoskale Strukturen in der südlichen Ostsee im Mai 1985, *Beitr. Z. Meeresk., 58,* 8–18, 1988.

Dowell, M. D., Optically active components and their relationship with mesoscale features in the Baltic coastal zone, Meereswiss. Berichte (Marine Science Reports), Gobex–Summary Report, IOW, *19*, 114–139, 1996.

Gade, M., and W. Alpers, Using ERS-2 SAR images for routine observation of marine pollution in European coastal waters, *Sci. Total Environ.*, *237–238*, 441–448, 1998.

Gade, M., O. Rud, and M. Ishii, Monitoring algae blooms in the Baltic Sea by using optical and microwave sensors, In Proc. IGARSS '98, 757–759, 1998.

Gidhagen, L., Coastal upwelling in the Baltic, SMHI Reports, RHO 37, Parts 1 and 2, Swedish Meteorological and Hydrological Institute, Norrköping, 1984.

Hansen, L., N. K. Hojerslev, and H. Sogaard, Temperature monitoring of the Danish marine environment and the Baltic Sea, Kobenhavns University Report 52, Copenhagen, Denmark, 77 pp, 1993.

Holligan, P. M., T. Aarup, S. B. Groom, and J. Aiken, The North Sea satellite colour atlas, *Cont. Shelf Res.*, *8/9*, 665–764, 1989.

Horstmann, U., Distribution patterns of temperature and water colour in the Baltic Sea as recorded in satellite images: Indicators for phytoplankton growth, *Ber. Inst. f. Meeresk.*, Kiel, *106*, 145 pp, 1983.

Horstmann, U., Satellite remote sensing for estimating coastal offshore transports, In *Coastal-Offshore Ecosystem Interactions*, edited by B.-O. Jansson, American Geophysical Union, Washington, D.C., 50–66, 1988.

Horstmann, U., H. van der Piepen, and K. W. Barrot, The influence of river water on the southeastern Baltic Sea as observed by Nimbus 7/CZCS imagery, *Ambio*, *15,* 286–289, 1986.

Horstmann, U., W. Koch, S. Lehner, and W. Rosenthal, Mesoscale wind fields and their variation retrieved from the synthetic aperture radar aboard ERS-1/2, In Proc. Fifth Int. Conf. Rem. Sens. Mar. Coastal Env., I, ERIM, Ann Arbor, 116–123, 1998.

Humborg, C., G. Nausch, T. Neumann, F. Pollehne, and N. Wasmund, The exceptional Oder flood in summer 1997—the fate of nutrients and particulate organic matter, *Dt. Hydrogr. Z.*, *50*, 169–181, 1998.

Kahru, M., Using satellites to monitor large-scale environmental change: A case study of cyanobacteria blooms in the Baltic Sea, In *Monitoring Algal Blooms*, edited by M. Kahru and C. W. Brown, Springer-Verlag, Berlin, 43–61, 1997.

Kahru, M., U. Horstmann, and O. Rud, Satellite detection of increased cyanobacteria blooms in the Baltic Sea: Natural fluctuations or ecosystem change?, *Ambio*, *23,* 469–472, 1994.

Kononen, K., Dynamics of the toxic cyanobacteria blooms in the Baltic Sea, *Finn. Mar. Res.*, *261*, 3–36, 1992.

Kononen, K., and J.-M. Leppänen, Patchiness, scales and controlling mechanisms of cyanobacterial blooms in the Baltic Sea: Application of a multiscale research strategy, In *Monitoring Algal Blooms*, edited by M. Kahru and C. W. Brown, Springer-Verlag, Berlin, 63–84, 1997.

Lehmann, A., Ein dreidimensionales baroklines wirbelauflösendes Modell der Ostsee. *Ber. Inst. f. Meeresk., Kiel, 179*, 176 pp, 1992.

Matthäus, W., G. Nausch, H. U. Lass, K. Nagel, and H. Siegel, Hydrographisch-chemische Zustandseinschätzung der Ostsee 1997, *Meereswiss. Ber., Warnemünde, 29*, 1–65, 1998.

Melsheimer, C., and M. Gade, Rain cells over the sea monitored by synthetic aperture and weather radars: A comparison, In Proc. Oceans '98, 1445–1448, 1998.

Mohrholz, V., M. Pastuszak, S. Sitek, K. Nagel, and H. U. Lass, The exceptional Oder flood in summer 1997—riverine mass and nutrient transport into the Pomeranian Bight, *Dt. Hydrogr. Z., 50*, 129–144, 1998.

Saydam, A. C., and I. Polat, Dust-influenced algae blooms, In Proc. SPIE Conf. Ocean Optics XIV, 1998.

Siegel, H., and M. Gerth, Distribution patterns of Oder discharge in the Pomeranian Bight during the exceptional flood in summer 1997, In Proc. Fifth Int. Conf. Rem. Sens. Marine Coastal Env., 481–485, 1998.

Siegel, H., and M. Gerth, Satellite-based studies of the Oder flood event in the southwestern Baltic Sea in summer 1997, *Remote Sensing Environ.*, in press, 2000.

Siegel, H., M. Gerth, R. Rudloff, and G. Tschersich, Dynamical features in the western Baltic Sea investigated using NOAA AVHRR data, *Dt. Hydrogr. Z., 3*, 191–209, 1994.

Siegel, H., M. Gerth, and T. Schmidt, Water exchange in the Pomeranian Bight investigated by satellite data and shipborne measurements, *Cont. Shelf Res., 16*, 1793–1817, 1996.

Siegel, H., M. Gerth, and T. Ohde, Case 1 and Case 2 algorithms for MOS-IRS and their application to different regions, In Proc. 2nd Workshop MOS-IRS Ocean Colour, 1998a.

Siegel, H., W. Matthäus, R. Bruhn, M. Gerth, G. Nausch, T. Neumann, and C. Pohl, The exceptional Oder flood in summer 1997—distribution of the Oder discharge in the Pomeranian Bight, *Dt. Hydrogr. Z, 50*, 145–167, 1998b.

Siegel, H., M. Gerth, and A. Mutzke, Dynamics of the Oder River plume in the southern Baltic Sea—satellite data and numerical modelling, *Cont. Shelf Res., 19*, 1143–1159, 1999a.

Siegel, H., M. Gerth, T. Neumann, and R. Doerffer, Case studies on phytoplankton blooms in coastal and open waters of the Baltic Sea using Coastal Zone Colour Scanner data, *Int. J. Remote Sensing, 20*, 1249–1264, 1999b.

Siegel, H., M. Gerth, R. Tiesel, and G. Tschersich, Seasonal and interannual variations in satellite-derived sea surface temperature of the Baltic Sea in the 1990s, *Dt. Hydrogr. Z., 51*, in press, 1999c.

Strübing, K., Fernerkundung von Meereis, *Promet, 3/4*, 114–120, 1990.

Strübing, K., Eiskartierung mit ERS-1 Radaraufnahmen—Möglichkeiten und Probleme, *DLR Mitteilungen, 95/04*, Tagungsband des 11, Nutzerseminars des DFD der DLR, 26–28, 1995.

Ulbricht, K. A., and D. Schmidt, Massenauftreten mariner Blaualgen in der Ostsee auf Satellitenaufnahmen erkannt, *DFVLR-Nachrichten, 22*, 913–915, 1977.

Herbert Siegel, Institut für Ostseeforschung, Warnemünde, Seestrasse 15, D-18119 Rostock, Germany. (email, herbert.siegel@io-warnemuende.de; fax, +49-381-519-7480)

Satellites, Oceanography and Society
edited by David Halpern

Chapter 14

Remote-sensing studies of seasonal variations of surface chlorophyll-*a* concentration in the Black Sea

Nikolay P. Nezlin
P. P. Shirshov Institute of Oceanology, Russian Academy of Sciences, Moscow

Abstract. Seasonal variations of Sea-viewing Wide Field-of-view Sensor (SeaWiFS) and Coastal Zone Color Scanner (CZCS) measurements of surface chlorophyll-*a* concentration in the Black Sea are described. For areas where the water depth is greater than 200 m, surface chlorophyll-*a* concentration was maximum in winter. SeaWiFS surface chlorophyll-*a* concentrations in 1997–1998 were larger than CZCS data in 1978–1986; the change seems to result from both eutrophication and long-term climatic oscillations. In 1997–1998, the maximum surface chlorophyll-*a* concentration occurred during autumn instead of winter–spring; this seems to result from Danube River runoff, which is maximum in August 1997; the possible mechanism of this influence is discussed.

1. Introduction

The present state of the Black Sea ecosystem should be considered disastrous (Mee 1992). It is important when discussing this ecosystem to distinguish between anthropogenic changes and variations resulting from climatic variations. Spaceborne observations can provide the means to do so. The precision of these data seems to be rather low, but data are regular and numerous. In contrast, oceanographic in-situ observations are insufficient in number.

The results of remote sensing from many satellites enable regular observations of the world ocean. The most important data, from the point of view of biology, is ocean color, which characterizes the concentration of plant pigments (primarily chlorophyll-*a*) in the surface layer. Pigment concentration is highly correlated with biomass of different ecological groups of plankton (Vinogradov et al. 1995, 1996a, 1996b, 1997, 1999).

A comprehensive array of ocean color observations was collected between 1978 and 1986 by the Coastal Zone Color Scanner (CZCS) radiometer (Hovis et al. 1980). A new era of remote-sensing observations of ocean color started in August 1997 with the launch of the OrbView-2 platform carrying the Sea-viewing Wide Field-of-view Sensor (SeaWiFS) radiometer. This paper presents the results of an analysis of CZCS and SeaWiFS data about seasonal variations of the ecosystem of the Black Sea, with emphasis on the productive winter period. The physical environment was characterized by sea surface temperature (SST) measured since 1981 by the National Oceanic and Atmospheric Administration (NOAA) Advanced Very-High Resolution Radiometer (AVHRR).

2. Black Sea Circulation

Several features of Black Sea bathymetry are shown in Figure 1. The abyss (depth >1500 m) is separated from the coastal margin by a steep continental slope, except in the northeast. A wide continental shelf (mean depth about 50 m) is between the Crimean peninsula and the west coast; the width decreases towards the south. The shelf of the southern coast abruptly terminates at 30°30'E (Sakarya River). Along the southern and eastern coast, the continental shelf is very narrow. The abyssal region or "open" region of the Black Sea occupies over one-half of the total area.

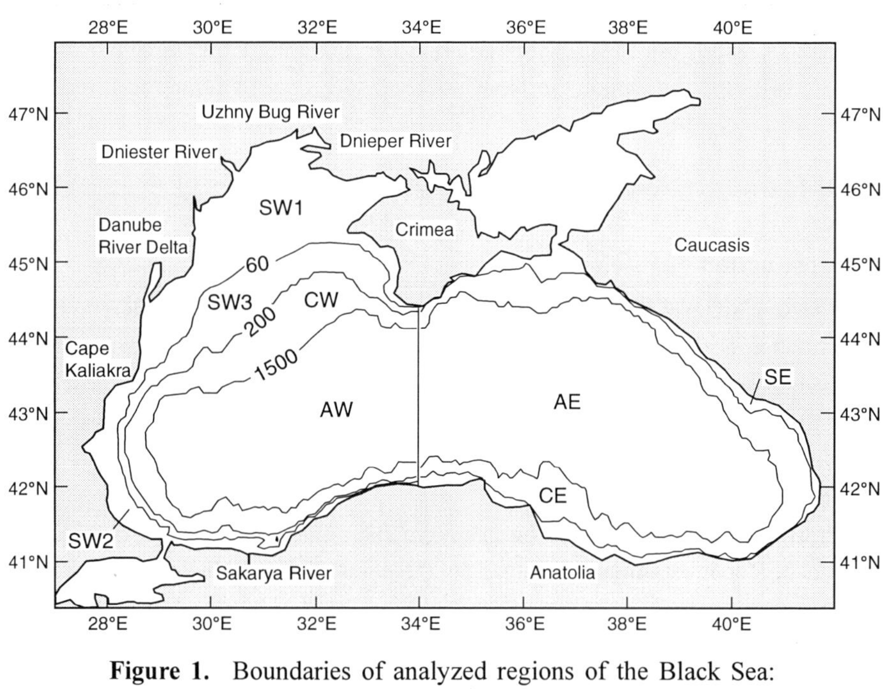

Figure 1. Boundaries of analyzed regions of the Black Sea: AW, western open; AE, eastern open; CW, western slope; CE, eastern slope; SE, eastern shelf; SW1, northwestern interior shelf; SW2, southwestern interior shelf; SW3, western outer shelf. Depth contours are in meters.

A basin-scale cyclonic boundary current, named Rim Current (Oguz et al. 1992) or Main Black Sea Current in Russian literature, is the predominant feature of Black Sea circulation (Figure 2). The Rim Current is less than 75-km wide and has an average speed of 20 cm s^{-1}. It separates the cyclonic circulation in the interior of the basin from the anticyclonic current in the narrow coastal zone. Upwelling occurs in the cyclonic gyres. In the anticyclonic gyres and along the periphery of the Rim Current, water sinks to remove terrigenic contaminants from the surface layer, thus isolating interior waters from the influence of the polluted coastal ecosystems. The difference between shelf- and deep-sea ecosystems is so large that these domains should be analyzed separately. Thus, the Black Sea is divided into open (>1500 m), slope (200–1500 m), and shelf (<200 m) areas.

The Black Sea has a pronounced two-layer haline stratification (Murray et al. 1991). The upper layer is formed by mixing of resident water with fresh river water. The thickness of the upper layer is about 150 m. The uppermost 30–40 m experiences seasonal temperature variations of 6–24°C. The salinity of the 30- to 40-m near-surface layer is typically 18.0 practical salinity units (psu) and varies seasonally by as much as 1.9 psu as a result of changes in evaporation, precipitation, runoff, etc. (Oguz et al. 1992). The lower part of the upper layer is occupied by cold intermediate water (CIW), whose temperature is < 8°C. Below 150 m, the water masses are remarkably stable. At 150 m the salinity reaches 21.0 psu.

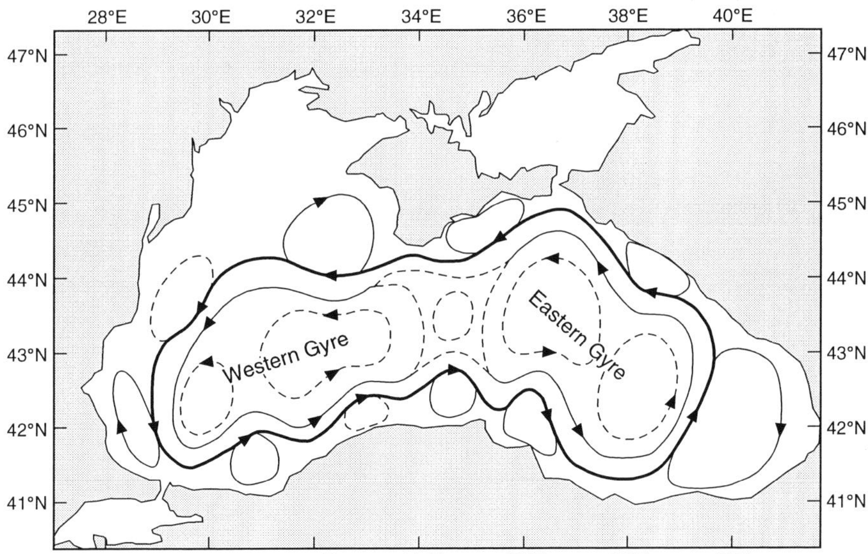

Figure 2. Schematic circulation of the Black Sea (after Oguz et al. 1993). Solid and dashed lines indicate quasi-permanent and recurrent features of the general circulation, respectively.

The depth of the main pycnocline fluctuates with space and time. During winter, the layer of maximum density gradient (main pycnocline) rises at the domes of cyclonic gyres to depths as shallow as 25 m, which is in the euphotic zone (Ovchinnikov and Popov 1987). This results in the winter phytoplankton bloom (Mikaelyan 1995). At the same time, the main pycnocline descends to 160–180 m in the convergent zone along the periphery of the basin at the shoreward boundary of the Rim Current and in the anti-cyclonic eddies in the southeast.

3. Data and Methods

The 0.1758° resolution of CZCS data (Feldman et al. 1989) was the same as the eight-day average nighttime AVHRR SST data. SeaWiFS was launched 1 August 1997, and data acquisition commenced 20 September 1997. SeaWiFS scans approximately 90% of the ocean every two days. The band-center wavelengths (in nm) are 412 (violet), 443 (blue), 490 and 510 (blue-green), 555 (green), 670 (red), and 765 and 865 (near infrared). In this study, global grids of 5.27', equal to 9.77 km latitude and 7.23–7.82 km longitude, were used.

The Black Sea was divided into the same eight regions (Figure 1) analyzed by Nezlin and Dyakonov (1998a, 1998b), distinguishing eastern and western parts, and open (or interior), slope, and shelf areas. The narrow eastern shelf was treated as one region; the wide western shelf was divided into three regions, with the boundary between the north-western and southwestern interior parts located near Cape Kaliakra.

CZCS and SeaWiFS data were analyzed using nonparametric statistical criteria involving 25th and 75th percentiles. The 25th percentile means 25% of the values are below the value, and 75th percentile means 75% are below. Figure 3 illustrates the boundaries of 50% of values.

Maps of SeaWiFS surface chlorophyll-*a* concentration (Figures 4a and 4b) were drawn from eight-day composites after the grids were smoothed with a 50-km radius cosine filter to remove small-scale variability. The eight-day averaged AVHRR SST anomalies (Figure 5) were computed relative to the 1981–1996 mean SST, also with a 50-km radius cosine filter. Only basic features of SST and chlorophyll-*a* distributions over the Black Sea were analyzed.

Danube River discharge in 1997 was obtained from the Danube Hydrometeorological Observatory of the Ukrainian State Committee of Hydrometeorology.

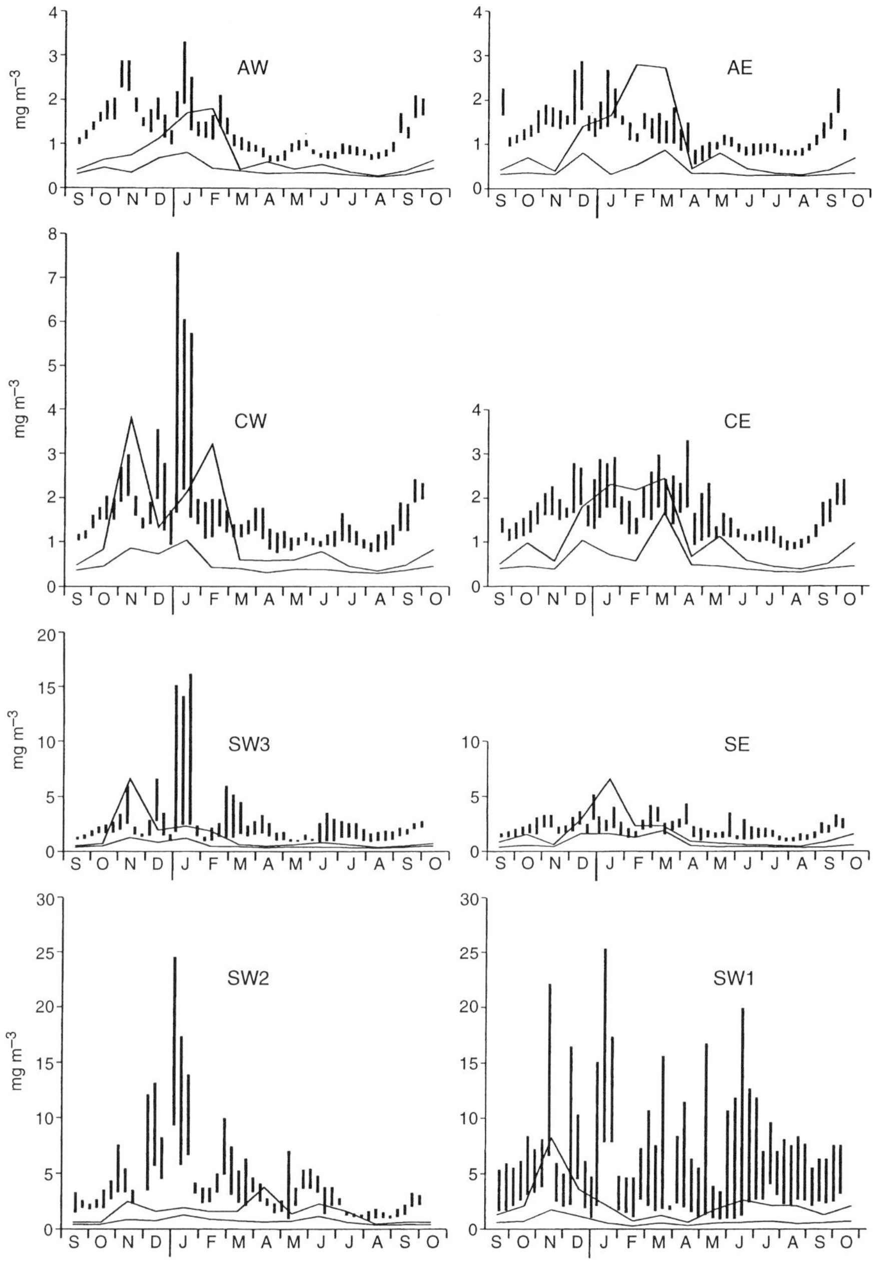

Figure 3. Pigment concentrations in eight areas of the Black Sea. Solid curves represent 25th and 75th percentiles of monthly mean CZCS data during 1978–86. Vertical lines represent 25th and 75th percentiles of eight-day mean SeaWiFS data September 1997–October 1998.

4. Results

4.1 Seasonal variation in surface chlorophyll-*a* concentration

In all regions of the Black Sea, excluding the most shallow SW1 and SW2 areas, monthly mean CZCS values (Figure 3) were lower during April–October, the "summer" period, compared to those occurring during November–March, the "winter" period (Nezlin 1997). SeaWiFS data were typically higher than CZCS data, especially during the warm season when SeaWiFS values were about two times higher. During 1997, high surface chlorophyll-*a* concentration appeared in the open and slope regions of the Black Sea in late September, about one month earlier than usual; hence, the duration of the 1997–1998 winter was longer and shifted to an earlier period compared with 1978–1986. The seasonal patterns in the western shelf regions SW1, SW2, and SW3 (Figure 3) differ from those observed in other regions. On the northwestern interior shelf (SW1), the surface chlorophyll-*a* values are always high, especially for SeaWiFS data. Pronounced chlorophyll-*a* peaks are evident on the southwestern (SW2) and outer-western shelves (SW3) during January 1998.

Three possible reasons are proposed to explain the higher values of surface chlorophyll-*a* concentration in 1997–1998 compared with those for 1978–1986. CZCS and SeaWiFS are different instruments. CZCS data resolution (~20 km) is half that of SeaWiFS (~10 km). In addition, SeaWiFS data are much more numerous than CZCS. During 1978–1986, there are some months when no CZCS data are available for the Black Sea because of clouds, especially during winter, or because the instrument was not operating. SeaWiFS data have good spatial and temporal coverage over each eight-day period. Differences between CZCS and SeaWiFS capabilities are more significant for coastal regions. Chlorophyll-*a* concentration is usually high in the narrow band along the coast, and this region was observed more often by SeaWiFS than it was by CZCS.

Especially high surface chlorophyll-*a* values occur during winter when satellite observations of ocean color and SST are absent more often than during summer because of the increased amount of clouds. However, it was for the summer period that SeaWiFS values regularly exceeded CZCS data for the open and slope regions (AW, AE, CW, CE). During summers, both CZCS and SeaWiFS collected sufficient data. Hence, instrument differences seem not to be the reason for the observed inconsistency in the magnitude of the chlorophyll-*a* data.

The observed increase in surface chlorophyll-*a* concentration in the Black Sea in 1997–1998, as compared with 1978–1986, could be representative of an increase in phytoplankton biomass or plant pigment concentration associated with eutrophication. A similar trend was detected at the beginning of the 1990s from in-situ chlorophyll-*a* data collected over the Black Sea (Vedernikov and Demidov 1993). Eutrophication advanced with the invasion of the newcomer ctenophore *Mnemiopsis leidyi* (Vinogradov et al.

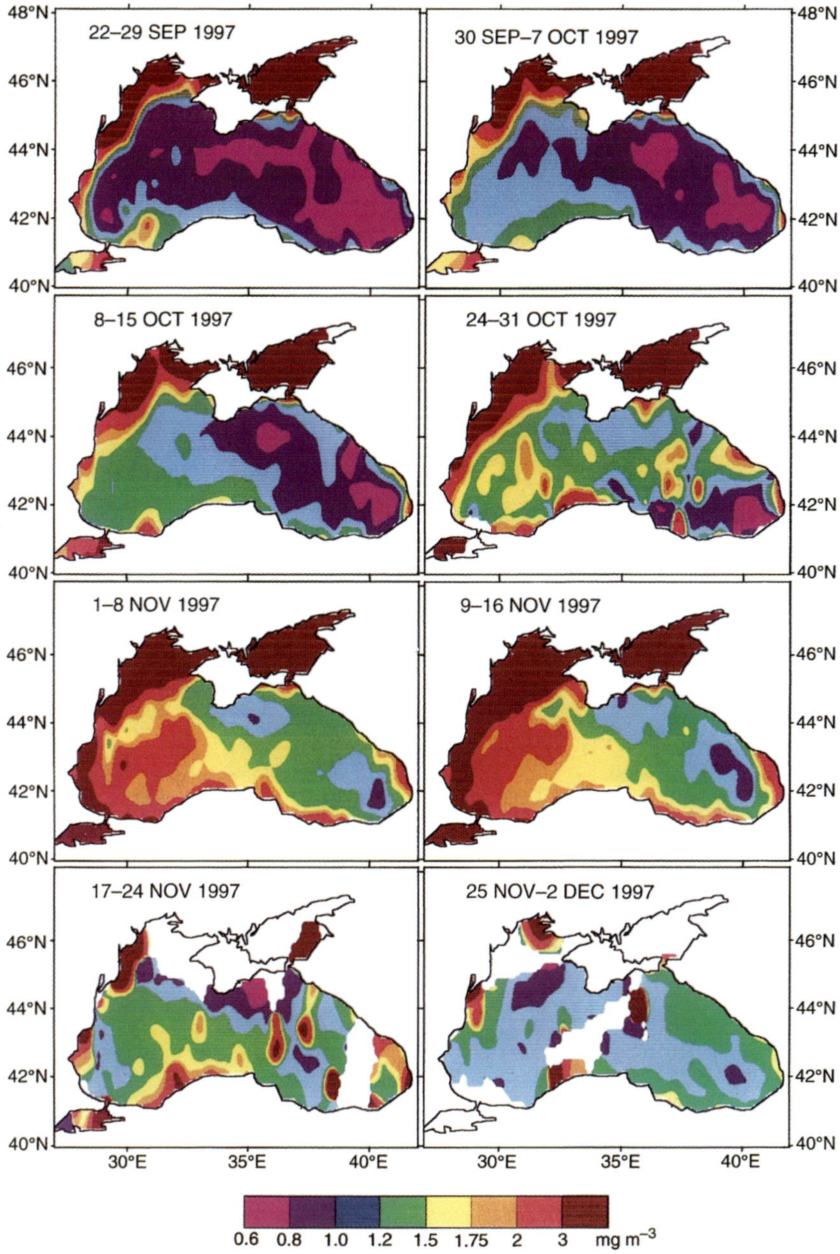

Figure 4a. SeaWiFS surface chlorophyll-*a* concentrations (mg m^{-3}) over the Black Sea from 22 September 1997 to 2 December 1997. From 16 to 23 October 1997, the number of observations was insufficient for mapping.

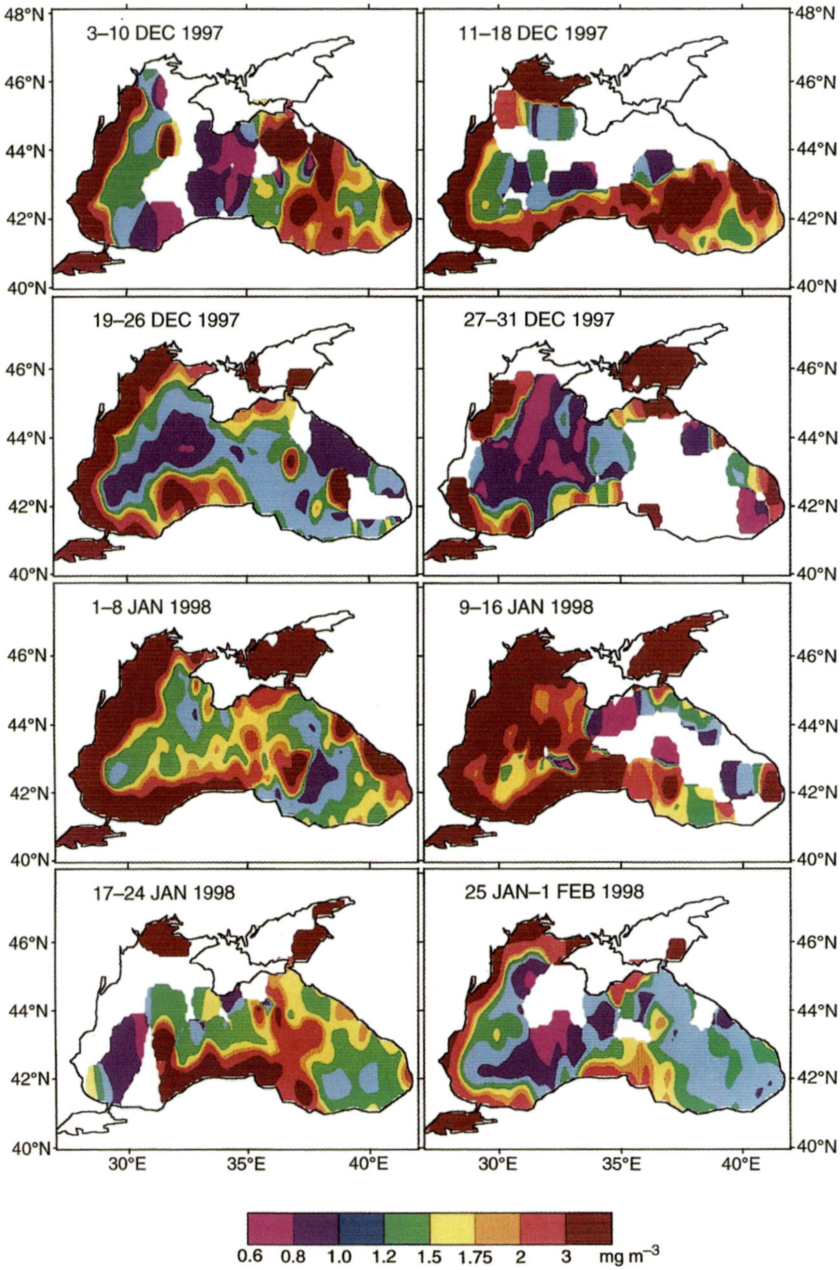

Figure 4b. SeaWiFS surface chlorophyll-*a* concentrations (mg m^{-3}) over the Black Sea from 3 December 1997 to 1 February 1998.

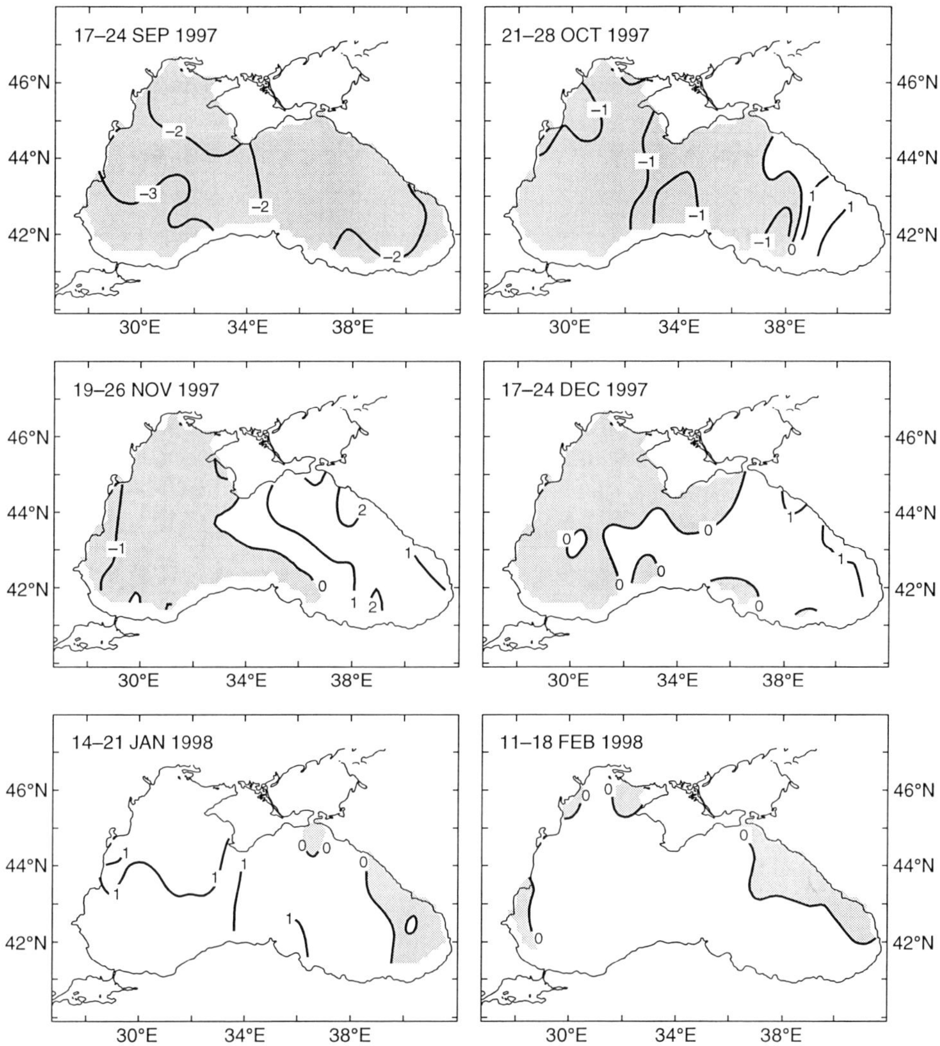

Figure 5. AVHRR eight-day averaged sea surface temperature anomalies (°C) over the Black Sea. Negative anomaly is shaded.

1992). This carnivore destroyed the bulk of phytophagous zooplankton to cause significant structural changes in the whole pelagic community, including an increase in phytoplankton biomass. Mikaelyan (1997) noted the gradual increase in phytoplankton biomass in the Black Sea from the 1970s to the middle 1990s. In-situ observations of chlorophyll-*a* concentrations in the Black Sea from 1990 to 1996 were summarized by Yilmaz et al. (1998). The difference between in-situ chlorophyll-*a* data before 1990 and from 1990 to 1996 agrees well with the difference between CZCS and SeaWiFS data.

Another reason for higher surface chlorophyll-*a* concentration in the Black Sea during 1997–1998 might be related to climatic oscillations. Mikaelyan (1997) suggested that oscillations in the phytoplankton biomass have a period of about 20 years. He noted that during the years when the winter average air temperature was low (circa 1950, 1970, 1990) compared with the climatic average, summer phytoplankton biomass was higher compared with other years. During cold winters, circulation in the Black Sea increases, resulting in increased penetration of nutrient-rich, deep water into the upper productive layer (Ovchinnikov and Popov 1987). During mild winters between extremely cold years (circa 1960, 1980), phytoplankton biomass in summer was lower. Mikaelyan (1997) had predicted low phytoplankton biomass from 1998 to 2000, which was not confirmed by our observations. The winter of 1997–1998 was relatively warm, but surface chlorophyll-*a* concentration values were higher than the average values during the 1978–1986 period.

Nezlin and Dyakonov (1998a, 1998b) found a correlation between higher CZCS chlorophyll-*a* concentrations and milder winters of 1980–1982. The less-intensive convective mixing of the water column during a mild winter caused the seasonal thermocline to be eroded later in autumn and formed earlier in spring. According to Sverdrup (1955), the formation of a seasonal thermocline in spring is the prerequisite condition for a spring bloom of phytoplankton in temperate regions. During winter, the destruction of the seasonal thermocline by intensive convection results in phytoplankton cells being transported from the thin euphotic layer into poorly illuminated deep layers, causing photosynthesis to slow, in spite of high nutrient concentrations. In spring, the intensity of the convective process decreases, causing the formation of the seasonal thermocline. The high concentration of nutrients in the euphotic layer yields a phytoplankton bloom. Dnsuring summer, the seasonal thermocline hinders penetration of nutrients into the upper productive layer. Hence, phytoplankton growth rate is low compared with the amount of grazing by herbivorous plankton. In autumn, partial (but not complete) destruction of the seasonal thermocline causes nutrient-rich subthermocline waters to upwell into the euphotic layer, resulting in an increased phytoplankton growth rate and biomass. Thus, it is tempting to speculate that during the mild winter of 1997–1998, erosion of the seasonal thermocline was slow to yield higher surface chlorophyll-*a* concentrations during the autumn period (Figure 3). This concept was confirmed by modeling experiments (Oguz et al. 1996, 1998).

4.2 Dynamics of chlorophyll-*a* concentration during the cold season

In autumn 1997, surface chlorophyll-*a* concentration increased earlier in the deep western region AW than in the eastern region AE. The dynamics of this phenomenon are evident from the chlorophyll-*a* distribution over the Black Sea (Figure 4), which was regularly high only in the zone where the Danube and other rivers (Dnieper, Dniester, and Uzhny Bug) flow into the Black Sea. Beginning at the end of September, high surface chlorophyll-*a* concentrations spread southward along the western coast, then are carried eastward by the cyclonic Rim Current to fill the western half of the Black Sea. This process was completed by the middle of November, when chlorophyll-*a* concentration decreased. During the first half of December, high concentrations occurred in the eastern part of the Black Sea. During the first half of January, high concentrations again occurred in the western region, with chlorophyll-*a*-rich waters again penetrating from the region of the Danube, both directly into the open sea and advected along the Anatolian coast by the Rim Current.

It is very likely that the development of phytoplankton in the Black Sea during the autumn and winter of 1997–1998 was influenced by the intensity of river discharge, especially from the Danube. The latter contributes nearly 70% of the total runoff into the Black Sea (Cociasu et al. 1997) and seems to be the main source of its eutrophication (Zaitsev and Mamaev 1997). In 1997, the Danube River runoff (7100 $m^3\ s^{-1}$) was higher than the long-term average values of 6290 $m^3\ s^{-1}$ (Nikiforov and Diaconu 1963) or 6550 $m^3\ s^{-1}$ (Stancík and Jovanovic 1988). The strongest contrast occurred in August (Figure 6), when the runoff intensity was about twice that of the climatic mean. The unusually high intensity of Danube River discharge in August 1997 seems to have caused the atypical autumn bloom of phytoplankton biomass. In September, the runoff intensity decreased to a typical autumn value and then had a typical winter increase. River dis-

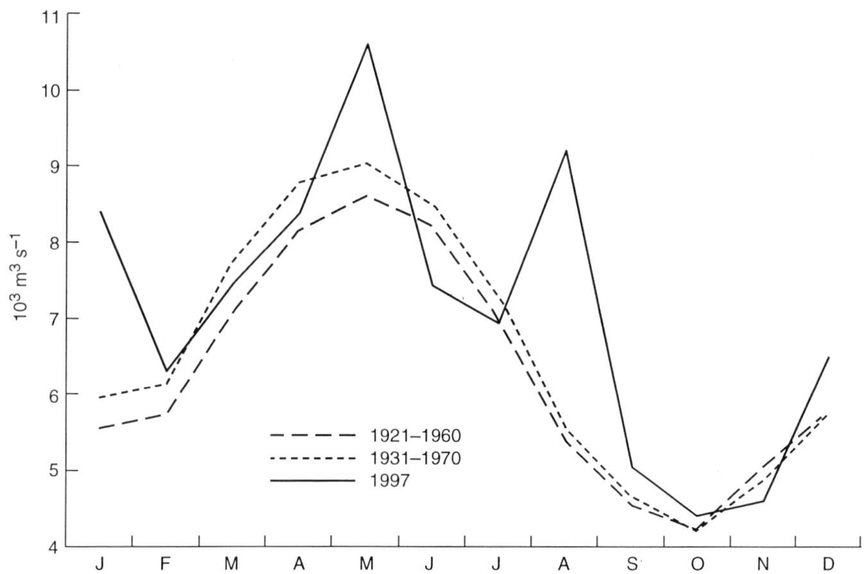

Figure 6. Intensity of Danube River runoff. Data for 1921–1960 are from Stancík and Jovanovic (1988); data for 1931–1970 are from Nikiforov and Diaconu (1963).

charge influences phytoplankton development in the Black Sea by pouring nutrients into the sea and by altering the buoyancy of the near-surface layer.

4.3 Danube River nutrient discharge

Surface chlorophyll-*a* concentrations in the coastal zones of the Black Sea are always higher compared with those of the deeper regions (Figure 3). Nutrient concentrations in the upper layer are also permanently high near the shore and seem not to limit phytoplankton growth.

The main nutrient transported by the Danube River is total inorganic nitrogen, whose values vary from 1000 to 3000 tons per day, with a maximum in April and a minimum in August (Cociasu et al. 1997). The minimum rate of primary production in the least productive open part of the Black Sea during the least productive period (summer) is about 300 mg carbon m^{-2} day^{-1} (Vedernikov and Demidov 1993, 1997; Stelmakh et al. 1998). The total area of the Black Sea is 4.2×10^5 km^2, and the area of the open part (>1500-m deep) is about 2.2×10^5 km^2. Using a carbon-to-nitrogen Redfield ratio of 6, the total nitrogen utilized by phytoplankton is about 2.1×10^4 tons day^{-1} over the Black Sea and 1.1×10^4 tons day^{-1} in the deep region. Thus, the maximum amount of nitrogen introduced by Danube River runoff could sustain <25% of the minimum value of primary production in the central Black Sea and <15% of production over the whole Black Sea. These amounts are insufficient to cause the variations in surface chlorophyll-*a* concentrations observed in 1997–1998. Similar calculations with phosphorus and silica resulted in lower values.

4.4 Black Sea temperature and salinity during the 1997–1998 winter

In September, the SST was below normal over the Black Sea and the anomaly reached –3°C in the western regions (Figure 5). In October, the SST anomaly had decreased to about –1°C, and in the eastern Black Sea it changed to positive values. The SST anomaly pattern in November was peculiar: SST was below normal in the western and southern Black Sea, and above normal in the eastern and northern parts. In December, the range of SST anomaly over the Black Sea narrowed to ±1°C. In January and February, the SST anomaly was insignificant over the Black Sea.

The significant negative SST anomaly in September resulted from very cold weather in which the monthly average air temperature anomaly was –3°C. The SST anomaly was maintained by the increased buoyancy created by desalination of the upper mixed layer, due to an atypical Danube River flood in August. This process strengthened the cyclonic circulation in the Black Sea. Cooler waters from the northwest were advected by the Rim Current along the Anatolian coast, and warmer waters from the southeast flowed along the Caucasian and Crimean coasts. In December this process weakened, and in January and February the surface waters of the Black Sea were slightly warmer than usual.

The positive SST anomalies in January and February indicated that the 1997–1998 winter was rather mild. Cyclonic circulation intensifies during winter (Oguz et al., 1992). During extremely cold winters the speed of the Rim Current is especially high. We hypothesize that in autumn 1997, the intensification of the Rim Current occurred earlier because of the extremely high Danube runoff in August. Oguz et al. (1995) showed that intensive river inflow into the northwestern shelf region also enhances the general circulation in the Black Sea. In autumn 1997, the Rim Current intensification was combined with lower intensity of upper-layer convective mixing due to rather mild winter air temperature, resulting in the early beginning of the winter phytoplankton bloom. This example illustrates that not only the winter air temperature, but the combination of temperature, river runoff, and other parameters must be taken into account in attempts to forecast the productivity of the Black Sea.

Acknowledgments. This work was supported in part by the RFBR Grant 97-0564513 and NATO Linkage Grant LG.97.1233. I would like to thank the SeaWiFS Project at the Goddard Space Flight Center for the production and distribution of the data used in this study. I thank V. Yu. Dyakonov (SIO RAS, Russia) for help in solving computer problems, and V. N. Morosov (Danube Hydrometeorological Observatory) and V. N. Mikhailov (Moscow State University) for supplying the data on the intensity of the 1997 Danube River discharge. I am grateful to A. I. Ginzburg and A. G. Kostianoy (SIO RAS) for a fruitful discussion on the hydrology of the Black Sea. Finally, thanks to two anonymous reviewers for a thorough analysis and significant improvements to this paper.

References

Cociasu, A., V. Diaconu, L. Popa, L. Buga, I. Nae, L. Dorogan, and V. Malciu, The nutrient stock of the Romanian shelf of the Black Sea during the last three decades, In *Sensitivity to Change: Black Sea, Baltic Sea and North Sea*, edited by E. Ozsoy and A. Mikaelyan, Kluwer Academic Publishers, Dordrecht, The Netherlands, 49–63, 1997.

Feldman, G. C., N. Curing, C. Ng, W. Esaias, C. R. McClain, J. Elrod, N. Maynard, D. Endres, R. Evans, J. Brown, S. Walsh, M. Carle, and G. Podesta, Ocean colour: Availability of the global data set, *Eos, Trans. Amer. Geophys. Un.*, *70*, 634–635, 1989.

Hovis, W. A., D. K. Clark, F. Anderson, R. W. Austin, W. H. Wilson, E. T. Baker, D. Ball, H. R. Gordon, J. L. Mueller, S. Y. El Sayed, B. Sturm, R. C. Wrigley, and C. S. Yentsch, Nimbus-7 Coastal Zone Color Scanner: System description and inertial imagery, *Science*, *210*, 60–63, 1980.

Mee, L. D., The Black Sea in crisis: a need for concerted international action, *Ambio, 21,* 278–286, 1992.

Mikaelyan, A.S., Winter bloom of the diatom *Nitzschia delicatula* in the open waters of the Black Sea, *Mar. Ecol. Prog. Ser., 129,* 241–251, 1995.

Mikaelyan, A. S., Longtime variability in phytoplankton communities in the open Black Sea in relation to environmental changes, In *Sensitivity to Change: Black Sea, Baltic Sea and North Sea,* edited by E. Ozsoy and A. Mikaelyan, Kluwer Academic Publishers, Dordrecht, The Netherlands, 105–116, 1997.

Murray, J. W., Z. Top, and E. Ozsoy, Hydrographic properties and ventilation of the Black Sea, *Deep-Sea Res.*, *38*, 663–689, 1991.

Nezlin, N. P., Seasonal variation of surface pigment distribution in the Black Sea on CZCS data, In *Sensitivity to Change: Black Sea, Baltic Sea and North Sea*, edited by E. Ozsoy and A. Mikaelyan, Kluwer Academic Publishers, Dordrecht, The Netherlands, 131–138, 1997.

Nezlin, N. P., and V. Yu. Dyakonov, Analysis of interannual variations of the surface chlorophyll concentration in the Black Sea from the data of CZCS radiometer, *Oceanology* (English translation), *38*, 636–641, 1998a.

Nezlin, N. P., and V. Yu. Dyakonov, Seasonal and interannual variations of surface chlorophyll concentration in the Black Sea on CZCS data, In *Ecosystem Modeling as a Management Tool for the Black Sea*, *1*, edited by L. I. Ivanov and T. Oguz, Kluwer Academic Publishers, Dordrecht, The Netherlands, 137–150, 1998b.

Nikiforov, Ya. D. and K. Diaconu, editors, *Hydrology of the Embouchment Area of Danube* (in Russian), Hidrometeoizdat, Moscow, 383 pp, 1963.

Oguz, T., H. Ducklow, P. Malanotte-Rizzoli, and J. W. Murray, Simulation of the Black Sea pelagic ecosystem by 1-D, vertically resolved, physical-biochemical models, *Fish. Oceanogr.*, *7*, *3/4*, 300–304, 1998.

Oguz, T., H. Ducklow, P. Malanotte-Rizzoli, S. Tugrul, N. P. Nezlin, and U. Unluata, Simulation of annual plankton productivity cycle in the Black Sea by a one-dimensional physical-biological model, *J. Geophys. Res.*, *101*, 16585–16599, 1996.

Oguz, T., V. S. Latun, M. A. Latif, V. V. Vladimirov, H. I. Sur, A .A. Markov, E. Ozsoy, B. B. Kotovshchikov, V. V. Eremeev, and U. Unluata, Circulation in the surface and intermediate layers of the Black Sea, *Deep-Sea Res.*, *40*, 1597–1612, 1993.

Oguz, T., P. E. La Violette, and U. Unluata, The upper layer circulation of the Black Sea: Its variability as inferred from hydrographic and satellite observations, *J. Geophys. Res.*, *97*, 12569–12584, 1992.

Oguz, T., P. Malanotte-Rizzoli, and D. Aubrey, Wind and thermohaline circulation of the Black Sea by yearly mean climatological forcing, *J. Geophys. Res.*, *100*, 6845–6863, 1995.

Ovchinnikov, I. M., and Yu. I. Popov, Evolution of the cold intermediate layer in the Black Sea, *Oceanology* (English translation), *27*, 555–560, 1987.

Stancík A., and S. Jovanovic, *Hydrology of the River Danube,* Priroda, Bratislava, 272 pp, 1988.

Stelmakh, L. V., O. A. Yunev, Z. Z. Finenko, V. I. Vedernikov, A. S. Bologa, and T. Ya. Churilova, Peculiarities of seasonal variability of primary production in the Black Sea, In *Ecosystem Modeling as a Management Tool for the Black Sea*, *1*, edited by L.I. Ivanov and T. Oguz, Kluwer Academic Publishers, Dordrecht, The Netherlands, 93–104, 1998.

Sverdrup H. U., The place of physical oceanography in oceanographic research, *J. Mar. Res.*, *14*, 287–294, 1955.

Vedernikov, V. I., and A. B. Demidov, Primary production and chlorophyll in deep regions of the Black Sea, *Oceanology* (English translation), *33*, 193–199, 1993.

Vedernikov, V. I., and A. B. Demidov, Vertical distribution of primary production and chlorophyll during different seasons in deep regions of the Black Sea, *Oceanology* (English translation), *37*, 376–384, 1997.

Vinogradov, M. E., V. V. Sapozhnikov, and E. A. Shushkina, *The Black Sea Ecosystem* (in Russian), Nauka, Moscow, 112 pp, 1992.

Vinogradov, M. E., E. A. Shushkina, O. V. Kopelevich, and S. V. Sheberstov, Photosynthetic productivity of the world ocean from satellite and expeditionary data, *Oceanology* (English translation), *36*, 531–540, 1996a.

Vinogradov, M. E., E. A. Shushkina, N. P. Nezlin, V. I. Vedernikov, and V. I. Gagarin, Correlation between different parameters of the ecosystem of the epipelagic zone of the world ocean, *Oceanology* (English translation), *39*, 54–63, 1999.

Vinogradov, M. E., E. A. Shushkina, and V. I. Vedernikov, The features of epipelagic ecosystems of the Pacific from satellite and expeditionary data: Primary production and its seasonal variations, *Oceanology* (English translation), *36*, 222–230, 1996b.

Vinogradov, M. E., E. A. Shushkina, V.I. Vedernikov, V. I. Gagarin, N. P. Nezlin, and S. V. Sheberstov, Characteristics of the Pacific epipelagic ecosystems based on the satellite and expedition data: Abiotic parameters, and production indices of phytoplankton, *Oceanology* (English translation), *35*, 208–217, 1995.

Vinogradov, M. E., E. A. Shushkina, V. I. Vedernikov, N. P. Nezlin, and V. I. Gagarin, Primary production and plankton stocks in the Pacific Ocean and their seasonal variation according to remote sensing and field observations, *Deep-Sea Res.*, *44*, 1979–2001, 1997.

Yilmaz, A., O. A. Yunev, V. I. Vedernikov, S. Moncheva, A. S. Bologa, A. Cociasu, and D. Ediger, Unusual temporal variations in the spatial distribution of chlorophyll-*a* in the Black Sea during 1990–1996, In *Ecosystem Modeling as a Management Tool for the Black Sea*, *1*, edited by L. I. Ivanov and T. Oguz, Kluwer Academic Publishers, Dordrecht, The Netherlands, 105–120, 1998.

Zaitsev, Yu., and V. Mamaev, *Marine Biological Diversity in the Black Sea: A Study of Change and Decline*, GEF Black Sea Environmental Programme, United Nations Publications, 208 pp, 1997.

Nikolay P. Nezlin, P. P. Shirshov Institute of Oceanology, Russian Academy of Sciences, 36 Nakhimovskiy Avenue, Moscow 117851, Russia. (email: nezlin@ecosys.sio.rssi.ru; fax +7-095-124-5983)

Satellites, Oceanography and Society
edited by David Halpern

Chapter 15

Remotely sensed coastal/deep-basin water exchange processes in the Black Sea surface layer

Anna I. Ginzburg and Andrey G. Kostianoy
P. P. Shirshov Institute of Oceanology, Russian Academy of Sciences, Moscow, Russia

Dmitry M. Soloviev and Sergei V. Stanichny
Marine Hydrophysical Institute, Sevastopol, Ukraine

Abstract. The role of mesoscale structures (eddies, dipoles, jets) in horizontal mixing and coastal/deep-basin water exchange in the Black Sea was investigated with National Oceanic and Atmospheric Administration (NOAA) Advanced Very-High Resolution Radiometer (AVHRR) imagery during 1993 and 1996–1998, together with relevant mean-daily meteorological data from seaports and available hydrographic data of different years. In summer 1993 two anticyclones with diameters of about 90 and 55 km co-existed without coalescence over the northwestern continental slope. Cyclones at the eddies' peripheries, entrained and ejected jets, filaments, and a pinched-off cyclone near Cape Hersones (44°35'N, 33°22'E) associated with wind-driven coastal upwelling contributed to the water exchange in the region. Four anticyclones about 50 km in diameter and associated cyclones at their peripheries were observed in the southeastern region in November 1996. Surface circulation in the region was considerably changed over several days because of the anticyclones' movements, and formations and disruptions of short-lived dipoles of anticyclones and associated cyclones at their peripheries. Near-shore anticyclonic eddies with diameters of 40–80 km and lifetimes up to one month, which form along the Caucasian coast and propagate with velocities up to 17 cm s^{-1} in the general direction of the Rim Current, can evolve into deep-sea eddies southwest of Novorossiisk. Offshore jets, up to about 200 km in length and associated with the anticyclones, are an effective mechanism of coastal/deep-basin water exchange in the northeastern region.

1. Introduction

The known schemes of the general circulation of the Black Sea (Neumann 1942; Bogatko et al. 1979; Ovchinnikov and Titov 1990; Oguz et al. 1993) are based on hydrographic measurements of different years and include a basin-scale cyclonic boundary current over the continental slope, cyclonic gyres in the basin interior, and quasi-stationary or recurrent anticyclonic eddies along the basin periphery. The schemes do not suggest coastal/deep-basin water exchange across the boundary current, named Rim Current by Oguz et al. (1992) or Main Black Sea Current in the Russian literature. However, early hydrobiological measurements of Mediterranean Sea plankton in Sevastopol Bay (Bogdanova and Shmeleva 1967) and polychaete larvae observed from Cape Sarych to the Anatolian coast (Kiseleva 1953) were indicative of the water exchange. Investigations of mechanisms of the exchange are of utmost importance for the Black Sea because of the poor ecological situation in this semi-enclosed basin.

Satellite imagery, which has been used for investigation of the Black Sea surface circulation since 1981 (Kazmin and Sklyarov 1982), has detected various nonstationary mesoscale features contributing to coastal/deep-basin water exchange. For example, in summer, Danube River water propagates along the western coast to the Bosphorus Strait with cyclonic vortices generated at the front between freshened coastal and saline offshore waters (Kazmin and Sklyarov 1982; Grishin 1993; Sur and Ilyin 1997). Few observations of complicated vortical structures, in particular dipoles, and jets were reported over the continental slope (Kazmin and Sklyarov 1982; Grishin 1993; Ginzburg 1994; Sur et al. 1994, 1996; Sur and Ilyin 1997). Mushroom-shaped flow extended about 160 km offshore from the coastal zone close to the Bosphorus Strait (Ginzburg 1995). Coastal eddies and filaments of upwelled coastal water occur along the Anatolian coast (Grishin et al. 1990; Oguz et al. 1992; Ginzburg 1994; Sur et al. 1994, 1996; Sur and Ilyin 1997). In summer, the meanders of the Rim Current are intense north of the Anatolian coast (Ginzburg 1994; Sur et al. 1994; Sur and Ilyin 1997) and south of the Kerch Strait (Sur and Ilyin 1997).

Spatial (less than 30 km) and temporal (several days to a month) scales of the nonstationary features (Figure 1) make it practically impossible to investigate their generation and evolution by hydrographic surveys. Satellite observations with high space-time resolutions are necessary. Ocean dynamics can vary considerably even in the same region during several days, and joint analysis of satellite data with in-situ hydrometeorological data is desirable, where possible, to understand the reasons for the variability. We present new results of analysis of visible and infrared (IR) images from the National Oceanic and Atmospheric Administration (NOAA) satellites with hydrometeorological data to describe the role of eddies and associated nonstationary flow patterns in horizontal mixing and coastal/deep-basin water exchange in the northwest, southeast, and northeast Black Sea.

2. Data

NOAA satellite images in 1993 and 1996–1998 were received in high-resolution picture transmission (HRPT) format by the Russian State Committee for Hydrometeorology in Moscow and by the Marine Hydrophysical Institute (MHI) in Sevastopol, respectively.

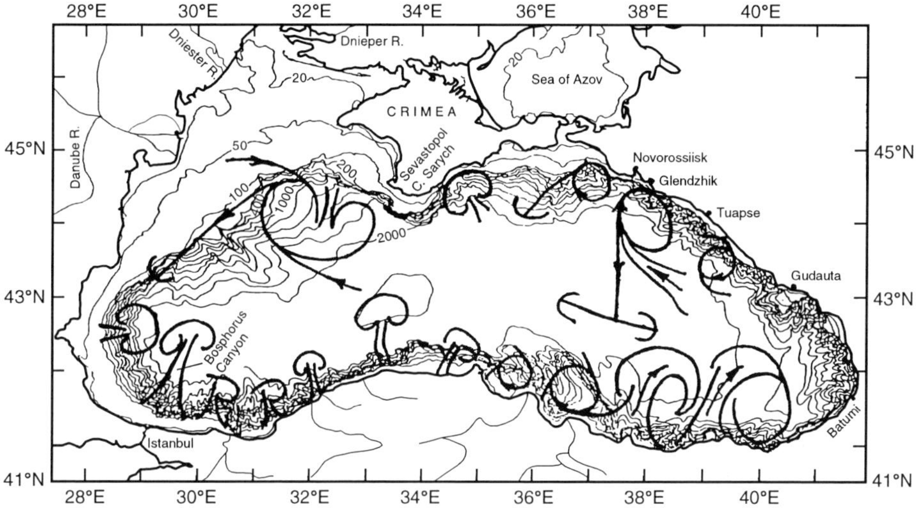

Figure 1. Schematic of coastal/deep-basin water exchanges in the Black Sea, superimposed on bathymetry with locations of sites mentioned in text.

The dataset included 27 IR images of the northwestern Black Sea (26 for April–August 1993 and 1 for June 1996), 12 visible-band images for June 1993, and 25 IR images of the eastern area (November 1996, September–November 1997, June 1998). A visible-band image was constructed from a linear combination of AVHRR channels 1 and 2, and represented the information about light diffusion associated with concentration of suspended matter. Atmospheric corrections were applied. Data were mapped on a Mercator projection with 1'-latitudinal and 1.5'-longitudinal resolutions. Sea surface temperature retrievals were computed with AVHRR channels 4 and 5, with a resolution of about 0.1°C.

Conductivity-temperature-depth (CTD) measurements were made over the northwestern continental slope on 16–17 June 1993 with Research Vessel *Akvanavt* of the P. P. Shirshov Institute of Oceanology. Also, mean-daily meteorological data were available from seaports in the Ukraine and Russia.

3. Mesoscale Structures in the Northwestern Region

The northwestern region is characterized by a very wide shelf, low salinity, high biological productivity, and pollutants due to discharges of large rivers (Danube, Dnieper, Dniester). Vortices and mushroom-shaped currents with horizontal scales of 15–50 km are frequently observed (Grishin and Subbotin 1992). However, the most interesting mesoscale features of the region are anticyclonic eddies up to 100 km in diameter, which occur over the continental slope to precondition water exchanges south of 45°N (Ginzburg 1994; Ginzburg et al. 1996).

The anticyclonic eddy west of Sevastopol, which has been repeatedly detected by hydrographic surveys (Neumann 1942; Bogatko et al. 1979), was embodied in the Oguz et al. (1993) surface circulation pattern as the "Sevastopol Eddy." However, satellite imagery shows that two closely spaced mesoscale anticyclones are frequently observed in the region (Figures 2a and 3). Anticyclone A1 (mean diameter of about 90 km) was observed in NOAA images on 19 and 27 April 1993 and 2 June–23 August 1993; although images for May were absent, the eddy observed in April and June–August was likely the same A1 eddy. Anticyclone A2 was seen from 9 June to 19 July 1993 (Figures 2 and 3); see also visible images for 10, 12, and 13 June and 18 July 1993 from NOAA-11 and Russian satellite Kosmos-1939 in Ginzburg (1994) and Ginzburg et al. (1996). From 9 June to 19 July, anticyclones A1 and A2 co-existed without coalescence within an area formed by an underwater ridge on the southwest and the 200-m isobath on the north and east. The largest displacements of the A1 and A2 centers were 90 km and 60 km, respectively (Figure 4), and the A2 eddy was always oriented to the northeast of A1 (Figure 4). Displacement velocities were not constant. In Julian days (JD) 160–161, both eddies remained practically immobile; however, during JD 162–166 both moved southwestward at about 16 cm s^{-1}. Directions of displacements of A1 and A2 also changed with time, with the same directions on JD 162–167 and 193–200, and opposite directions on JD 169–174 and 183–193. Therefore, the distance between A1 and A2 centers varied from a minimum 100 km on JD 168 (previous day to that shown in Figure 2a) to a maximum 140 km on JD 174 (Figure 3). Interaction of A1 and A2, even when they were far apart, is indicated in Figure 3 by the narrow jet.

Directions of A1 and A2 displacements were not always associated with the southwestward direction of the Rim Current. The eddy diameter decreased as it approached the 200-m isobath. Absence of A2 on 2 August 1993 (Figure 5b) suggests collapse of the eddy or its coalescence with A1. In hydrographic data recorded during September–October 1993, Georgiev et al. (1994) found a 100-km diameter anticyclone located 80 km southwest from the A1 position on 23 August (JD 235 in Figure 4). It is quite possible that this anticyclone was A1 shifted in the direction of the Rim Current.

One to three short-lived (up to a week), small-scale (about 10-km diameter) cyclones sporadically formed at the A1 and A2 peripheries. Generation of the small cyclones was not correlated with wind direction or wind speed, which rarely exceeded 5 m s^{-1}. However, on 30 July 1993 the northwesterly wind strengthened to 10 m s^{-1}, and we speculate that it created an intense jet and 90-km diameter cyclone (B1 in Figure 5c) at the A1 eastern periphery. The horizontal scale of the A1-B1 dipole was about 200 km (Ginzburg 1994).

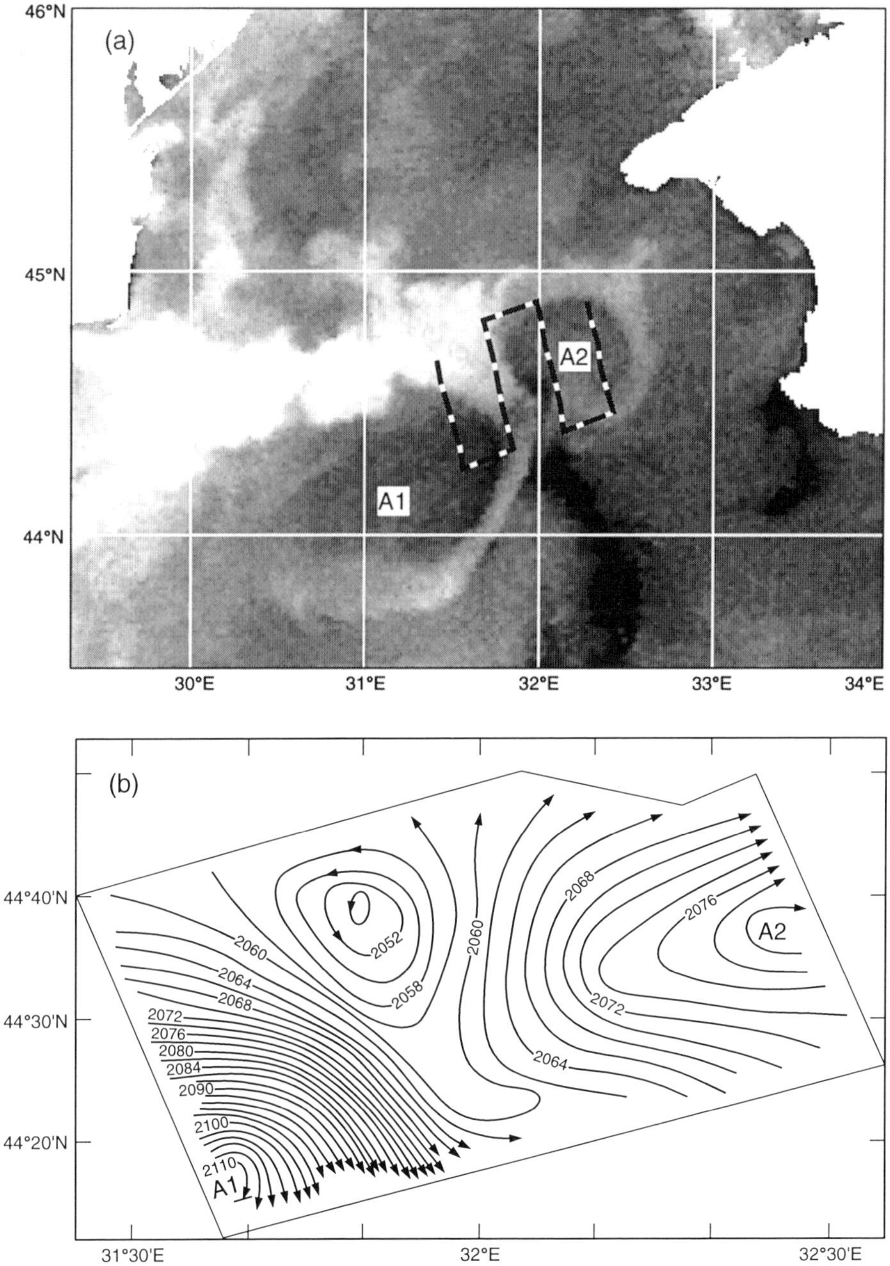

Figure 2. (a) NOAA visible image on 18 June 1993 with superimposed positions of CTD stations made during R/V *Akvanavt* survey on 16–17 June 1993. (b) Dynamic topography map (mm) at 5 dbar relative to 500 dbar. A1 and A2 mark centers of anticyclones.

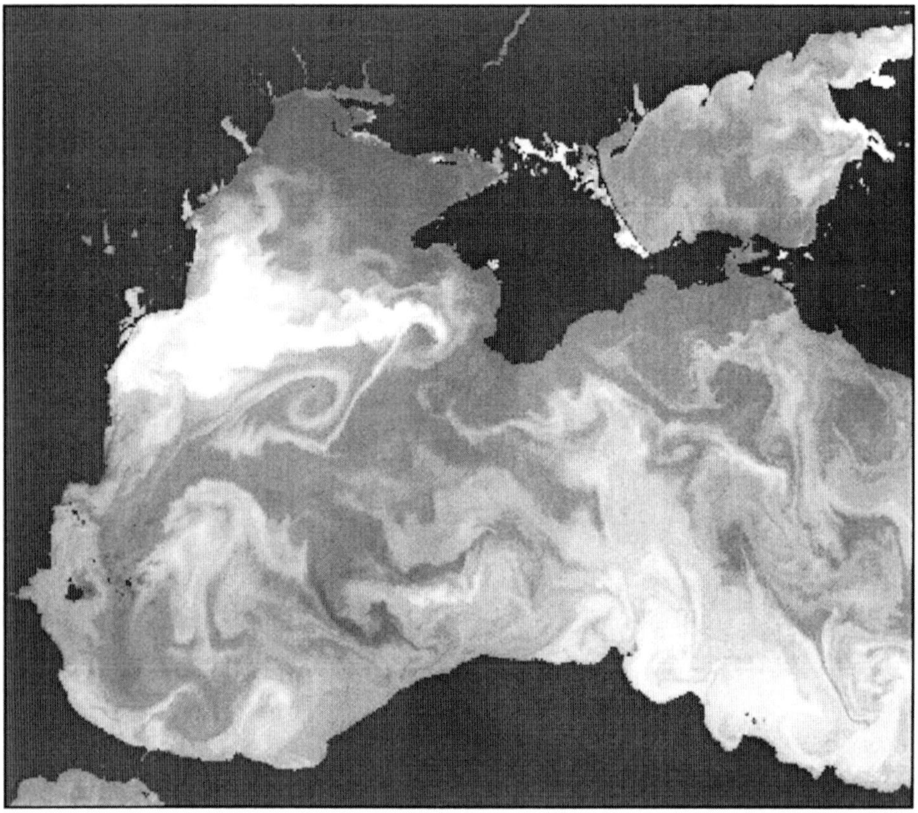

Figure 3. NOAA visible image on 23 June 1993.

The A1 eddy influenced the area 43°–45°N, 29°–33°E, which was much larger than the A1 area because of the generation of associated cyclones, entrainment of water, and ejection of jets. An example of water entrainment by A1 and A2 is seen in Figure 3, which shows turbid water from the Danube delta region along the northern peripheries of A1 and A2 and almost reaching to the Crimean shore (Ginzburg 1994). The added contribution to coastal/deep-basin water exchange was associated with anticyclones A1 and A2 and a cyclone detected by the hydrographic survey on 16–17 June 1993 (Figures 2a and 2b). The A1-A2 system and the short-lived (several days) cyclone were traced to 300-m depth. The current in the upper 150 m between A1 and the cyclone was southeastward with a speed more than 30 cm s^{-1}, and reversed flow at the northern periphery of the cyclone had a core at 150 m and a speed of about 10 cm s^{-1} (Ginzburg et al. 1998a).

Coastal upwelling occurred with favorable winds, irrespective of shoreline or bottom topography orientation (Figures 5 and 6). Wind speed in summer 1993 was rarely above 5 m s^{-1}. Upwelling occurred most often along Tendrovskaya Spit (Figure 5c) because of

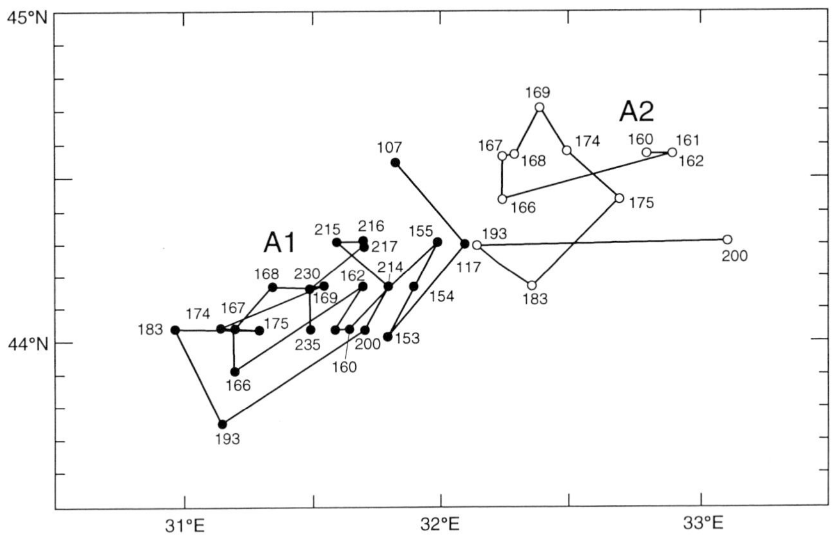

Figure 4. Positions of centers of anticyclones A1 (solid dot) and A2 (open circle) during April–August 1993. Numbers are dates in Julian day.

northwesterly winds. Offshore transport of upwelled water beyond the upwelling front was by short-lived (one to a few days) transversal filaments about 40 km long (longer near capes) and 4–14 km wide, separated by distances from 6 to 38 km. Sometimes the offshore end of the filaments terminated in vortices or vortex dipoles. The filament length considerably exceeded the 12-km baroclinic Rossby radius of deformation near Tendrovskaya Spit (Ginzburg et al. 1997). On 23 June 1993, cold water was carried in a filament more than 150 km from Cape Hersones (Figure 6), and the filament moved along the southeastern peripheries of A1 and A2. Maximum speed of filaments estimated from correlation of successive IR images on 2 and 3 August 1993 was 35 cm s^{-1}. Filament formation time did not exceed 24 hours, and temperature contrast relative to surrounding water was 1.6–2.1°C. Minimum temperature of upwelled water was 4–7°C colder than that beyond the upwelling zone and corresponded to depths less than 20–30 m, which was within or just below the seasonal thermocline (Ginzburg et al. 1997). Shallowness of seasonal thermocline and pycnocline aided the occurrence of upwelling under light winds.

The NOAA IR images in Figure 5 describe the generation and evolution of another nonstationary element of circulation associated with coastal upwelling—a pinched-off cyclone near Cape Hersones. On 30 July 1993 (Figure 5a), upwelling between Cape Hersones and Cape Sarych was created by steady 5 m s^{-1} northwesterly winds during the previous two days. Width of the upwelling zone was about 10 km, and offshore transport

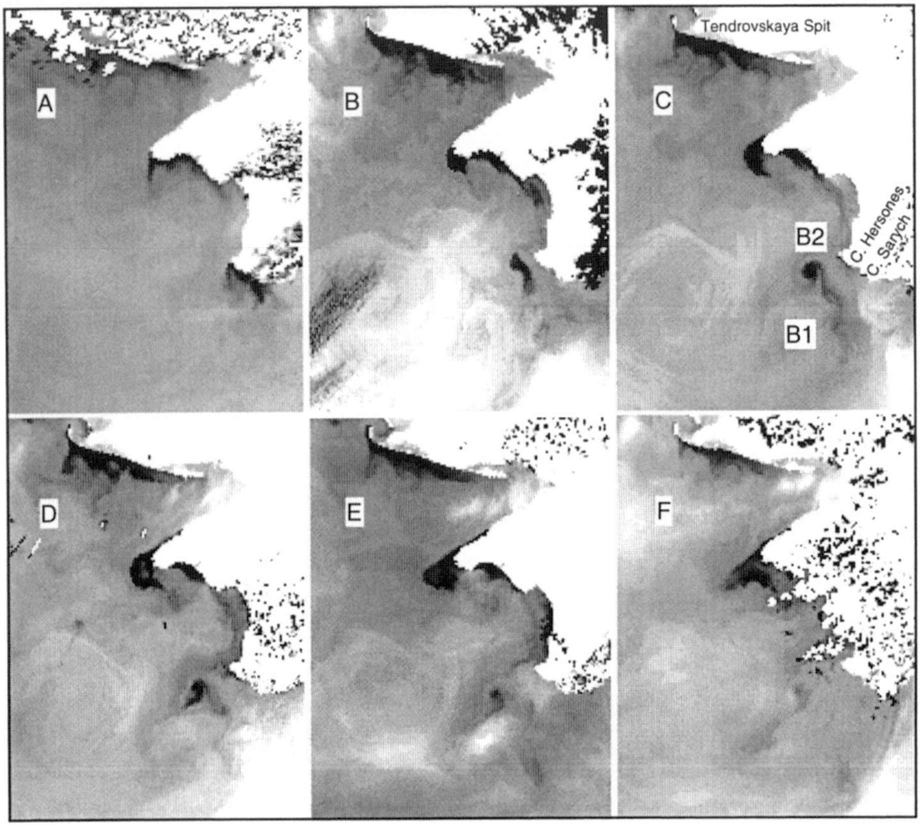

Figure 5. NOAA IR images in 1993 on (a) 30 July, (b–f) 2, 3, 4, 5, and 6 August, respectively. Darker (lighter) tone corresponds to colder (warmer) water.

of cold water by southward filaments was traced offshore for about 20 km. Surface water temperature changed from 11.3°C near shore on 30 July (Figure 7a) to 16°C at the terminal of the filament from Cape Hersones. Three days later, after wind speed increased on 30 July to 10 m s^{-1} and decreased to 2–3 m s^{-1} on succeeding days, upwelling was absent, and the cold-water patch separated from the coast (Figure 5b). The 10-km wide patch expanded in its head part to 17 km. Temperature within the patch gradually increased from 17.3°C at the center of the head part to 18.5°C in the tail part. One day later the patch evolved into cyclonic eddy B2 (Figure 5c), with 18-km diameter (Figure 7b) and uniform temperature. In succeeding days the temperature gradually increased (Figure 7a) and the eddy diameter decreased to 7 km on 6 August (Figure 7b). The B2 cyclone moved southwestward across isobaths with a mean speed of about 9 cm s^{-1}. By

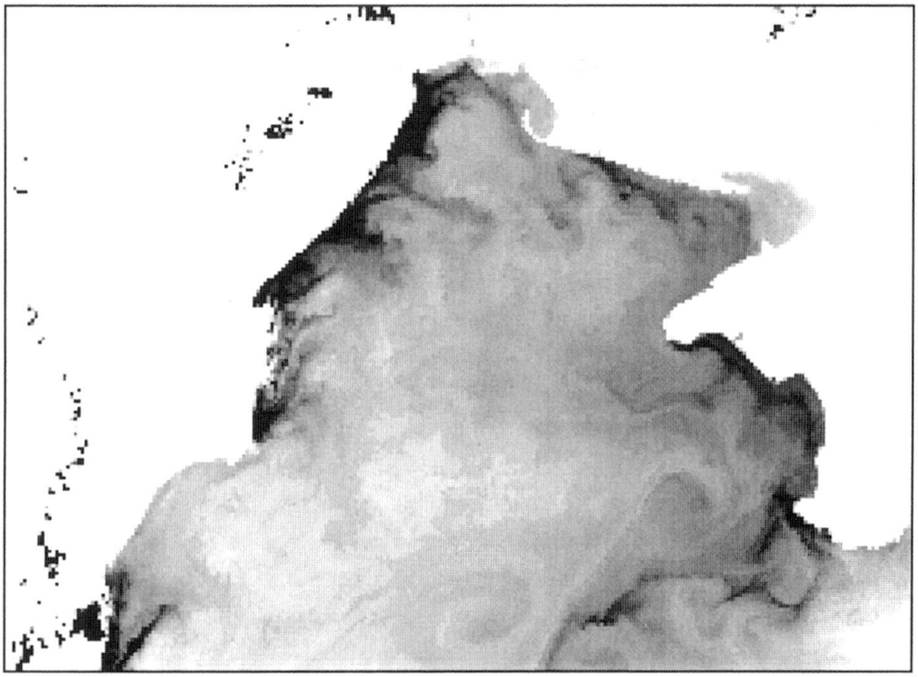

Figure 6. NOAA IR image on 23 June 1993 with darker (lighter) tone corresponding to colder (warmer) water.

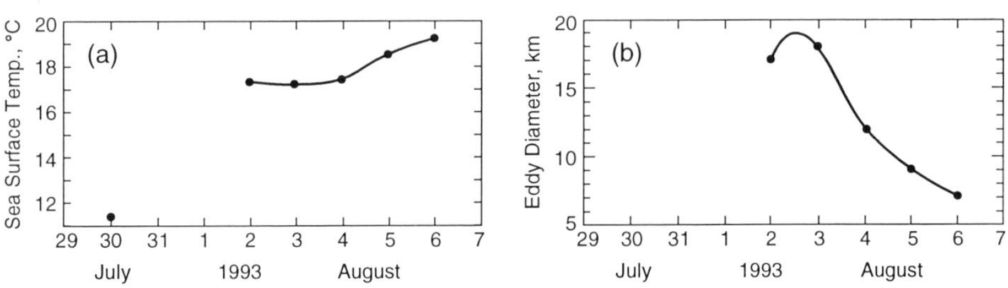

Figure 7. Time variations of (a) sea surface temperature and (b) diameter of pinched-off cyclone.

6 August 1993 it was about 50 km from Cape Hersones (Figure 5f). In addition, B2 water combined with water along the periphery of cyclone B1 to influence the water properties as far as 80 km from Cape Hersones.

4. Mesoscale Dynamics in the Southeastern Region

A large anticyclonic eddy has repeatedly been detected with hydrographic data in the southeastern Black Sea (Bibik 1964; Bogatko et al. 1979; Oguz et al. 1993); this persistent circulation feature was named "Batumi Eddy" by Oguz et al. (1993). However, satellite imagery in the region revealed considerable mesoscale variability associated with one or several eddies existing at the same time (Ginzburg et al. 1998c). Figure 8 shows four anticyclonic eddies with diameters of 45–50 km: A1, in the open sea, forms a quasi-symmetrical dipole with cyclone C1 at its southeastern periphery; A2, also located in the deep basin; and two coastal anticyclones, A3 and A4, which have cyclones on their peripheries. In particular, A3 had three associated cyclones. A similar system of three dipoles with one common anticyclone and three cyclones placed at 120° from each other was observed for the first time in a laboratory experiment by Fedorov et al. (1989). Quasi-symmetrical dipoles of the A1–C1 type were previously observed on a satellite image of the southeastern basin (Kazmin and Sklyarov 1982; Fedorov and Ginsburg 1992), and appear to be typical of the regional surface circulation.

During 13–18 November 1996, A4 practically stayed at the same place, whereas A2 and A3 moved southward and northwestward, respectively, with a mean speed of 8.5 cm s^{-1}. A1, after displacement southwestward at 18 cm s^{-1} during 13–15 November,

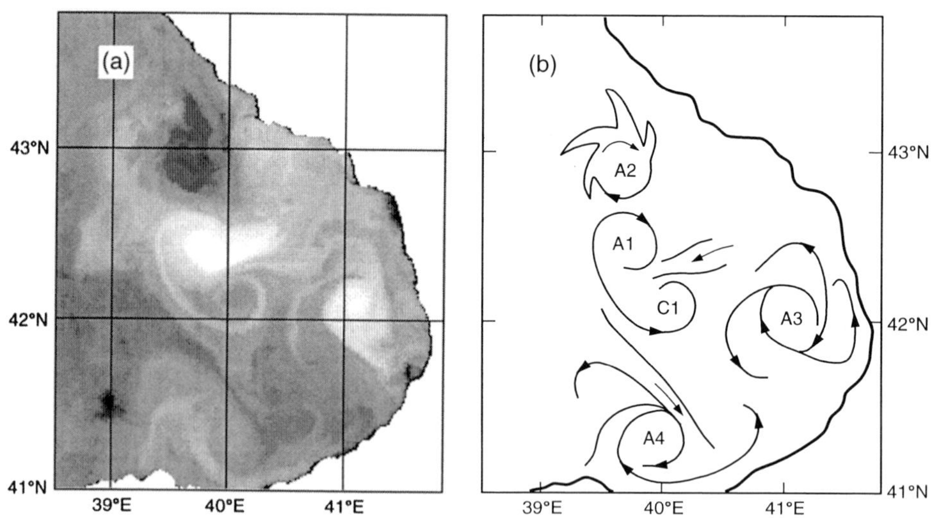

Figure 8. Surface circulation in the southeastern Black Sea on 14 November 1996 with (a) NOAA IR image with darker (lighter) tone corresponding to colder (warmer) water, and (b) corresponding scheme.

moved at 27 cm s^{-1} along the periphery of A2, supposedly under the influence of A2, and disrupted dipole A1–C1. Nonstationary dipoles of A4 and its periphery cyclones were continuously formed. Absence of satellite data after 18 November 1996 did not allow further monitoring. It is evident that local surface circulation can be considerably changed over several days because of anticyclones' movements, disruptions of dipoles, and formations of new ones.

Anticyclone diameters in the southeastern region vary from 45–50 km (Figure 8) to about 100 km. For example, in June 1998 (not shown), an anticyclone completed a chain of anticyclonic eddies formed off capes of the Anatolian coast, with diameters increasing to the east (shown schematically in Figure 1). Eddy size was not related to the seasonal cycle.

Dipoles, even short-lived ones, are a very effective mechanism of horizontal mixing in the surface layer. Entrainment of flow by a pair of eddies of opposite sign (e.g., dipole A1–C1 in Figure 8) transported warm water from Gudauta Bank to the open sea (Figure 8a). A narrow jet between A4 and a cyclone on its eastern periphery transported cold water from the deep basin to the southern coastal zone near 40°20'E (Figure 8). Fedorov and Ginsburg (1992) observed offshore transport of coastal water rich in suspended matter and onshore flow of clear, open-sea water that was associated with three dipoles in the southeastern region.

5. Eddies and Jets in the Northeastern Region

A characteristic feature of the regional circulation is near-shore anticyclonic eddies, which are formed within anticyclonic meanders of the Rim Current and move with speeds of 4–6 cm s^{-1} in the same direction as the Rim Current (Krivosheya et al. 1998a, 1998b). A near-shore anticyclonic eddy with 40-km diameter (named NAE-1) formed on 7–8 September 1997 west of Novorossiisk, in the region of shelf/slope widening (Figure 9a). NAE-1 progressively moved away from the coast, approximately southward during 7–18 September (Figures 9a and 9b), northwestward on 18–23 September, and then southwestward for 23 September–8 October. Its translation speed was uneven: about 8.4 cm s^{-1} between 12 and 17 September, 15 cm s^{-1} in the following six days, 4.5 cm s^{-1} from 24 September to 5 October, and 8.5 cm s^{-1} between 5 and 8 October. Total displacement of NAE-1 was about 115 km southwestward, with mean speed of about 4.5 cm s^{-1}. The diameter increased from 40 km to about 75 km. The NAE-1 lifetime was greater than one month, but absence of cloudless images did not allow further monitoring of NAE-1. No near-shore anticyclone eddy other than NAE-1 formed in the region between 7 September and 23 September 1997, whereas after 24 September anticyclonic eddies began to appear in rapid succession. Figure 9c shows a chain of three near-shore anticyclonic eddies, each about 40 km in diameter, between NAE-1 and Gelendzhik, and a meander of the Rim Current near Tuapse that manifested itself as a new NAE one day

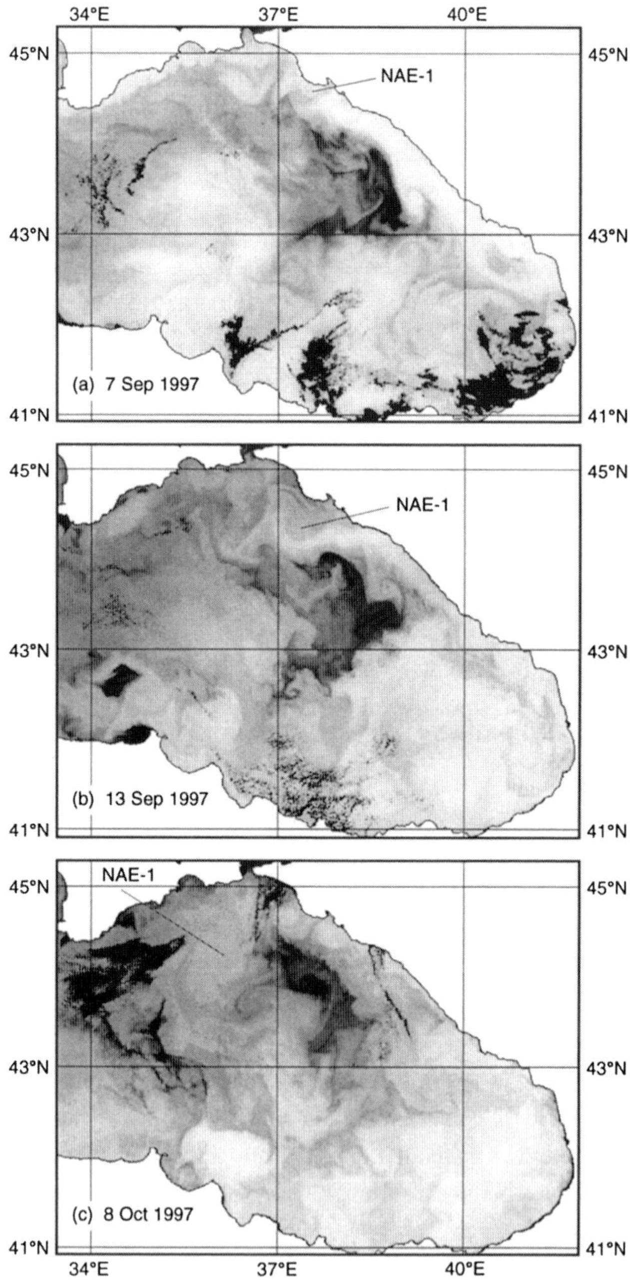

Figure 9. NOAA IR images on (a) 7 September, (b) 13 September, and (c) 8 October 1997 with darker (lighter) tone corresponding to colder (warmer) water.

later. All near-shore anticyclonic eddies moved northwestward along isobaths with speeds of 9–17 cm s^{-1}. However, not all near-shore anticyclonic eddies had non-zero translation speed. For example, an 80-km diameter near-shore anticyclonic eddy remained at the same place south of Gelendzhik (Figure 3) for at least 11 days (Ginzburg 1994).

Offshore jets are frequently observed with near-shore anticyclonic eddies. For example, a southwestward warm water jet about 85-km long occurred at the NAE-1 periphery (Figures 9a and 9b). A similar offshore jet, about 200 km in length, is seen in Figure 3. Evidently, the jets are an effective mechanism of coastal/deep-basin water exchange in the area.

6. Conclusions

Analysis of satellite data revealed that one-to-several anticyclonic mesoscale eddies exist simultaneously in the northwestern, southeastern, and northeastern Black Sea. In particular, two closely spaced mesoscale anticyclones co-existed without coalescence over the northwestern continental slope for about 1.5 months in summer 1993 (Figures 2–4). Different dynamical situations were observed in different years in the same season, e.g., four anticyclonic eddies in the southeastern region in November 1996 (Figure 8) and only one in November 1997 (Figure 9).

Translation speeds of anticyclones in all three regions reached 16 cm s^{-1}. High speed of movement and close proximity of anticyclones, as in the southeastern region in November 1996, can produce an aliasing error in the interpretation of hydrographic data recorded over several days with inadequate spatial resolution.

Water exchange between the coastal zone and deep basin in the Black Sea is determined, in large part, by processes associated with evolution of anticyclonic eddies, such as entrainment of coastal and open sea waters and generation of additional cyclones and jets at the peripheries. An added contribution to water exchange in the northeastern region is the separation of near-shore anticyclones near Novorossiisk and their evolution into deep-sea eddies. In addition, in the northwestern region, filaments of wind-driven coastal upwelling produce horizontal mixing, transport of cold water for offshore distances of several tens of kilometers and considerably greater distance near capes, and due to entrainment by mesoscale eddies beyond the upwelling zone. Characteristics of filaments in the northwestern region (temperature contrast relative to surrounding water, formation time, velocity, marked excess of length compared to Rossby radius of deformation) agree with observations of transversal filaments in other coastal zones of the ocean (Fedorov and Ginsburg 1992; Kostianoy and Zatsepin 1996; Kostianoy 1996).

Further analysis of satellite and in-situ data is desirable to understand regularities of generation of pinched-off cyclones near Cape Hersones during coastal upwelling (Figure 5).

Acknowledgments. This study was supported by the Russian Foundation for Basic Research (grants NN 98-05-64715 and 99-05-65528) and by the European Community grant INCO-Copernicus NIC15-CT96-0111.

References

Bibik, V. A., Peculiarities of water dynamics in the southeastern Black Sea and distribution of oceanographic elements, *Trudy of AzCherNIRO*, *23*, 23–31, 1964 (in Russian).

Bogatko, O. N., S. G. Boguslavsky, Yu. M. Beljakov, and R. E. Ivanov, Surface currents in the Black Sea, In Kompleksnye Issledovaniya Chernogo morya, Marine Hydrophysical Institute, Sevastopol, Ukraine, 25–33, 1979 (in Russian).

Bogdanova, A. K., and A. A. Shmeleva, Hydrological conditions of Mediterranean plankton types penetration in the Black Sea, In Dinamika vod I voprosy gidrohimii Chernogo Morya, Naukova Dumka, Kiev, 156–166, 1967.

Fedorov, K. N., and A. I. Ginsburg, *The Near-Surface Layer of the Ocean*, VSP, Utrecht, The Netherlands, 259 pp, 1992.

Fedorov, K. N., A. I. Ginsburg, and A. G. Kostianoy, Modeling of "mushroom-like" currents (vortex dipoles) in laboratory tank with rotating homogeneous and stratified fluids, In *Mesoscale/Synoptic Coherent Structures in Geophysical Turbulence*, edited by J. Nihoul and B. Jamart, Elsevier Science, New York, 15–24, 1989.

Georgiev, V. T., S. A. Gerasimov, and Yu. I. Popov, Hydrodynamic state of open waters of the northern Black Sea, In Issledovanie Ekosistemy Chernogo Morya, *1*, Ukraine National Center of Ecology of the Sea, Odessa, 18–24, 1994 (in Russian).

Ginzburg, A. I., Horizontal exchange processes in the near-surface layer of the Black Sea, *Issledovanie Zemli iz Kosmosa*, 75–83 (2), 1994 (in Russian); *Earth Observ. Remote Sensing*, *12*, 190–202, 1994 (in English).

Ginzburg, A. I., On nonstationary jet-like currents in the southwestern Black Sea, *Issledovanie Zemli iz Kosmosa*, 10–16 (4), 1995 (in Russian); *Earth Observ. Remote Sensing*, *13*, 519–528, 1995 (in English).

Ginzburg, A. I., A. G. Kostianoy, D. M. Soloviev, and S. V. Stanichny, Evolution of anticyclonic eddies in the northwestern Black Sea, *Issledovanie Zemli iz Kosmosa*, 67–76 (4), 1996 (in Russian); *Earth Observ. Remote Sensing*, *14*, 599–611, 1996 (in English).

Ginzburg, A. I., A. G. Kostianoy, D. M. Soloviev, and S. V. Stanichny, Coastal upwelling in the northwestern Black Sea, *Issledovanie Zemli iz Kosmosa*, 61–72 (6), 1997 (in Russian).

Ginzburg, A. I., E. A. Kontar, A. G. Kostianoy, V. G. Krivosheya, D. M. Soloviev, S. V. Stanichny, and S. Yu. Laptev, System of mesoscale eddies over the slope in the northwestern Black Sea in summer 1993 (satellite and ship data), *Okeanologiya*, *38*, 56–63, 1998a (in Russian).

Ginzburg, A. I., A. G. Kostianoy, D. M. Soloviev, and S. V. Stanichny, Cyclonic eddies associated with upwelling off the southwestern Crimea, *Issledovanie Zemli iz Kosmosa*, 83–88 (3), 1998b (in Russian).

Ginzburg, A. I., A. G. Kostianoy, D. M. Soloviev, and S. V. Stanichny, Variability of eddy picture in the southeastern Black Sea, *Issledovanie Zemli iz Kosmosa*, 3–15 (6), 1998c (in Russian).

Grishin, G. A., Satellite and ship-borne observations of hydrological fronts in the Black Sea and Mediterranean Sea, *Issledovanie Zemli iz Kosmosa*, 76–88 (5), 1993 (in Russian).

Grishin, G. A., I. G. Makeev, and S. V. Motyzhev, Observations of circulation in the western Black Sea by remote sensing, *Morskoy Gidrofizicheskiy Zhurnal*, 54–62 (2), 1990 (in Russian).

Grishin, G. A., and A. A. Subbotin, Study of the vortex dipole in the Black Sea using spaceborne data and shipboard observations, *Issledovanie Zemli iz Kosmosa*, 56–64 (5), 1992 (in Russian).

Kazmin A. S., and V. E. Sklyarov, Some peculiarities of the Black Sea circulation inferred from the Meteor satellite, *Issledovanie Zemli iz Kosmosa*, 42–49 (6), 1982 (in Russian).

Kiseleva, M. I., Polychaets larvae of the Black Sea, Avtoreferat dissertatzii (candidat biologicheskih nauk), Zoological Institute, Acad. Sci. USSR, Leningrad, 12 pp, 1953 (in Russian).

Kostianoy, A. G., Investigation of the Sicilian upwelling on the base of satellite data, Tech. Rep., Stazione Oceanografica CNR, La Spezia, Italy, 99 pp, 1996.

Kostianoy, A. G., and A. G. Zatsepin, The West African coastal upwelling filaments and cross-frontal water exchange conditioned by them, *J. Mar. Sys.*, *7*, 2–4, 349–359, 1996.

Krivosheya, V. G., F. Nyffeler, V. G. Yakubenko, I. M. Ovchinnikov, R. D. Kos'yan, and E. A. Kontar, Experimental studies of eddy structures within the Rim Current zone in the northeastern part of the Black Sea, In *Ecosystem Modeling as a Management Tool for the Black Sea*, *2*, edited by L. I. Ivanov and T. Oguz, Kluwer Academic Publishers, Dordrecht, The Netherlands, 131–144, 1998a.

Krivosheya, V. G., I. M. Ovchinnikov, V. B. Titov, V. G. Yakubenko, A. Yu. Skirta, Meandering of the Main Black Sea Current and eddies formation in the northeastern part of the Black Sea in summer 1994, *Okeanologiya*, *38*, 546–553, 1998b (in Russian).

Neumann, G., Die absolute topografie des physikalischen Meeresniveaus und die Oberflachenstromungen des Schwarzen Meeres, *Ann. Hydrogr. Maritimen Meteorol.*, *70*, 265–282, 1942.

Oguz T., V. S. Latun, M. A. Latif, V. V. Vladimirov, H. I. Sur, A. A. Markov, E. Ozsoy, V. V. Kotovshchikov, V. V. Eremeev, and U. Unluata, Circulation in the surface and intermediate layers of the Black Sea, *Deep-Sea Res.*, *40*, 1597–1612, 1993.

Oguz T., P. E. La Violette, and U. Unluata, The upper layer circulation of the Black Sea: Its variability as inferred from hydrographic and satellite observations, *J. Geophys. Res.*, *97*, 12569–12584, 1992.

Ovchinnikov, I. M., and V. B. Titov, Anticyclonic vorticity of currents in the Black Sea coastal zone, *Doklady Akad. Sci.*, *314*, 1236–1239, 1990 (in Russian).

Sur, H. I., and Yu. P. Ilyin, Evolution of satellite derived mesoscale thermal patterns in the Black Sea, *Prog. Oceanogr.*, *39*, 109–151, 1997.

Sur, H. I., E. Ozsoy, and U. Unluata, Boundary current instabilities, upwelling, shelf mixing and eutrophication processes in the Black Sea, *Prog. Oceanogr.*, *33*, 249–302, 1994.

Sur, H. I., E. Ozsoy, Yu. P. Ilyin, and U. Unluata, Coastal/deep ocean interactions in the Black Sea and their ecological/environmental impacts, *J. Mar. Sys.*, *7*, 293–320, 1996.

Anna Ginzburg, P. P. Shirshov Institute of Oceanology, Russian Academy of Sciences, Moscow, Russia. (email, akfedorov@glasnet.ru; fax, +7-095-124-5983)

Satellites, Oceanography and Society
edited by David Halpern

Chapter 16

Satellite-derived flow characteristics of the Caspian Sea

Halil İ. Sur
Institute of Marine Sciences and Management, University of Istanbul, Istanbul, Turkey

Emin Özsoy
Institute of Marine Sciences, Middle East Technical University, Erdemli, Turkey

Rashit Ibrayev
Institute of Numerical Mathematics, Russian Academy of Sciences, Moscow, Russia

Abstract. Mesoscale dynamics of the Caspian Sea are analyzed with satellite data to capture rapid submesoscale motions not sufficiently resolved by in-situ measurements. A combination of Sea-viewing Wide Field-of-view Sensor (SeaWiFS) ocean color data and Advanced Very-High Resolution Radiometer (AVHRR) sea surface temperature (SST) measurements provide a novel evaluation of marine biogeochemical and physical processes and their forcing mechanisms. The main feature of the Caspian Sea is higher pigment concentration with respect to other basins in the Mediterranean region. SeaWiFS detects riverine sediments introduced into the Caspian Sea by the Ural, Terek, Volga, and other rivers, and planktonic flora created in the Caspian Sea by nutrient-rich river runoff. The influence of river flow into the north-northwestern shelf area is especially evident after springtime flooding. Runoff from the Volga River has a major impact on the biomass in the northern Caspian Sea. Eddies and river plumes in coastal waters transport materials and momentum into the Caspian's northern and middle basins. In winter, the cyclonic (counterclockwise) circulation leads to much higher SST in the eastern part of the Caspian than in the west. In summer, wind-induced upwelling yields a pronounced decrease in temperature and higher biomass in the upper layer of the eastern Caspian Sea. In spring, cold water is formed over the entire northern Caspian Sea.

1. Introduction

The new capability to measure the ocean surface at high spatial wavenumber with imaging sensors onboard satellites has motivated detailed observations of complex oceanic processes. Although the temperature of the ocean skin (about 20 micrometers) measured by infrared sensors, such as the National Oceanic and Atmospheric Administration (NOAA) Advanced Very-High Resolution Radiometer (AVHRR), is frequently used to identify surface flow features, diurnal solar heating, evaporation, and other factors temporarily mask the true value of surface temperature, adversely affecting correlations with the ocean's underlying dynamical features (Deschamps and Frouin 1984). Thus, the sea surface temperature (SST) pattern does not necessarily correspond to the ocean current pattern. In addition, the heat stored by the surface mixed layer can be uniform in relatively large areas of the ocean, and therefore many flow features may have a color signature, even when they do not appear on infrared imagery (Ahlnas et al. 1987).

The Sea-viewing Wide Field-of-view Sensor (SeaWiFS) on the SeaStar satellite collects data from a depth range (approximately a few tens of meters) comparable to the attenuation distance of visible light in the ocean and may give a better indication of surface circulation than AVHRR SST. SeaWiFS is a modern instrument designed to assess biological productivity in large areas of the ocean. However, phytoplankton pigments are not constant tracers and often contain "nutrient enrichment" effects produced by vertical mixing, density discontinuities, or upwelling of nutrients into the euphotic zone for plankton growth (Yentsch 1984). Regions of increased fertility are typically the boundaries of ocean currents or eddies. Observations of spatial distributions of plankton and suspended matter, which are usually concentrated in shelf waters, reveal circulation patterns and interactions between waters on the shelf and over the deep sea. Consequently, the combination of SST and ocean color images is very useful for studying ocean conditions.

Satellite data are used to identify a number of important dynamical processes in the Caspian Sea. Wind-induced upwelling events occur along the eastern periphery of the Caspian Sea. Significant amounts of fresh water, nutrients, and sediments enter the Caspian Sea from the Volga, Ural, and Terek Rivers. Riverine input is advected along the coast by a boundary current, which interacts with coastline and bathymetry to produce eddies, and is being partially injected into the sea's interior as a result of turbulent exchanges. Satellite data have proven useful in similar environments where rapid changes and turbulent features cannot be sufficiently resolved by in-situ data (Sur et al. 1994, 1996, 1997).

Cloud-free NOAA AVHRR SST and SeaStar SeaWiFS ocean color images are used to infer regional flow dynamics of the Caspian Sea. All image processing was done at the Institute of Marine Sciences, Middle East Technical University, and at the Institute of Marine Sciences and Management, University of Istanbul.

2. Oceanography of the Caspian Sea

The Caspian Sea is the world's largest inland water body with no outlet to the ocean. Its surface area and volume are about 4×10^5 km^2 and 7.8×10^4 km^3. The north-south length is about 1200 km and the east-west width is approximately 310 km. The Caspian Sea is divided into the northern, middle, and southern parts (Figure 1). The northern part is very shallow, averaging about 5–6 m in depth; its maximum depth is 20 m. The 788-m maximum depth of the middle Caspian Sea is located in the Derbent Depression near the western coast. The western slope of this depression is much steeper than the eastern slope. The southern Caspian Sea is separated from the middle by the Apsheron Ridge, which is the continuation of the Apsheron Peninsula. Water depths along the ridge are less than 180 m. The deepest part of the southern Caspian Sea is 1025 m (Voropayev 1997) and is located east of the Kura River delta. Several ridges rising to 500 m occur in the southern depression. Kara Bogaz Gol, a large gulf on the eastern shore, is an extensive evaporite basin that is below the level of the Caspian Sea.

The salinity range of the Caspian Sea is 3–13 practical salinity units (psu), increasing from north to south. Temperature almost totally determines the density stratification because the salinity of the Caspian Sea is relatively homogeneous. The Volga River supplies 82% of the annual volume of inflow to the Caspian Sea. Riverine water accounts for about 80% of the annual water input. Because the Volga supplies about 65% of all water to the Caspian, sea level variations are highly dependent on Volga River discharge, which peaks in May–June, when about 35% of the annual flow occurs (Figure 2).

The Caspian Sea surface circulation consists of cyclonic (counterclockwise) eddies in the southern, middle, and northern regions (Terziev et al. 1992) (Figure 3). The mesoscale eddy features have a seasonal evolution (Trukhchev et al. 1995; Rashit et al. 1999). Caspian Sea currents are considered to be mainly wind-generated. The northern Caspian Sea has strong easterly winds in autumn and winter, and weak southerly winds in summer. The middle and southern Caspian Sea has northerly-northwesterly winds during summer, with easterly-southeasterly winds in winter. Currents on the northern shallow shelf are especially influenced by wind (Bondarenko 1993) and, in addition, are strongly influenced by the spatial gradient of buoyancy produced by river discharge. The inflows of the Volga and Ural Rivers are responsible for the southwestward-southward current in the near-delta region. In the coastal regions of the middle and southern Caspian Sea, currents correlate with wind direction and are typically toward the northwest, north, southeast, and south. Easterly currents are also observed near the east coast. Along the western coast of the middle Caspian Sea, the prevailing currents are southeastward and southward. Other types of currents, such as baroclinic currents and seiches, also play an important role in local circulation patterns. In summer, upwelling occurs along the eastern coast of the middle Caspian Sea (Kosarev and Yablonskaya 1994).

Sea level has had large fluctuations this century to impact two key natural resources: oil and natural gas, and caviar-producing sturgeon. In the early 1930s, the Caspian's sea level began to decline, dropping 3 m by the mid-1970s. However, beginning in 1978, sea level began to rise, increasing about 2.5 m in the past 20 years. Flooding in the coastal

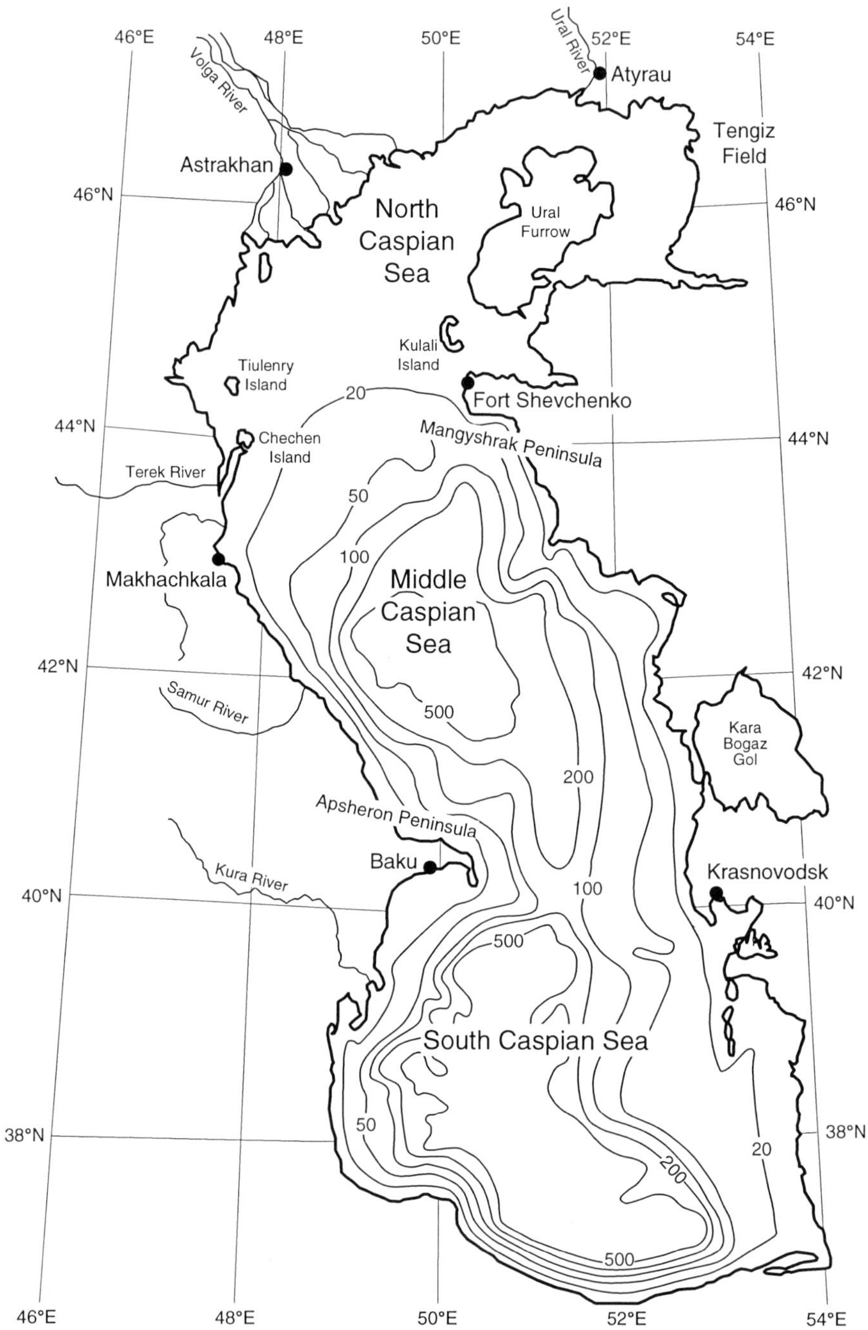

Figure 1. Bathymetry and location map of the Caspian Sea (depth in meters).

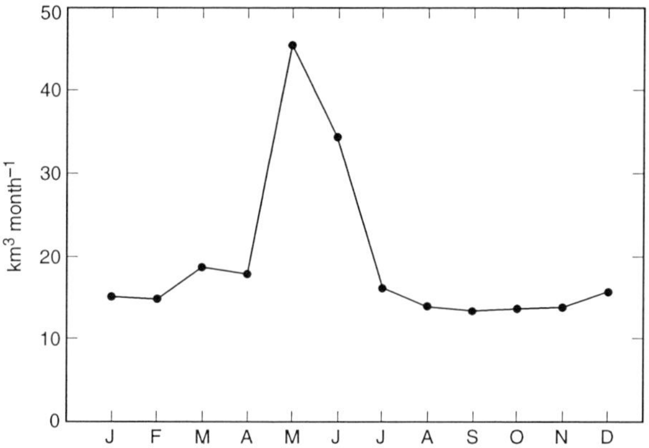

Figure 2. Long-term measurements of monthly average Volga River discharge (after Baidin and Kosarev 1986).

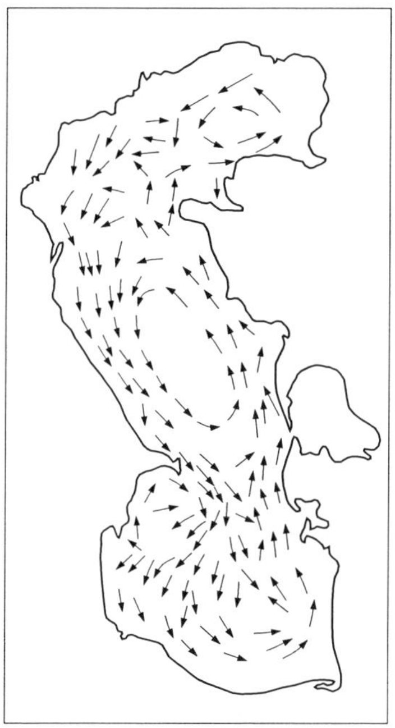

Figure 3. Sea surface currents of the Caspian Sea based on indirect methods of current measurement (floats, bottles, etc.) (after Lednev 1943).

zone damaged buildings, roads, beaches, oil wells, and much farm land. A higher sea level of the Caspian could lead to an increase in environmental problems for the sea's ecosystems as more and more oil is washed into the sea (Glantz and Zonn 1997).

3. Results

There are significant SST differences between the northern and southern regions. In winter the SST gradient is about 0.8°C per degree latitude; in summer the SST is uniform from south to north. Ice forms in mid-November, lasting 5–6 months (Rodionov 1994), and covers almost the entire shallow northern Caspian Sea. During severe winters, shallow bays freeze and shorelines are icy along the eastern coast. Along the western coast, drifting ice is found as far south as the Apsheron Peninsula.

Cold water flows southward along the west coast and warm water intrudes northward from the southern Caspian to the middle Caspian in the cyclonic system of water circulation (Figure 4a). The zonal width of the southward intrusion of cold water along the western continental shelf decreases towards the south (Figure 4a), and is parallel to the decreasing width of the shelf. The offshore boundary of cold water follows a constant depth contour between 20 and 50 m along the entire western shelf of the middle Caspian Sea. Blending of cold and warm waters proceeds along the coast to the Apsheron Ridge, where cold water mixes efficiently with warm water. The cold water tongue diminishes

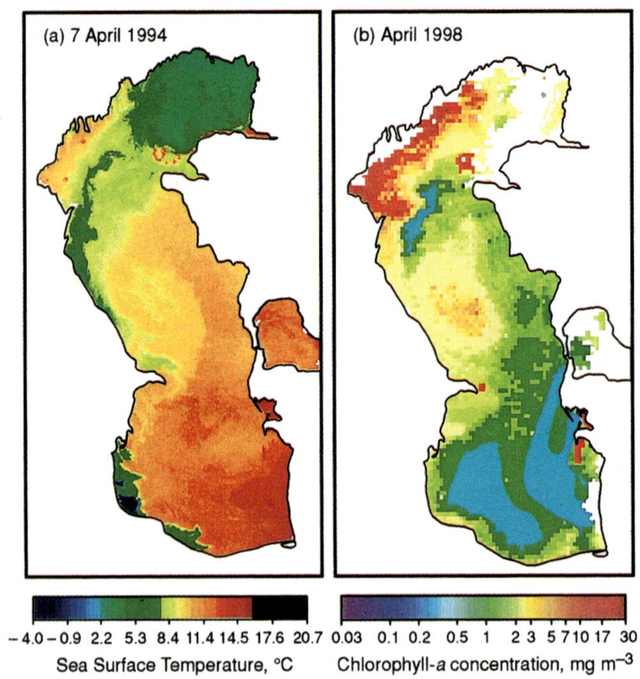

Figure 4. Development of flow along the northwestern and western coasts of the Caspian Sea shown in (a) AVHRR SST on 7 April 1994 and (b) monthly mean SeaWiFS chlorophyll-*a* concentration.

beyond the Apsheron Peninsula. In addition, the cyclonic circulation in the middle Caspian Sea shows marked differences in SST between the eastern and western parts (Figure 4a).

High phytoplankton concentration was linked to the southward transport of cold nutrient-rich water from the northwest shelf (Figure 4b). When the cold water reaches the coastline north of the Apsheron Peninsula, it flows eastward, becomes partially separated from the coast, and is then trapped in an elongated cyclonic (counterclockwise) eddy over the deepest region of the middle Caspian Sea. This meridionally stretched cyclonic eddy is in agreement with the circulation pattern shown in Figure 3. SeaWiFS data (Figure 4b) shows a spectacular phytoplankton bloom within this eddy. The nutrients to drive the phytoplankton production associated with the eddy originate from river discharges in the northern region. River water forms the trophic regimes in the northern part of the sea, which are the main areas for juvenile sturgeon. The increase in biological productivity, specifically the blue-green algae *Cyanophyta*, develops in summer after the Volga River discharge is at its peak and spreads over half of the northern Caspian Sea.

Southward winds along the east coast of the Caspian Sea produce persistent upwelling of cold water in summer. Cold-water patches initially form as eddies (Figure 5a), and a week later (Figure 5b) the eddies have been stretched in the offshore direction, taking the

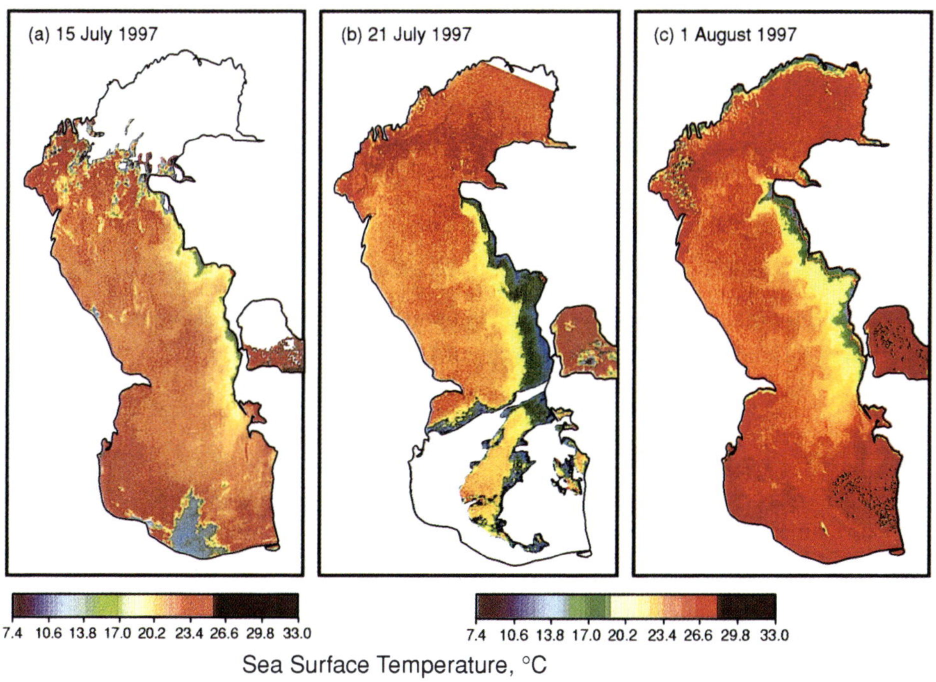

Figure 5. AVHRR SST on (a) 15 July, (b) 21 July, and (c) 1 August 1997.

form of filaments. Growing filaments extend offshore (Figure 5c). Nutrients injected by coastal upwelling into the euphotic zone result in higher phytoplankton biomass and productivity along the eastern coast.

4. Summary and Conclusions

A limited study of satellite imagery indicates a cyclonic flow pattern in the Caspian Sea. Effects of the Volga River along the western coast of the northern and middle Caspian Sea were identified. Riverine materials are transposed by the coastal current and distributed along the coast and across the frontal region by turbulent motions. Flow separation caused by the Apsheron Peninsula is also evident, and influences transport as well as vertical motion along the coast. Southward wind stress along the eastern coast produces upwelling, giving rise to seasonal changes in phytoplankton concentrations along the eastern coast. The observed flow patterns were similar to those occurring when sea level was lower (Figure 3), suggesting that sea level rise in the past 20 years has not affected flow patterns.

The large size of the Caspian Sea requires spaceborne sensors to monitor SST, phytoplankton pigment concentration, wind vectors, and other variables at adequate time and space scales to identify environmental conditions. Most environmental processes in the Caspian Sea are strongly influenced by interactions between the current and coastal boundaries, freshwater runoff, and bottom topography. Atmospheric forcing by wind and by heat and fresh water fluxes is very important and leads to distinct effects on both surface and deep-sea phenomena. Scientific questions about natural and anthropogenic variations of inland-drainage seas (e.g., Caspian, Great Salt Lake, Chad Lake) require remote-sensing data for comparative studies.

Acknowledgments. This study was funded in part by the Middle East Technical University and the Research Fund of the University of Istanbul.

References

Ahlnas, K., T. C. Royer, and T. H. George, Multiple dipole eddies in the Alaska Coastal Current detected with Landsat Thematic Mapper data, *J. Geophys. Res.*, *92*, 13041–13047, 1987.

Baidin, S. S., and A. N. Kosarev, editors, *The Caspian Sea Hydrology and Hydrochemistry*, Nauka, Moscow, 262 pp, 1986.

Bondarenko, A. L., *Currents of the Caspian Sea and Formation of Salinity Field of the Waters of the North Caspian Sea*, Nauka, Moscow, 122 pp, 1993.

Deschamps, P. Y., and R. Frouin, Large diurnal heating of the sea surface observed by the HCMR Experiment, *J. Phys. Oceanogr.*, *14*, 177–184, 1984.

Glantz, M. H., The regionalization of climate-related environmental problems, In *Scientific, Environmental, and Political Issues in the Circum-Caspian Region*, edited by M. H. Glantz and I. S. Zonn, Kluwer Academic Publishers, Dordrecht, 3–10, 1997.

Kosarev, A. N., and E. A. Yablonskaya, *The Caspian Sea*, SPB Academic Publishing, Moscow, 259 pp, 1994.

Lednev, V. A., *Currents of the Northern and Middle Caspian Basins*, Marine Transport Publishing, Moscow, 97 pp, 1943.

Rashit, A.I., E. Özsoy, A. S. Sarkisyan, and H. İ. Sur, Seasonal variability of the Caspian Sea circulation driven by climatic wind stress and river discharge, *J. Phys. Oceanogr.*, submitted, 1999.

Rodionov S. N., *Global and Regional Climate Interaction: The Caspian Sea Experience*, Kluwer Academic Publishers, Dordrecht, 241 pp, 1994.

Sur, H. İ., E. Özsoy, and Ü. Ünlüata, Boundary current instabilities, upwelling, shelf mixing, and eutrophication processes in the Black Sea, *Prog. Oceanogr.*, *33*, 249–302, 1994.

Sur, H. İ., E. Özsoy, Y. P. Ilyin, and Ü. Ünlüata, Coastal/deep ocean interactions in the Black Sea and their ecological/environmental impacts, *J. Mar. Systems*, *7*, 293–320, 1996.

Sur, H. İ., and Y. P. Ilyin, Evolution of satellite derived mesoscale thermal patterns in the Black Sea, *Prog. Oceanogr.*, *39*, 109–151, 1997.

Terziev, F. S., A. N. Kosarev, and A. A. Kerimov, editors, *Hydrometeorology and Hydrochemistry of Seas. Volume VI: Caspian Sea: Issue 1: Hydrometeorological Conditions*, Hydrometeoizdat, St. Petersburg, 359 pp, 1992.

Trukhchev D., A. Kosarev, D. Ivanova, and V. Tuzhilkin, Numerical analysis of the general circulation in the Caspian Sea, *Comptes Rendus de l'Académie Bulgare des Sciences,* Sofia, *48*, 35–38, 1995.

Voropayev, G. V., The problem of the Caspian Sea level forecast and its control for the purpose of management optimization, In *Scientific, Environmental, and Political Issues in the Circum-Caspian Region*, edited by M. H. Glantz and I. S. Zonn, Kluwer Academic Publishers, Dordrecht, 105–117, 1997.

Yentsch, C. S., Satellite representation of features of ocean circulation indicated by CZCS colorimetry, In *Remote Sensing of Shelf Sea Hydrodynamics*, edited by J. C. J. Nihoul, Elsevier, New York, 336–354, 1984.

Halil İbrahim Sur, Institute of Marine Sciences and Management, University of Istanbul, Istanbul, Turkey. (email, sur@istanbul.edu.tr; fax, +90-212-526-8433)

Satellites, Oceanography and Society
edited by David Halpern

Chapter 17

Analyzing the 1993–1998 interannual variability of NCEP model ocean simulations: The contribution of TOPEX/Poseidon observations

Richard W. Reynolds

National Climatic Data Center, National Oceanic and Atmospheric Administration, Camp Springs, Maryland

David Behringer, Ming Ji, Ants Leetmaa, and Christophe Maes

National Centers for Environmental Prediction, National Oceanic and Atmospheric Administration, Camp Springs, Maryland

Femke Vossepoel

Institute for Earth-Oriented Space Research, Delft University, Delft, The Netherlands

Yan Xue

National Centers for Environmental Prediction, National Oceanic and Atmospheric Administration, Camp Springs, Maryland

Abstract. The National Centers for Environmental Prediction climate forecast system for El Niño Southern Oscillation (ENSO) includes a coupled ocean and atmosphere general circulation model. A critical element for a skillful forecast of ENSO is accurate initialization of the tropical Pacific Ocean component of the coupled system. For accurate ocean analyses, assimilation of both in-situ and satellite data is required to correct ocean model biases and to better capture sea surface temperature (SST) variability. Our results suggest that Topography Experiment (TOPEX)/Poseidon (T/P) sea level data have a strong potential for improving ocean analyses and coupled forecasts, in the same way that assimilation of T/P data improved model sea level. However, our present assimilation scheme corrects only temperature; T/P data can, therefore, only influence temperature. In the western tropical Pacific, salinity is an important contribution to sea level. Thus, T/P data need to be correctly partitioned between temperature and salinity for more accurate ocean analyses.

1. Introduction

During the last decade, coupled atmosphere-ocean models have been successfully used to produce forecasts of the El Niño Southern Oscillation (ENSO) phenomenon. The coupled model used at the National Oceanic and Atmospheric Administration (NOAA) National Centers for Environmental Prediction (NCEP) gave one of the better forecasts of the 1997–1998 warm episode (Barnston et al. 1999; Latif et al. 1998). NCEP treats the forecast as a deterministic initial value problem for the tropical Pacific within the limit of the predictability (about one year). An essential component of the NCEP climate forecast system is a variational ocean data assimilation system (Behringer et al. 1998). This system assimilates real-time in-situ and satellite observations into the NOAA Geophysical Fluid Dynamics Laboratory (GFDL) modular ocean model (MOM), configured for the Pacific to produce realistic ocean initial conditions for the coupled model forecast. The coupled model, which includes a low-resolution version of the NCEP atmospheric medium forecast model, produces forecasts of tropical sea surface temperature (SST) for up to one year (Ji et al. 1998).

Satellite data may play an important role in prediction. Chen et al. (1999) used satellite surface winds to initialize the Cane-Zebiak coupled model (Zebiak and Cane 1987) and found much improved predictions for the 1997–1998 warm episode. In addition, Chen et al. (1998) found similar improvements in predicting the 1997–1998 El Niño by assimilating Topography Experiment (TOPEX)/Poseidon, named T/P, data to initialize the Cane-Zebiak coupled model.

Assimilation of subsurface temperature measurements into an ocean general circulation model has a significant impact on the model simulations (Halpern and Ji 1993; Halpern et al. 1998) and, subsequently, on the outcome of coupled model predictions (Rosati et al. 1996; Ji and Leetmaa 1996). As we will show, additional assimilation of satellite altimetry data also has a clear impact. However, errors in data and bias in the model-forcing system make it difficult to assess the impact of satellite altimetry data on forecast accuracy. In the present paper, we focus on the impact of assimilation of satellite altimeter data to produce accurate ocean analyses and initial conditions for a coupled ocean-atmosphere general circulation model. This is a first step toward a comprehensive assessment of the impact of assimilation of satellite observations for interannual climate prediction.

2. Satellite Data

Three satellite data products that affect ocean model analyses are wind stress, SST, and sea level. The wind stress field used to force the ocean model is obtained from the NCEP operational atmospheric analyses (Kanamitsu et al. 1991). The NCEP atmospheric model assimilates satellite winds with other satellite and in-situ data. The SST field is produced by an analysis of SST retrievals measured by the Advanced Very-High Resolution Radiometer (AVHRR) on NOAA polar-orbiting spacecraft and in-situ SST data (Reynolds and Smith 1994). The T/P data product created by Cheney et al. (1994) is used.

To determine which type of satellite data should be the most important for ENSO prediction, we examined SST, sea level, and wind stress data in the tropical Pacific for the 1964–1999 period. The wind stress field was created from the Florida State University (FSU) data product (Goldenberg and O'Brien 1981). For 1964–1981 and 1982–1999 the Smith et al. (1995) and Reynolds and Smith (1994) SST data products, respectively, were used. Sea level was computed with the ocean general circulation model forced with FSU winds without assimilation of ocean data. Xue et al. (2000) showed that the skill of a Markov model prediction of SST up to a year was much higher with inclusion of sea level data. This result suggests that T/P data should be the most important source of satellite data for ENSO prediction. Additional reasons to emphasize T/P data are the relatively modest impact of satellite wind data on the NCEP tropical surface wind stress data product (Atlas and Hoffman 2000) and the modest effect of satellite SST data on the assimilation system. Although satellite SST data can locally improve the SST analysis in regions without in-situ data, SST measurements from the Tropical Atmosphere Ocean (TAO) moored array (McPhaden 1993) have good coverage within 8° of the equator in the Pacific.

3. Results

The ocean model used in the NCEP coupled model covers the Pacific Ocean (45°S–55°N) and has a three-dimensional variational assimilation system. The model is forced by wind fields from the NCEP operational atmospheric forecast system with the Oberhuber (1998) monthly climatological heat fluxes. In this paper, the assimilation system corrects only temperature; modification of the assimilation scheme to correct temperature and salinity is discussed elsewhere (Vossepoel and Behringer 2000). To evaluate the impact of T/P data, two datasets are prepared for assimilation. The first set of data used for assimilation consists of SST analyses (Reynolds and Smith 1994) and subsurface temperature measurements recorded with expendable bathythermographs (XBT) from ships and temperature sensors suspended below TAO buoys. The second set of data includes the previous dataset plus surface dynamic heights computed from T/P data. The two datasets and the corresponding ocean model analyses are named NCEP-XBT and NCEP-TP, respectively (Ji et al. 2000).

A satellite altimeter does not directly measure surface dynamic topography. To compute dynamic topography from T/P data, geoid topography must be subtracted. However, the geoid topography is not known well enough to be used. Thus, only the time variable part of T/P data is assimilated; a three-year (1993–1995) mean was used. Assimilation of T/P data also requires an estimate of the model sea level deviation relative to the same three-year time interval, which was computed from NCEP-XBT analyses for 1993–1995.

In addition, the data assimilation system is configured to assimilate XBT and TAO temperature measurements between the surface and 720 m, where most seasonal-to-interannual variability occurs (Picaut et al. 1995).

An overview of the importance of data assimilation in the tropical Pacific is displayed in Figure 1. From 20°S to 20°N, the standard deviations of model sea level computed without data assimilation (Figure 1b) tend to be lower than those computed from T/P data (Figure 1a), particularly in the western Pacific. The NCEP-XBT (Figure 1c) simulated sea level was in better agreement with T/P data (Figure 1a) compared to simulations made without data assimilation (Figure 1b). From 10°S to 10°N, the largest differences between NCEP-XBT model sea level (Figure 1c) and sea level computed without data assimilation (Figure 1b) was at 6°S, 160°E. With assimilation of T/P data, the NCEP-TP simulated sea level variations (Figure 1d) had very good agreement with T/P data

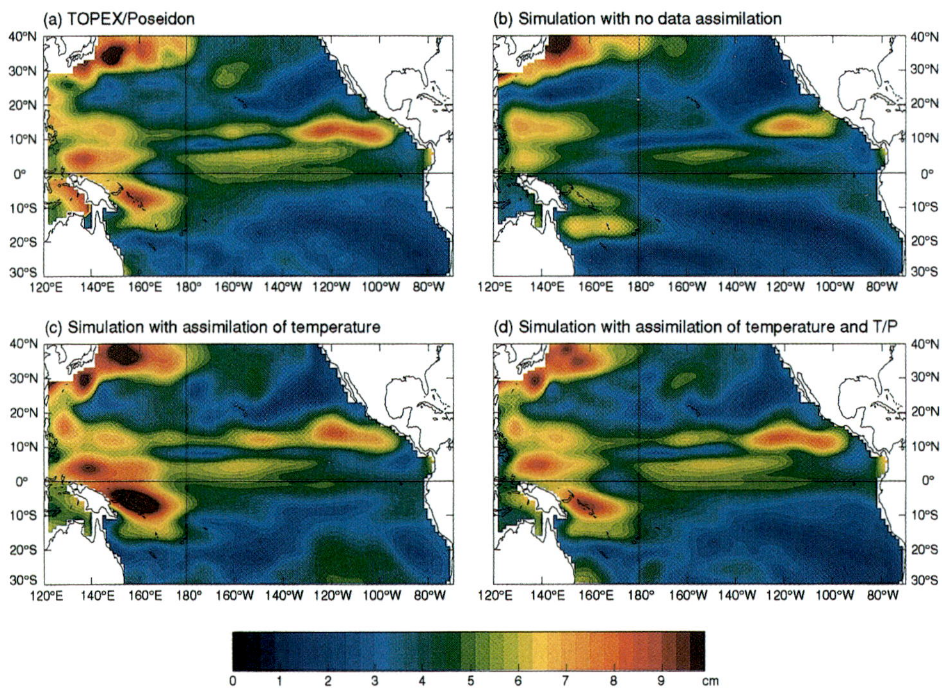

Figure 1. Standard deviation (cm) of monthly sea level from 1993 to 1996 from (a) T/P data, (b) ocean model simulation without data assimilation, (c) ocean model simulation with assimilation of temperature measured at the surface and at depth (named NCEP-XBT), and (d) ocean model simulation with assimilation of temperature measured at the surface (SST) and at depth (XBT, TAO), and with assimilation of T/P data (named NCEP-TP).

(Figure 1a); slight differences (discussed below) were likely related to an interannual salinity variation, which was not correctly represented in the model.

To evaluate the accuracy of the NCEP-XBT and NCEP-TP simulations, model sea level was computed at the location of four tide gauge stations in the western tropical Pacific where the accuracy of the model simulation is critical for coupled forecasts (Ji et al. 1998). The four stations selected were: Kapingamarangi (1.1°N, 155°E), Nauru (0.5°S, 167°E), Tarawa (1.4°N, 173°E), and Kanton (2.8°S, 172°W). The tide gauge data were not assimilated into the model and were thus independent data for validation. Monthly tide gauge data were computed relative to their respective 1993–1995 means. Figure 2 shows that, during 1993–1995, month-to-month variability of both simulations tends to agree with each other and with tide gauge data. However, starting in late 1995, the NCEP-TP analyses tended to follow the tide gauge data while the NCEP-XBT analyses showed a positive bias relative to both the NCEP-TP analyses and tide gauge data.

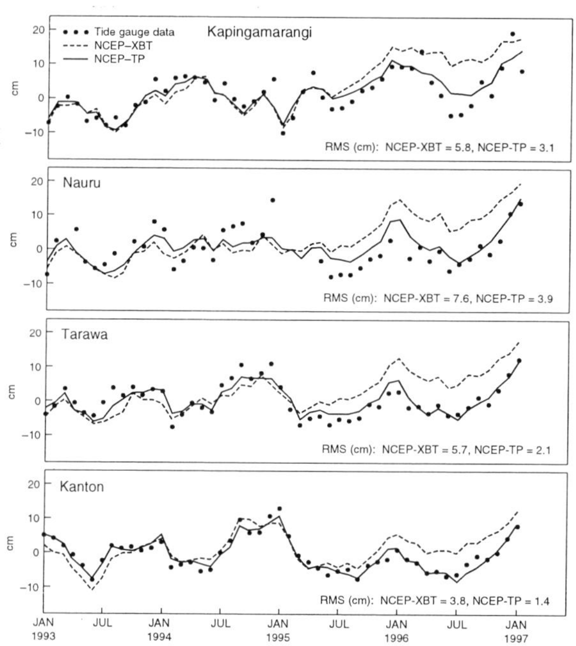

Figure 2. Monthly mean sea level measured at four tide gauges, and NCEP-XBT and NCEP-TP simulations of sea level. Curves have been normalized by removing their respective 1993–1995 means. The ordinate is in centimeters; rms differences between simulated and measured sea levels are shown in the lower right-hand corner of each panel. Tide gauge data are located at Kapingamarangi (1.1°N, 155°E), Nauru (0.5°S, 167°E), Tarawa (1.4°N, 173°E), and Kanton (2.8°S, 172°W).

The T/P and tide gauges are measuring the same variable and, therefore, it is not surprising that the NCEP-TP simulations agree better with the tide gauge data than the NCEP-XBT simulations. The comparisons suggest that the NCEP-XBT analyses are in error in the equatorial western Pacific in 1996. The average difference between the root-mean-square (rms) differences of NCEP-XBT and the tide gauges was about 5.7 cm, which exceeds the observational error for both T/P and tide gauges, but which is smaller than the 10- to 15-cm variability associated with ENSO.

It is useful to look at the differences between NCEP-XBT and NCEP-TP simulations in relation to the input temperature data. For this purpose, we selected the TAO site at 0°, 180°, where almost continuous SST and subsurface temperature were measured during 1993–1996. These data and all other real-time TAO temperature data are assimilated in both simulations. The data assimilation scheme produces a weighted combination of all measurements and the first-guess analysis; therefore, discrepancies between the simulation and observations exist. Comparison of the NCEP-XBT and NCEP-TP simulated subsurface temperatures with the TAO observations (Figure 3) shows similar differences between TAO data and each simulation of temperature. However, during 1996, the rms temperature differences relative to TAO are generally larger for NCEP-TP than for the NCEP-XBT simulations. The additional assimilation of T/P data yields cooling of the upper ocean during 1996 relative to TAO data to correctly lower sea level height, as described in Figure 2. Thus, assimilation of T/P data improves simulation of sea level but introduces larger errors in subsurface temperature.

4. Salinity

Our assimilation technique assumed the barotropic part of tropical sea level variations can be neglected compared to the baroclinic part, in accord with results reported by Chao and Fu (1995) and Picaut et al. (1995). We also assumed that the contribution of salinity to surface dynamic height was negligible compared to temperature. We now believe that the contribution of the interannual salinity variation cannot be ignored, particularly in the western tropical Pacific.

The magnitude of the impact of salinity variations was determined with conductivity, temperature, and depth (CTD) data (Ando and McPhaden 1997), which were used to compute the 1984–1997 mean vertical profile of salinity. Surface dynamic heights relative to 1000 m were calculated from each measured temperature and salinity profiles, and from each observed temperature profile and the 1984–1997 mean salinity profile. The difference between the two estimates of time-varying dynamic height is representative of the impact of salinity variations. The largest difference in surface dynamic height computed with a mean salinity profile and with time-varying salinity profile data was found in the western tropical Pacific. An example is shown along 165°E (Figure 4), where typical

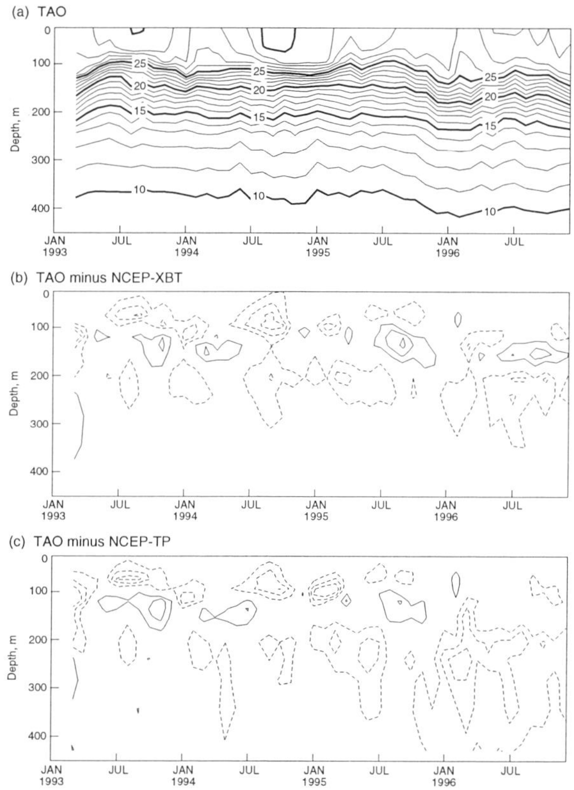

Figure 3. (top) Monthly mean TAO temperature measurements at 0°, 180°; the contour interval is 1°C. Monthly temperature differences relative to TAO data for (middle) NCEP-XBT and (bottom) NCEP-TP simulations; the contour interval is 0.5°C. Negative differences are shown with dashed lines and positive differences are shown with solid lines.

differences are several dyn cm and reach ±8 dyn cm (Maes 1998). Had observations of time-varying salinity profiles been continuously assimilated into the ocean general circulation model, similar to the assimilation procedure for subsurface temperature data, the NCEP-XBT sea level simulations could have had better agreement with T/P data.

Two methods were recently developed at NCEP to determine subsurface salinity. Vossepoel et al. (1999) estimated a vertical distribution of salinity from the observed temperature, a climatological temperature-salinity relation, and sea surface salinity (SSS). Maes (1999) and Maes et al. (2000) used CTD data to generate empirical orthogonal functions (EOF) of temperature and salinity, and estimated subsurface salinity from the EOFs in combination with T/P and SSS data.

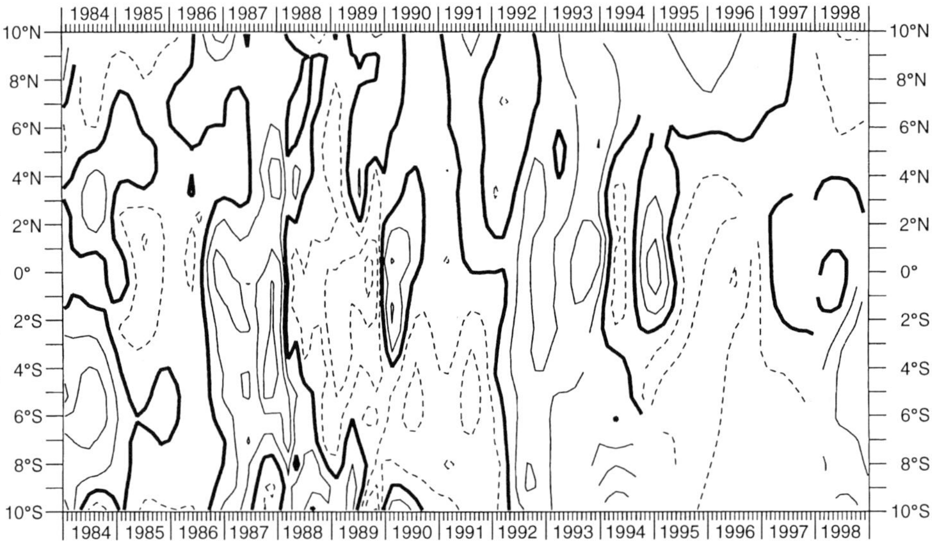

Figure 4. Difference in surface dynamic height relative to 1000 m, $\Delta D|_{1000}^{0}$, along 165°E, defined as $\Delta D|_{1000}^{0}$ computed with observed temperature and observed salinity minus $\Delta D|_{1000}^{0}$ computed with observed temperature and 1984–1997 mean salinity. Contour interval is 2 dyn cm. Negative differences are shown with dashed lines and positive differences are shown with solid lines. The zero contour is shown as a thick line.

5. Concluding Remarks

In the NCEP ocean model data assimilation method described in this paper, T/P data influenced only subsurface temperature. The assimilation scheme is being modified in order for both temperature and salinity to be affected by T/P data. Progress has been made to improve ocean model simulations with a bivariate assimilation technique (Vossepoel and Behringer 2000).

Results indicated that SSS data were useful in estimating subsurface salinity. Lagerloef (2000) and Le Vine et al. (2000) are optimistic regarding future prospects for satellite SSS measurements, which would augment T/P data to produce better ocean simulations and, hopefully, better forecasts.

References

Ando, K., and M. McPhaden, Variability of surface-layer hydrography in the tropical Pacific Ocean, *J. Geophys. Res.*, *102*, 23063–23078, 1997.

Atlas, R., and R. N. Hoffman, The use of satellite surface wind data to improve weather analysis and forecasting, In *Satellites, Oceanography and Society*, edited by D. Halpern, Elsevier Science Publishers, New York, this book, 2000.

Barnston, A. G., M. H. Glantz, and Y. He, Predictive skill of statistical and dynamical climate models in SST forecasts during the 1997-98 El Niño episode and the 1998 La Niña onset, *Bull. Amer. Meteorol. Soc.*, *80*, 217–243, 1999.

Behringer, D. W., M. Ji, and A. Leetmaa, An improved coupled model for ENSO prediction and implications for ocean initialization. Part I: The ocean data assimilation system, *Mon. Wea. Rev.*, *126*, 1013–1021, 1998.

Chen, D., M. A. Cane, and S. E. Zebiak, The impact of sea level data assimilation on the Lamont model prediction of the 1997-98 El Niño, *Geophys. Res. Lett.*, *25*, 2837–2840, 1998.

Chen, D., M. A. Cane, and S. E. Zebiak, The impact of NSCAT winds on predicting the 1997–98 El Niño: A case study with the Lamont model, *J. Geophys. Res.*, *104*, 11321–11327, 1999.

Cheney, R., L. Miller, R. Agreen, N. Doyle, and J. Lillibridge, TOPEX/Poseidon: The 2-cm solution, *J. Geophys. Res.*, *99*, 24555–24564, 1994.

Chao, Y., and L.-L. Fu, A comparison between the TOPEX/Poseidon data and global ocean general circulation model during 1992–1993, *J. Geophys. Res.*, *100*, 24965–24976, 1995.

Goldenberg, S. B., and J. J. O'Brien, Time and space variability of tropical Pacific wind stress, *Mon. Wea. Rev.*, *109*, 1190–1207, 1981.

Halpern, D., and M. Ji, An evaluation of the National Meteorological Center weekly hindcast of upper-ocean temperature along the eastern Pacific equator in January 1992, *J. Climate*, *6*, 1221–1226, 1993.

Halpern, D., M. Ji, A. Leetmaa, and R. W. Reynolds, Influence of assimilation of subsurface temperature measurements on simulations of Equatorial Undercurrent and South Equatorial Current along the Pacific equator, *J. Atmos. Oceanic Tech.*, *15*, 1471–1477, 1998.

Ji, M., D. W. Behringer, and A. Leetmaa, An improved coupled model for ENSO prediction and implications for ocean initialization. Part II: The coupled model, *Mon. Wea. Rev.*, *126*, 1021–1034, 1998.

Ji, M., and A. Leetmaa, Impact of data assimilation on ocean initialization and El Niño prediction, *Mon. Wea. Rev.*, *125*, 742–753, 1996.

Ji, M., R. W. Reynolds, and D. W. Behringer, Use of TOPEX/Poseidon sea level data of ocean analyses and ENSO prediction: Some early results, *J. Climate*, *13*, 216–231, 2000.

Kanamitsu, M., J. C. Albert, K. A. Campana, P. M. Caplan, D. G. Deaven, M. Iredell, B. Katz, H.-L. Pan, and G. H. White, Recent changes implemented into the global forecast system at NMC, *Wea. Forecasting*, *6*, 425–435, 1991.

Latif, M., D. Anderson, T. Barnett, M. Cane, R. Kleeman, A. Leetmaa, J. O'Brien, A. Rosati, and E. Schneider, A review of the predictability and prediction of ENSO, *J. Geophys. Res.*, *103*, 14375–14393, 1998.

Lagerloef, G. S. E., Recent progress towards satellite measurements of the global sea surface salinity field, In *Satellites, Oceanography and Society*, edited by D. Halpern, Elsevier Science Publishers, New York, this book, 2000.

Le Vine D. M., J. B. Zaitzeff, E. J. D'Sa, J. L. Miller, C. Swift, and M. Goodberlet, Sea-surface salinity: Toward an operational remote-sensing system, In *Satellites, Oceanography and Society*, edited by D. Halpern, Elsevier Science Publishers, New York, this book, 2000.

McPhaden, M. J., TOGA-TAO and the 1991–93 El Niño-Southern Oscillation event, *J. Oceanogr.*, *6*, 36–44, 1993.

Maes, C., Estimating the influence of salinity on sea level anomaly in the ocean, *Geophys. Res. Lett.*, *25*, 3551–3554, 1998.

Maes, C., A note on the vertical scales of temperature and salinity and their signature in dynamic height in the western Pacific Ocean: Implications for data assimilation, *J. Geophys. Res.*, *104*, 11037–11048, 1999.

Maes, C., D. W. Behringer, R. W. Reynolds, and M. Ji, Retrospective analysis of the salinity variability in the western tropical Pacific Ocean using an indirect minimization approach, *J. Atmos. Oceanic Tech.*, in press, 2000.

Oberhuber, J. M., An atlas based on the "COADS" data set: The budgets of heat, buoyancy and turbulent kinetic energy at the surface of the global ocean, Rep. No. 15, Max-Plank-Institut für Meteorologie, Hamburg, 20 pp, 1998.

Picaut, J., A. J. Busalacchi, M. J. McPhaden, L. Gourdeau, F. I. Gonzalez, and E. C. Hackert, Open-ocean validation of TOPEX/Poseidon sea level in the western equatorial Pacific, *J. Geophys. Res.*, *100*, 25109–25127, 1995.

Reynolds, R. W., and T. M. Smith, Improved global sea surface temperature analyses, *J. Climate*, *7*, 929–948, 1994.

Rosati, A., K. Miyakoda, and R. Gudgel, The impact of ocean initial conditions on ENSO forecasting with a coupled model, *Mon. Wea. Rev.*, *125*, 754–772, 1996.

Smith, T. M., R. W. Reynolds, R. E. Livezey, and D. C. Stokes, Reconstruction of historical sea surface temperatures using empirical orthogonal functions, *J. Climate*, *9*, 1403–1420, 1996.

Vossepoel, F. C., and D. W. Behringer, Impact of sea level assimilation on salinity variability in the western equatorial Pacific, *J. Phys. Oceanogr.*, in press, 2000.

Vossepoel, F. C., R. W. Reynolds, and L. Miller, The use of sea level observations to estimate salinity variability in the tropical Pacific, *J. Atmos. Oceanic Tech.*, *16*, 1401–1415, 1999.

Xue, Y., A. Leetmaa, and M. Ji, ENSO prediction with Markov models: The impact of sea level, *J. Climate*, in press, 2000.

Zebiak, S. E., and M. A. Cane, A model El Niño-Southern Oscillation, *Mon. Wea. Rev.*, *115*, 2262–2278, 1987.

Richard W. Reynolds, National Climatic Data Center, National Oceanic and Atmospheric Administration, 5200 Auth Road, Camp Springs, MD 20746, U.S.A. (email, richard.w.reynolds@noaa.gov; fax, +1-301-763-8125)

Satellites, Oceanography and Society
edited by David Halpern

Chapter 18

Recent progress toward satellite measurements of the global sea surface salinity field

Gary S. E. Lagerloef
Earth and Space Research, Seattle, Washington

Abstract. With present and incipient technology, it appears to be possible to map the global surface salinity field on climatological scales from satellites. Measurements would be made with passive microwave radiometry at 1–3 GHz (L- and S-bands), which are frequencies also used for soil moisture measurements. There may be additional scientific value for sea ice observations. Studies are being done to assess the scientific importance and applications of global surface salinity measurements, and to evaluate the potential accuracy and spatio-temporal resolution with present and near-future technologies. Three principal scientific objectives have been identified as: (1) Improving seasonal to interannual (e.g., El Niño and La Niña) climate predictions; (2) Improving ocean rainfall estimates and global hydrologic budgets; and (3) Monitoring large-scale salinity events. The assimilation of surface salinity data into ocean and climate models provides an important tool for addressing these scientific issues. Technical problems include radiometer systems and calibration, ambient corrections, algorithms, satellite systems, and orbit mechanics. Several radiometer designs are being considered. Retrieval simulations for one sensor, including the major sensor and geophysical error sources, predict errors of about 0.1 practical salinity units (psu, equivalent to grams of salt per kilogram seawater or parts per thousand, 0/00) in equatorial areas and about 0.2 psu at high latitudes, for 100-km and monthly averages. This assumes mostly uncorrelated error and could be further reduced by larger area averages. Such observations represent a significant advance in global ocean remote-sensing capability for climate research.

1. Introduction

The principles, as well as the technical challenges, of satellite remote sensing of sea surface salinity (SSS) using low microwave frequencies (1–3 GHz) have been recognized for more than two decades (Lagerloef et al. 1995; Swift and McIntosh 1983). A Salinity Sea Ice Working Group (SSIWG) (www.esr.org/ssiwg/mainssiwg.html) has been organized to evaluate the technical and scientific merit of SSS remote sensing, as well as the added benefit to sea ice measurements in this frequency band. The scientific merit of such a sensor is not a stand-alone issue, because the same (1–3 GHz) microwave frequencies are used to measure terrestrial soil moisture (Jackson et al. 1982), and the radiometric signal for soil moisture is much stronger than for SSS (Swift 1993).

The primary issue for salinity is whether the gap between scientific requirements and technical capabilities is sufficiently narrow to merit a satellite mission in the near future. One of the major technical obstacles for working in this microwave band is the need for a very large antenna to obtain the desired degree of spatial resolution. Footprint size is inversely proportional to antenna aperture size; a 5- to 10-m aperture is required for an approximately 20- to 40-km footprint, depending on orbit altitude and view angle. Such systems are becoming affordable, and the scientific need is sufficiently compelling that a union of scientific interests—soil moisture, SSS, and sea ice—can support one or more missions to address the global surface hydrology.

2. Why Measure Sea Surface Salinity From Space?

Salinity is emerging as an important parameter for climate observations (Schmitt 1998), but in-situ observations of surface salinity remain very sparse (Lagerloef et al. 1995). Satellites offer the advantage of systematic global surface sampling that will complement the autonomous, vertically profiling buoys proposed for the Climate Variability and Predictability (CLIVAR) Program. The SSIWG is evaluating what fundamental or highly significant scientific contributions would be made by measuring global SSS from satellite, and what is required in terms of spatio-temporal resolution and accuracy. Three broad, primary scientific objectives for SSS remote sensing have been identified (Lagerloef 1998): (1) Improving seasonal to interannual El Niño climate predictions involves using SSS data to initialize coupled climate forecast models (Reynolds et al. 1998) and to model the role of freshwater flux in the formation and maintenance of both barrier layers and mixed-layer heat budgets in the tropics (Lukas and Lindstrom 1991); (2) Precipitation and, to a lesser extent, evaporation over the ocean are still poorly understood, and are related to the hydrologic budget and to latent heating of the overlying atmosphere. The "ocean rain gauge," which is the measurement of temporal, advective, and diffusive variations of salinity, shows considerable promise in reducing uncertainties in the surface freshwater flux on climatic time scales; (3) Large-scale salinity events, such as ice melt, major river runoff, and monsoons, affect ocean circulation. In the Nordic Seas (Dickson 1988), SSS variations influence the rate of oceanic convection and poleward heat transport. High-latitude SSS measurements will be technically challenging because of the accuracy needed, and the relatively weak radiometric signature at low sea temperature (Section 3). Salinity variations are generally much stronger in the coastal ocean and mar-

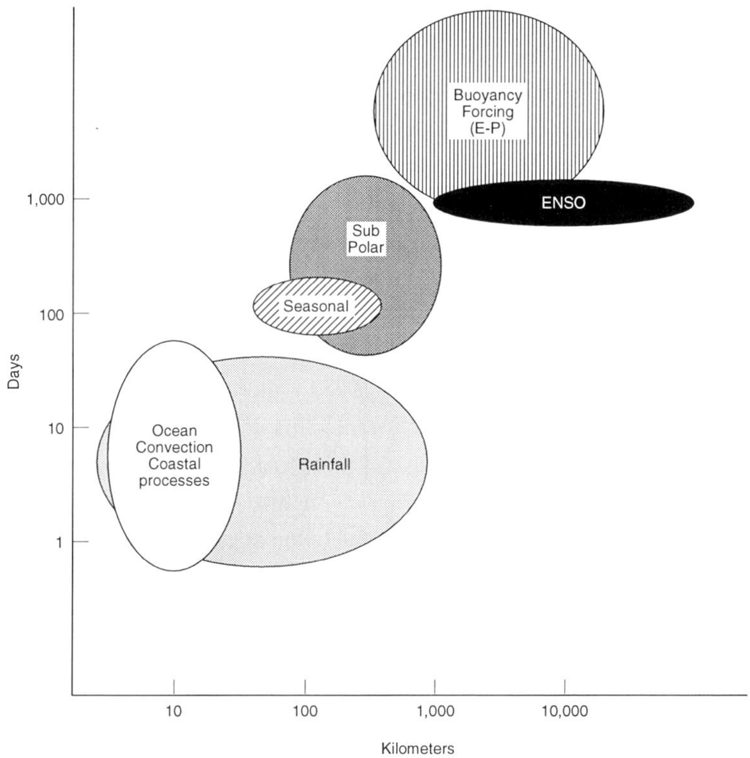

Figure 1. A schematic of the principal SSS phenomena of scientific interest according to length and time scales (Lagerloef 1998).

ginal seas than in the open ocean, but the large footprint size (~20–40 km) will limit nearshore applications of the data. Many larger marginal seas, such as the East China Sea, Bay of Bengal, Gulf of Mexico and Coral Sea/Gulf of Papua, have a SSS range of sufficient magnitude to be adequately measured.

Figure 1 illustrates a schematic frequency-wavenumber diagram of the scientific phenomena and the space and time scales to be resolved. One of the major factors that has restrained development of salinity remote sensing is that the radiometric signal is very small. The signal and noise are approximately equal for individual measurements at the smallest spatial scale. Some degree of space-time filtering will be necessary to recover signals of climatological interest (Lagerloef et al. 1995). It is apparent that no single accuracy requirement is appropriate for the range of scientific problems to be addressed. For seasonal-to-interannual phenomena, especially relevant to El Niño and La Niña prediction, a salinity satellite mission could provide useful data in near-real-time to the forecast models, as well as a more precise delayed-mode research product. For the El Niño problem, SSS data would have a length of greater than 100 km, weekly time scales, and a resolution of 0.05–1.0 psu. A coarse resolution of, say, 200 km × 500 km over five-day averages made available within several days of measurement could provide adequate SSS data for El Niño forecast models. The delayed-mode research data might be more smoothed in time (30 days), better resolved in space (100 or 200 km), and merged with in-situ data for better calibrations and the removal of biases.

Meeting the requirement for the surface freshwater flux problem, the SSS requirements are approximately 0.05 psu, 2° × 2° and monthly resolution in low-to-mid latitudes (Lagerloef 1998). This would resolve a monthly freshwater flux of about 5 cm in the largest precipitation zones (roughly 3–5 m/yr). Larger area averages would reduce uncertainty even further. Schmitt (1998) has suggested a more demanding 0.01 psu requirement, which would resolve a freshwater buoyancy flux equivalent to about 10 Wm^{-2} heat flux. Near-real-time products might be useful for the El Niño problem. Numerical modeling impact studies may be the most appropriate method to address these issues.

In high latitudes, SSS will be more restricted to the delayed-mode research data product. Ice melt or convective overturning are in the order of 100 km and 0.01 to 0.1 psu. Useful near-real-time measurements of the high-latitude phenomena are unlikely. However, for long time scale climatic events, research quality data will require co-analysis with in-situ data and filtering satellite data to further reduce errors. Averaging more than 500 km × 500 km and about 100 days may lead to sufficient accuracy for examining large-scale changes on seasonal-to-decadal scales related to ice-water conversions, such as the Great Salinity Anomaly (Dickson 1988) or the breakup of Antarctic ice shelves.

Assimilating SSS data into ocean and climate models will be an important development for addressing the above scientific questions. The Global Ocean Data Assimilation Experiment (GODAE) (http://www.BoM.GOV.AU/bmrc/mrlr/nrs/oopc/godae/homepage.html), has established SSS measurement requirements: optimal (200 km, 10-day sampling, 0.1 psu error) and threshold (500 km, 10-day sampling, 1.0 psu error).

3. Salinity Remote Sensing

Technical limitations are dictated in part by the principles of salinity remote sensing. The relation between radiometer brightness temperature (T_b) and surface temperature (SST) is given by the emissivity, $\in$:

$$T_b = \in (SST)$$

Emissivity is governed in part by the electrical conductivity of sea water, hence salinity. The effect increases at lower microwave frequencies and is large enough at 1–3 GHz that SSS is detectable (Swift and McIntosh 1983; Klein and Swift 1977). A near-linear relation exists between SSS and T_b, with other factors affecting emissivity held constant, so that the relation can be expressed as:

$$\Delta SSS = \Delta T_b A(SST, f, \sigma, p)$$

where A is a proportionality for a given SST, radio frequency (f), incidence angle (σ), and polarization state ($p = H$ or V). Figure 2 shows the cases for $f = 1.41$ GHz, the primary channel for SSS, with $\sigma = 0$, and either H or V polarization (same for zero incidence). The left panel shows the relation between T_b and SST for various contours of SSS. The zone between 32 and 37 psu represents the range for open ocean conditions, remote from coastal runoff. There is a relatively small T_b change of a few K in this SSS range. The curvature indicates that the sensitivity to SST is also very small, and is indeed zero at certain temperatures. It remains necessary to know SST to retrieve SSS, as illustrated in the right panel, which shows T_b versus SSS for various contours of SST. The signals are stronger at warmer SST, such that A ≈ 1.5 to A ≈ 2 in temperate and tropical open ocean conditions and A≈ 3 to A ≈ 4 in polar regions. Thus, every 0.1 K T_b uncertainty yields 0.15 to 0.2 psu uncertainty in low to mid latitudes, and 0.3 to 0.4 psu in high latitudes. Useful salinity accuracy requires that sensor precision and ambient corrections have uncertainties of 0.1 K or less. The highest resolution data will remain noisy. An assumption made repeatedly here, and one that requires further verification, is that residual errors will be sufficiently uncorrelated that lower frequency and wavenumber features will not be contaminated. Accordingly, we will need to rely heavily on spatio-temporal averaging, or other filtering, to resolve the larger scales and lower frequencies of SSS variability on climate scales.

The capability to measure salinity with remote sensing has been demonstrated by aircraft experiments. Miller et al. (1998) produced the first image of SSS in the coastal zone with an airborne salinity mapper. Variations over a range of several psu were clearly resolved. Available ground truth indicated a retrieval accuracy of about 1 psu.

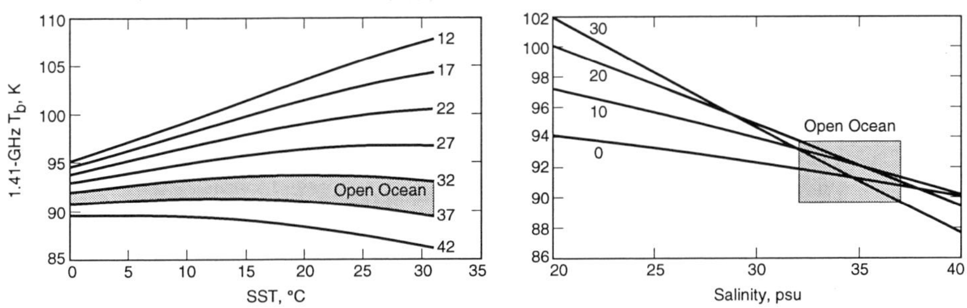

Figure 2. Left: Radiometric brightness temperature (T_b) as a function of SST for salinity, in psu. Right: T_b as a function of SSS for various SST.

4. Candidate Satellite Systems to Measure Salinity

4.1 Soil Moisture Ocean Salinity

This satellite mission proposal (http://www-sv.cict.fr/cesbio/smos) was approved by the European Space Agency (ESA). Soil Moisture Ocean Salinity (SMOS) is a merger of Microwave Imaging Radiometer with Aperture Synthesis (MIRAS), an ESA technical development program, and Radiometrie Appliquée la Measure de la Salinité et de l'Eau du Sol (RAMSES), a Centre National d'Études Spatiales (CNES) mission proposal. The sensor is a Y-shaped array, two-dimensional interferometric, 1.4-GHz, dual-polarization radiometer (Figure 3). The orbit will be 6 a.m./p.m. helio-synchronous at an altitude of 757 km, and the mission has a planned three- to five-year duration. SMOS is designed to

Figure 3. Artist's conception of the SMOS satellite (courtesy Yann Kerr, 1999).

address terrestrial hydrology and ocean science, as the name implies. The flexible interferometric imaging geometry allows for higher spatial resolution imaging over land and larger scale multi-beam averaging over the ocean to reduce measurement noise. The ground resolutions will vary from about 30–90 km within the field of view. Ocean retrieval and image reconstruction algorithms will be quite complicated and the ultimate accuracies require further simulation studies to be resolved. The anticipated launch year is 2004 or 2005.

4.2 Ocean Salinity Soil Moisture Integrated Radiometric Imaging System

The Ocean Salinity Soil Moisture Integrated Radiometric Imaging System (OSIRIS) is a large mesh antenna design under development at the Jet Propulsion Laboratory. OSIRIS is designed with the primary objective of obtaining ocean salinity retrievals, as well as soil moisture, with the highest possible measurement accuracy using current technology. OSIRIS intends to use a high-gain, real-aperture antenna to attain a low-noise, well-calibrated, multi-channel radiometer with the necessary ancillary measurements to correct important sources of geophysical error, such as SST, surface roughness, and ionosphere (Njoku et al. 1999). Large-aperture mesh antenna designs have been developed by aerospace companies for microwave telecommunications, are lightweight, and can be adapted to passive microwave radiometry. OSIRIS includes an approximately 6-m diameter, conical-scanning antenna, and constant incidence angle viewing geometry that will allow forward and backward beams to be averaged with a spot resolution of about 40 km (Figure 4). The baseline radiometer will measure both 1.4 and 2.7 GHz (L- and S-band) with horizontal and vertical polarizations and horizontal-vertical cross-polarizations (Stokes-3), using a flexible design that makes it relatively inexpensive to add channels. An optional L-band radar for wind and sea state correction is also being evaluated. Antenna beam efficiency is expected to exceed 95%. The flexible bandwidth over the ocean can be employed to reduce the root-mean-square (rms) noise per pixel to about 0.1 K (about 0.2–0.3 psu salinity error). Simulation studies of the retrieval accuracy of this system, including geophysical corrections, show that accumulated errors over 100-km scales for one-month averages are about 0.1 psu in the tropics and 0.1–0.2 psu globally (Figure 5). Errors with long correlation scales (1600 km) are modeled in these simulations and special attention to radiometer calibration stability will be necessary to reduce errors further.

4.3 Hydrostar

Hydrostar is a thinned array, one-dimensional, aperture synthesis design applying real aperture in the along-track and synthetic aperture in the cross-track directions (Figure 6). The mission design has been in development at the Goddard Space Flight Center for several years in collaboration with the University of Michigan, and is based on the Electroni-

Figure 4. Artist's conception of the Hydrostar satellite (courtesy T. England, 1999).

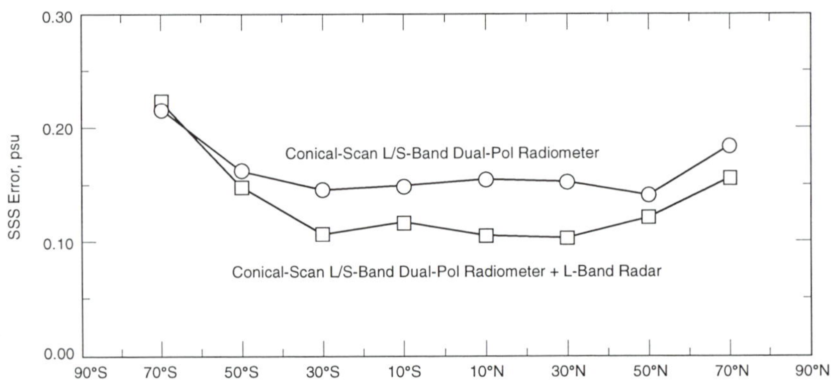

Figure 5. Zonally averaged root-mean-square SSS error based on simulated OSIRIS SSS retrievals averaged over all longitudes for 4 weeks (courtesy Simon Yueh, 1999).

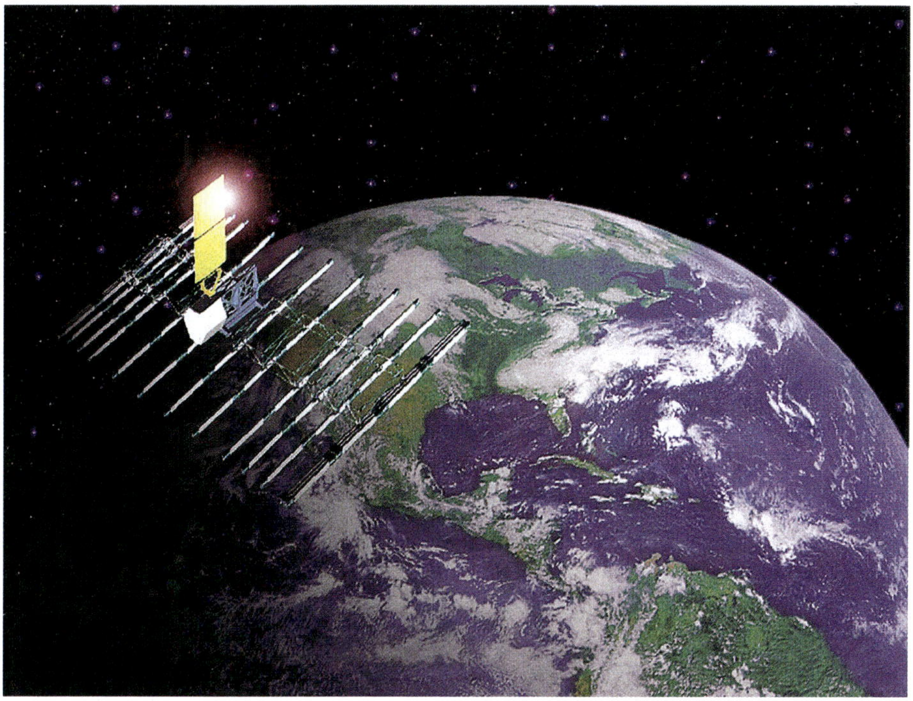

Figure 6. Artist's conception of the Hydrostar satellite (courtesy T. England, 1999).

cally Scanned Thinned Array Radiometer (ESTAR) concept (Swift 1993). This is a cross-track scan, single-channel (L-band, H-pol) system with about 30- to 40-km resolution, and would have somewhat more limited salinity measuring accuracy than the systems described above. Accordingly, the primary science objective is soil moisture, with salinity as an experimental measurement.

5. Sources of Salinity Retrieval Error

A number of technical challenges must yet be solved to ensure research quality satellite SSS data. Several uncertainties are related to geophysical corrections which must be modeled and derived from ancillary observations. Key parameters are outlined (see Lagerloef 1998; Lagerloef et al. 1995; Swift and McIntosh 1983 for a more complete discussion): (1) Sea surface temperature will require ancillary data or multiple radiometer channels for simultaneous SSS and SST retrievals; (2) Surface roughness and foam (wind stress) are not well known at L and S microwave frequencies and there is a need for further experimental studies. The effect is expected to be about 0.1 K per m s^{-1} wind speed, but may be larger. Ancillary wind or sea roughness measurements or multiple radiometer and radar channels for simultaneous SSS and roughness retrieval will be required; (3) Ionospheric Faraday rotation influences the polarization vector and the effect increases with lower frequency. It can be mitigated by avoiding solar mid-day orbits and making vertical and horizontal polarized measurements, and a measurement of the third Stokes parameter; (4) Rain distorts microwave radiation, and the effect can be significant during heavy rain. The rain effect on T_b is about 0.01 RH, where R is rain rate (mm/hr) and H is rain-layer depth (km). The effect is about 0.1 K for a 10-mm/hr, 1-km thick rain layer. It is important to make corrections for this effect, rather than flagging and rejecting rain-contaminated data, because of the importance of retrieving SSS within the tropical precipitation zones; (5) The effects of atmospheric water vapor and clouds are nearly negligible at these frequencies; (6) Atmospheric absorption is nearly a constant offset with minor variations that can be corrected with sea level pressure data; (7) Sun glint will be mitigated by choosing an orbit avoiding high sun angles; (8) Galactic noise will be flagged when galactic core radiation reflected off the sea surface will be in the radiometer view; (9) Some isolated radio interference may be encountered in the 1.413-GHz hydrogen absorption band, which is safeguarded for radio astronomy.

6. Summary and Conclusions

Various technologies for ocean remote sensing by satellite have progressed over the past two decades. Routine global observations of sea level, SST, ocean color, surface winds, sea ice, and rainfall are now commonplace. Ocean surface salinity may be added to the list of spaceborne observations within the next five years. With radiometers now being designed, research quality SSS data can be obtained for climate studies on seasonal-to-interannual time scales. The problem of relatively low signal-to-noise ratio from the sensor can be mostly overcome by spatial and temporal filtering. Other systematic and random errors due to sensor design and geophysical corrections must also be addressed rigorously. Simulation studies indicate that the rms errors from an optimally designed sensor could be as small as 0.1 psu in the tropics and subtropics and about 0.2 psu in higher latitude on monthly and 100-km grid scales.

References

Dickson, R. R., R. Meincke, S.-A. Malmberg, and J. J. Lee, The 'Great Salinity Anomaly' in the Northern North Atlantic, 1968–1982, *Prog. Oceanogr.*, *20*, 103–151, 1988.

Jackson, T. J., T. J. Schmugge, and J. R. Wang, Passive microwave remote sensing of soil moisture under vegetation canopies, *Water Resources Res.*, *18*, 1137–1142, 1982.

Klein, L. A., and C. T. Swift, An improved model for the dielectric constant of sea water at microwave frequencies, *IEEE Trans. Antennas and Prop.*, *AP-25*, 104–111, 1977.

Lagerloef, G., C. Swift, and D. LeVine, Sea surface salinity: The next remote sensing challenge, *J. Oceanogr.*, *8*, 44–50, 1995.

Lagerloef, G., Final report of the first workshop, Salinity Sea Ice Working Group (SSIWG), preliminary assessment of the scientific and technical merits for salinity remote sensing from satellite (http://www.esr.org/lagerloef/ssiwg/ssiwgrep1.v-2.html), revised 16 December 1998.

Lukas, R., and E. Lindstrom, The mixed layer of the western equatorial Pacific Ocean, *J. Geophys. Res.*, *96*, 3343–3357, 1991.

Miller, J., M. Goodberlet, and J. Zaitzeff, Airborne salinity mapper makes debut in coastal zone, *Eos, Trans. Am. Geophys. Un.*, *79*, 173 and 176–177, 1998.

Njoku, E., W. J. Wilson, S. H. Yueh, and Y. Rahmat-Samii, A large antenna microwave radiometer-scatterometer concept for high-resolution surface sensing, *IEEE Trans. Geosci. Remote Sensing*, submitted.

Reynolds, R., M. Ji, and A. Leetmaa, Use of salinity to improve ocean modeling, *Physics and Chemistry of the Earth*, *23*, 545–555, 1998.

Schmitt, R. W., GOSAMOR, A program for Global Ocean SAlinity MonitORing, a proposed contribution to CLIVAR (http://www.bom.gov.au/bmrc/mrlr/nrs/oopc/godae/gosamor.htm), March 1998.

Swift, C. T., and R. E. McIntosh, Considerations for microwave remote sensing of ocean-surface salinity, *IEEE Trans. Geosci. Remote Sensing*, *GE-21*, 480–491, 1983.

Swift, C. T., ESTAR—The electronically scanned thinned array radiometer for remote-sensing measurement of soil moisture and ocean salinity, NASA Technical Memorandum 4523, Goddard Space Flight Center, Greenbelt, MD, 40 pp, 1993.

Gary S. E. Lagerloef, Earth and Space Research, 1910 Fairview Avenue East, Suite 102, Seattle, WA 98102, U.S.A. (email, lagerloef@esr.org; fax, +1-206-726-0524)

Satellites, Oceanography and Society
edited by David Halpern

Chapter 19

Sea surface salinity: Toward an operational remote-sensing system

David M. Le Vine

Goddard Space Flight Center, National Aeronautics and Space Administration, Greenbelt, Maryland

James B. Zaitzeff and Eurico J. D'Sa

National Environmental Satellite, Data, and Information Service, National Oceanic and Atmospheric Administration, Washington, D.C.

Jerry L. Miller

Naval Research Laboratory, Stennis Space Center, Mississippi

Calvin Swift

University of Massachusetts, Amherst, Massachusetts

Mark Goodberlet

Quadrant Engineering, Amherst, Massachusetts

Abstract. Salinity is an important component of biological and physical processes in the ocean, but its distribution throughout the ocean is poorly known. Passive microwave sensors are capable of measuring sea surface salinity, and recent advances in technology suggest that such measurements may soon be feasible on a global basis from Earth orbit. Aircraft-mounted instruments have been developed to validate the technology and address the problem of developing robust algorithms for remote sensing of sea surface salinity. Two aircraft instruments, the Scanning Low Frequency Microwave Radiometer (SLFMR) and the Electronically Scanned Thinned Array Radiometer (ESTAR), have successfully demonstrated remote sensing of salinity in the coastal ocean. Sensors for monitoring salinity from space that utilize the thinned array technology demonstrated by ESTAR have recently been proposed to the National Aeronautics and Space Administration and the European Space Agency.

1. Introduction

Salinity plays an important role in the ocean's biological and physical processes. On continental shelves, river runoff creates large salinity variations, which in turn control density-driven coastal currents. In the coastal zone, salinity gradients can easily be ten times more influential than temperature gradients in determining the density field that drives low-frequency circulation. This circulation redistributes biological and chemical substances, including pollution, along and across the shelf, as well as vertically within the water column. In the open ocean, the thermohaline component of ocean circulation involves the full volume of the ocean and is capable of moving vast amounts of heat and water.

Changes in salinity are caused by events occurring at the surface, such as evaporation, precipitation, melting ice, and river discharge. Hence, an examination of salinity at the surface is a first step in understanding its effects on ocean density and associated circulation. Sea surface salinity (SSS) is also important for modeling energy exchange at the surface. For example, air-sea heat fluxes in equatorial regions are thought to be modulated by the near-surface salinity structure (Reynolds et al. 2000). Thus, salinity may play a key role in El Niño events and will be needed for accurate prediction of such events.

The mean SSS distribution for the global ocean (Levitus et al. 1994) is shown in Figure 1, which represents data compiled from many individual ship measurements during the past 125 years. The smooth contours hide the fact that even using data collected over this long period leaves large portions of the ocean poorly sampled (Lagerloef et al. 1995). Satellite monitoring to provide global maps of salinity would help resolve large-scale features of the salinity field and provide new information on its variability with time, which is not practical to obtain with in-situ measurement techniques.

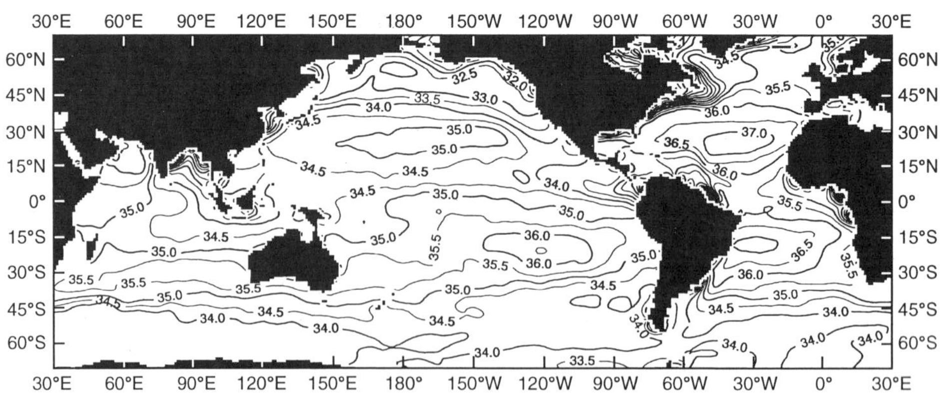

Figure 1. Average sea surface salinity from 70°S to 70°N, compiled by Levitus et al. (1994) from many individual ship measurements, mostly during 1900–1999.

Sea surface salinity can be detected remotely with passive microwave sensors, and it may be possible to make such a measurement from space (Blume et al. 1978; Swift and MacIntosh 1983). Changes in salinity modulate the natural emission of electromagnetic radiation from the ocean's surface. This phenomenon is illustrated in Figure 2, which shows the amplitude of the microwave signature, expressed in brightness temperature, as a function of salinity and sea surface temperature (SST) (Blume et al. 1978; Klein and Swift 1977). Figure 2 shows that the microwave signature is more sensitive to changes in salinity as the frequency decreases, and as a result the measurement of SSS is most effective at the low frequency end of the microwave spectrum. For remote sensing from space, the choice of frequency is limited by both man-made noise and by the effects of the ionosphere. A 27-MHz band at 1.413 GHz (L-band), set aside for passive use, is the window most often identified for remote sensing of SSS. Other choices, such as the frequency bands allocated for passive use at 2.65 GHz (S-band) and 0.8 GHz (P-band), have disadvantages of less bandwidth and more noise, and have either less sensitivity to SSS (S-band) or more problems with the ionosphere (P-band).

The physics of remote sensing of SSS at L-band have been demonstrated with airborne radiometers, and an observation has even been made from space. Observations of SSS differences between fresh water and sea water were made with an L-band radiometer on Skylab (Lerner and Hollinger 1977). Simultaneous measurement with L- and S-band radiometers eliminates the need for coincident SST data, as has been demonstrated with airborne instruments (Thomann 1976; Blume et al. 1978). A sensor system for global SSS measurements from space obtaining SST data from an alternative source was proposed more than 15 years ago (Swift and MacIntosh 1983). Measurements with modern aircraft instruments have demonstrated that SSS can be retrieved with scientifically useful accuracy of about 1 practical salinity unit (psu) in coastal areas, using L-band measurement combined with SST data from another source (Miller et al. 1998; Le Vine et al. 1998). These instruments provide the basis for contemporary prospects for remote sensing from space.

2. Aircraft Remote Sensors

2.1 Scanning Low Frequency Microwave Radiometer

The Scanning Low Frequency Microwave Radiometer (SLFMR) represents new technology for remote sensing of sea surface salinity (Miller et al. 1998). The instrument was designed specifically for mapping coastal salinity and to achieve both accuracy and large swath width (Miller et al. 1998; Goodberlet et al. 1997). The SLFMR bandwidth and frequency are 25 MHz and 1.413 GHz, respectively. The instrument has a single antenna that is scanned across track to form an image. The antenna beamwidth is about 16° and has six across-track positions at ± 7°, ± 22°, and ± 39° (Figure 3). The radiometric accuracy is about 0.4 K, with an integration time of 1.0 s. Although the SLFMR can be operated from

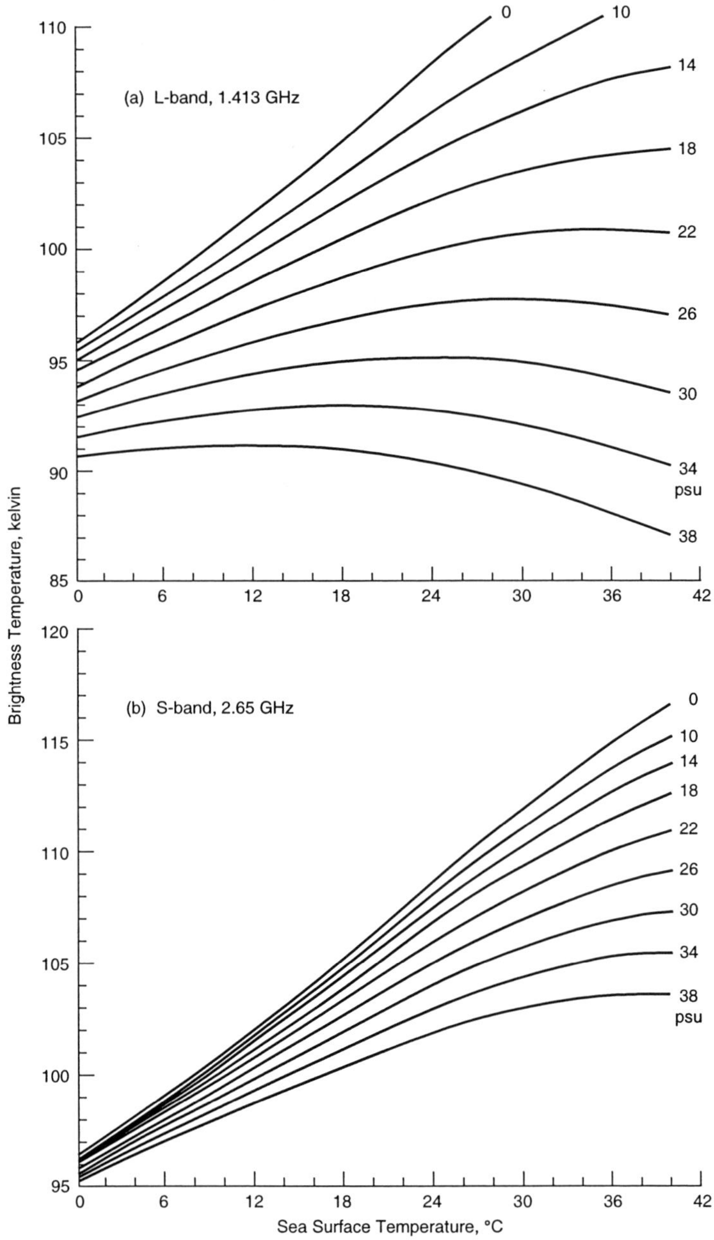

Figure 2. Microwave brightness temperature as a function of SSS and SST for observations at (a) L-band, 1.413 GHz and (b) S-band, 2.65 GHz (after Blume et al. 1978; Klein and Swift 1977).

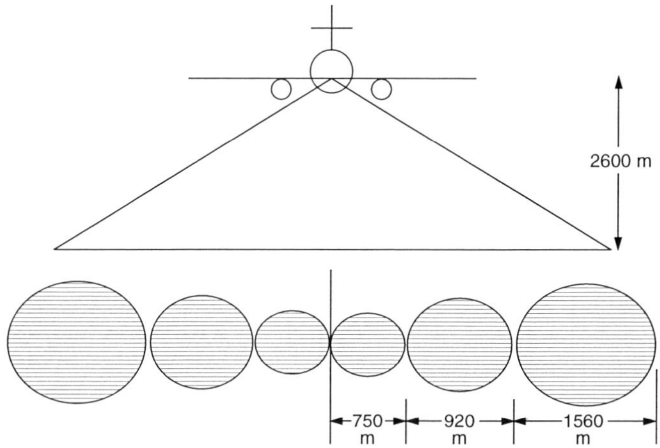

Figure 3. Remote-sensing geometry and beamwidth of the SLFMR.

both large and small aircraft, it was designed specifically for measurements from small aircraft. The SLFMR was integrated onto the NASA P-3 aircraft in August 1999 for a series of salinity measuring experiments in the Atlantic Ocean, off the coast of Delaware.

Figure 4 is an example of measurements with the SLFMR at the mouth of the Chesapeake Bay in September 1996 (Miller et al. 1998). The measurements were made at an altitude of about 2.6 km with a resolution of about 1 km. The large change in salinity between the two images is attributed to runoff from Hurricane Fran, which passed over the area on 5–8 September. Comparison of SLFMR data with ship measurements in Florida Bay during experiments in October 1997 indicates an accuracy per pixel of about 1.5 psu for the SLFMR; however, pixels in the SLFMR image overlap and averaging can be done to reduce noise. The SSS accuracy in Figure 4 is about 0.5 psu.

A second-generation SLFMR is currently being built to measure SSS. The goal is an accuracy of 0.5 psu, assuming a maximum SST error of 0.5°C. At this accuracy, precise knowledge of the incidence angle of the radiometer beams is important, and special consideration is being given to monitoring aircraft orientation. The galactic background radiation at 1.4 GHz must also be included in the retrieval algorithm. The most uncertain source of error in the retrieval of SSS is the effect of surface roughness, which is dependent on wind speed and fetch as well as other parameters affecting the surface. For example, surfactants also modify surface roughness. Research is needed to determine the magnitude of the error and the parameters needed to model it.

2.2 Electronically Scanned Thinned Array Radiometer

The Electronically Scanned Thinned Array Radiometer (ESTAR) is a prototype of new technology for passive microwave remote sensing from space. This is an interfero-

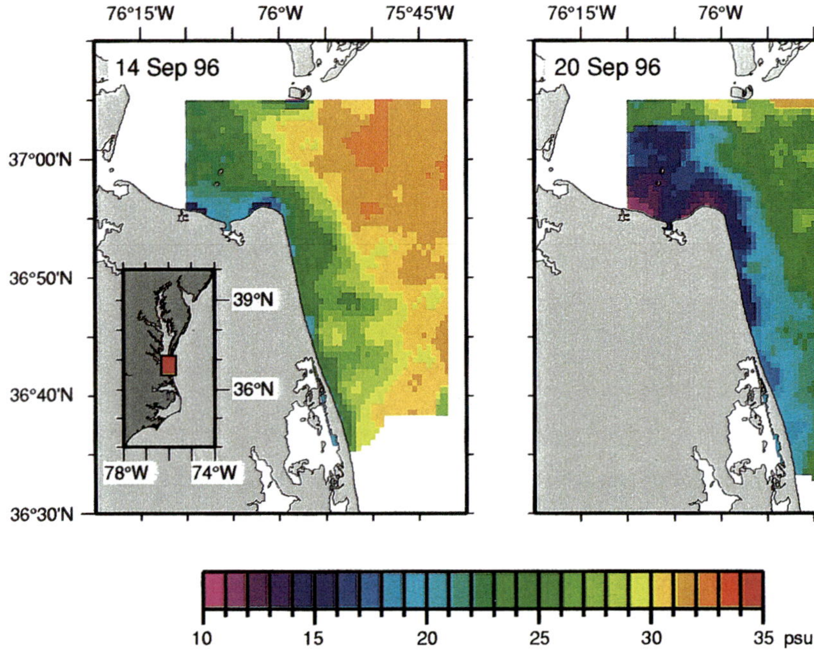

Figure 4. Airborne SLFMR measurements of sea surface salinity of the Chesapeake Bay plume on (right) 14 September 1996 and (left) 20 September 1996.

metric technique called aperture synthesis, which is employed to reduce the size of the antenna aperture needed in space (Le Vine et al. 1989; Swift et al. 1993). Antenna aperture is an important issue for monitoring SSS from space because at the long wavelengths needed to monitor SSS, large antennas are needed to achieve adequate spatial resolution (Le Vine et al. 1989).

Aperture synthesis for remote sensing from space is similar, in principle, to Earth rotation synthesis practiced in radio astronomy (Thompson et al. 1986). In this technique, signals from pairs of antennas with many different spacings are measured. Each spacing yields a sample point in the Fourier transform of the brightness temperature distribution of the scene, and a map of the brightness temperature is obtained in software after the measurements have been made by inverting the transform (Le Vine et al. 1994). With this technique, the resolution of a single large antenna aperture can be obtained with relatively few small antennas arranged in a manner to provide many combinations of antenna pairs, each with different separation distance. ESTAR is an aircraft instrument built to develop this technique for remote sensing.

ESTAR is a hybrid of a real and a synthetic aperture radiometer (Figure 5). Long "stick" antennas oriented with the long axis in the direction of motion produce a narrow fan-like beam with good resolution along track and no resolution across track. Across-track resolution is obtained using aperture synthesis. To implement aperture synthesis, the voltage from antenna pairs with different spacing are multiplied together and averaged (Le Vine et al. 1994). ESTAR has five antennas spaced in integer multiples of one-half wavelength; from this arrangement it is possible to form seven independent baselines (Le Vine et al. 1990; Le Vine et al. 1994).

ESTAR has demonstrated that images can be made with accuracies required for both the measurement of soil moisture (Le Vine et al. 1994; Jackson and Le Vine 1996) and

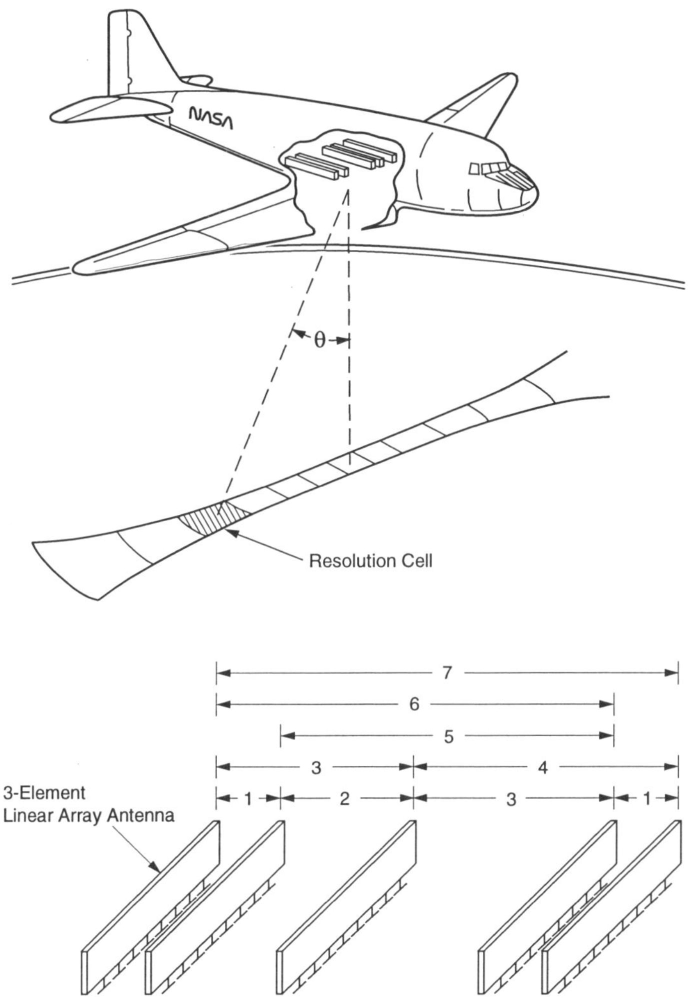

Figure 5. (top) Imaging geometry and (bottom) antenna configuration of the ESTAR radiometer. Dimensions on the bottom are in units of one-half wavelength (~10.5 cm).

SSS in coastal regimes (Le Vine et al. 1998). Among its accomplishments are SSS measurements made during the Delaware Coastal Current experiment (Le Vine et al. 1998) to study the relatively low-saline outflow originating in Delaware Bay (Munchow and Garvine 1993). In these experiments, good agreement was found between shipboard thermosalinograph measurements of salinity (Figure 6a) and ESTAR measurements (Figure 6b). The SSS difference between ESTAR and ship data was about 1 psu (Figure 6c). Larger differences occurred in the northwest, and may have been the result of the time difference between the two measurements: The ship required nearly two days to map this region, whereas the ESTAR map required less than two hours (Le Vine et al. 1998).

3. Proposals for Measuring Sea Surface Salinity from Space

The success of recent aircraft measurements of SSS, as well as the increasing awareness of the important role of salinity in ocean processes, has led to proposals for measuring salinity from space. Several of these are described briefly in the following sections.

3.1 Hydrostar

The success of ESTAR has generated research on the development of synthetic aperture technology for remote sensing from space. In the United States, a proposal has been submitted to fly a spaceborne instrument patterned after ESTAR. This instrument, called Hydrostar, is designed to measure soil moisture, with the measurement of sea surface salinity being a research project. Like ESTAR, it employs aperture synthesis in the across-track dimension and operates at L-band (1.413 GHz) with horizontal polarization.

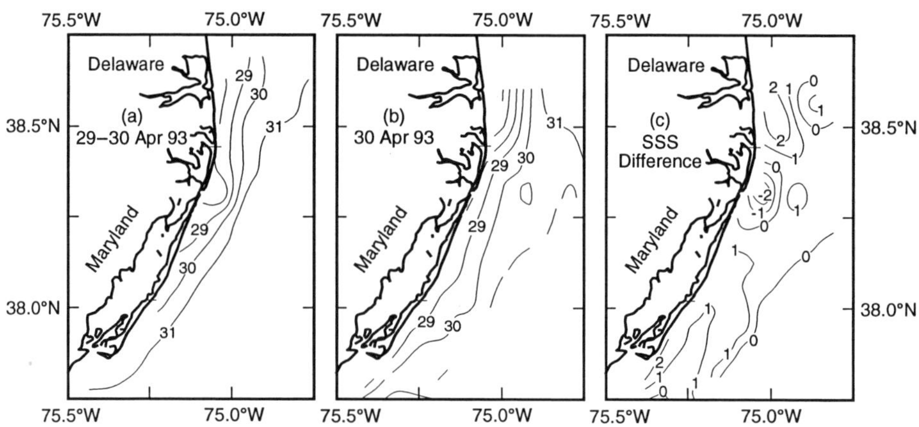

Figure 6. Sea surface salinity of the outflow from Delaware Bay; (a) shipboard thermosalinograph data; (b) airborne ESTAR radiometer data; (c) difference between ship and aircraft data.

The soil moisture-driven science objectives of the Hydrostar mission call for a revisit time of three days or less. The 670-km altitude sun-synchronous orbit (observations at constant local time) has a 6 a.m. equatorial crossing and achieves the desired revisit time by processing data in the across-track dimension to ± 35 degrees. In its normal operation mode the integration time per image is 0.5 s, and internal calibration is in the form of a stream of short pulses from the reference noise diode. Absolute calibration is accomplished using data in a manner similar to that developed for ESTAR (Le Vine et al. 1994).

Hydrostar employs 16 antenna sticks, each 5.8-m long, arranged in a minimum redundancy array with a 9.5-m maximum across-track separation (Figure 7). The minimum antenna spacing is one-half wavelength, and the array has 90 independent antenna spacings, which are integer multiples of half a wavelength. Each antenna is a rectangular waveguide stick with 36 slots cut in the narrow wall of the waveguide and inclined to pro-

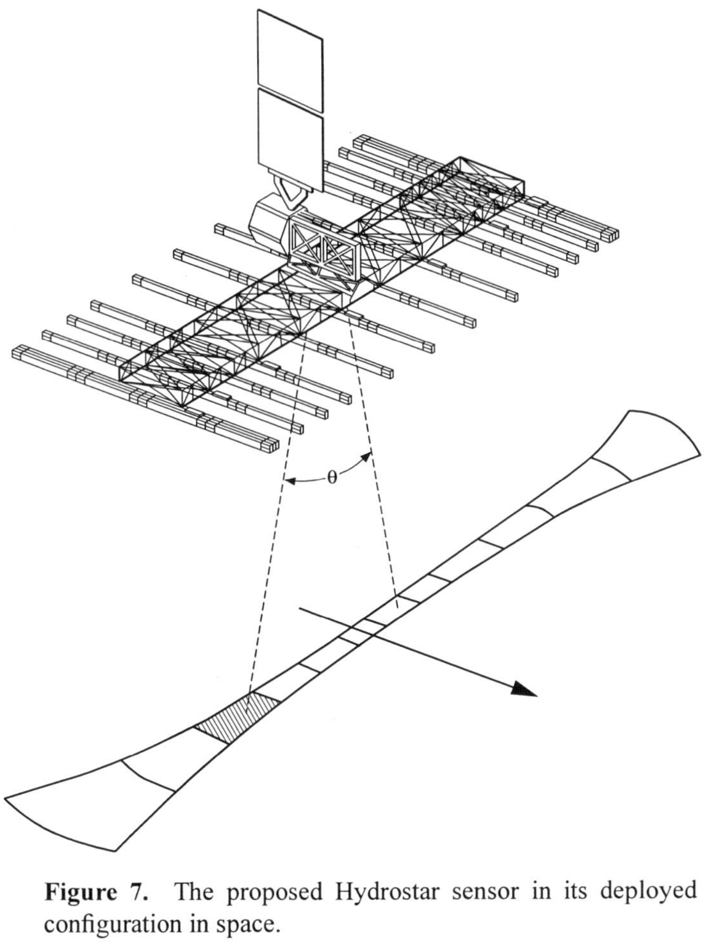

Figure 7. The proposed Hydrostar sensor in its deployed configuration in space.

duce horizontal polarization. Each stick has an along-track beam width of 2.3° at band center that corresponds to a footprint size of 27 km at nadir. Operation over the desired bandwidth is obtained by subdividing each antenna stick into four resonant subarrays of nine slots each, coupled together with a flexible combining network. The antennas are fabricated in composite to minimize weight and maximize rigidity and thermal stability. The Hydrostar brightness temperature sensitivity is 1 K in each 0.5 s image, with 3 K accuracy. The 9.5-m across-track distance of the stick array was chosen so that across-track resolution matched along-track resolution at the swath edge. At nadir, the across-track resolution has a peak-to-null beam width of 1.2°, corresponding to 14-km resolution.

3.2 Soil Moisture Ocean Salinity mission

The European Space Agency (ESA) has developed an aircraft instrument called Microwave Imaging Radiometer with Aperture Synthesis (MIRAS) that employs aperture synthesis in two dimensions (Martin-Neira and Goutoule 1997). This instrument operates at L-band and C-band and employs dual-polarized antennas arranged in a "Y" configuration. The "Y" is tilted forward at about 40°, yielding a swath with a relatively constant incidence angle. A proposal to build a spacecraft version of MIRAS has recently been selected for funding by ESA. This instrument, called Soil Moisture Ocean Salinity (SMOS), operates at only one frequency, L-band (1.413 GHz), and employs aperture synthesis in two dimensions using antennas in the "Y" configuration (Figure 8). The 760-km sun-synchronous orbit will have a 6 a.m. equatorial crossing time. The instrument is dual-polarized and is tilted forward approximately 21°. The arms of SMOS are about 4.5 m long and accommodate 27 antennas spaced approximately 0.8 wavelengths apart. The SMOS brightness temperature sensitivity will be 0.8–2.2 K, with 3 K accuracy. The sensor will obtain footprint resolution of 30–90 km, depending on the location of the pixel in the field of view.

3.3 Ocean Salinity Soil Moisture Integrated Radiometer-radar Imaging System

The Ocean Salinity Soil Moisture Integrated Radiometer-radar Imaging System (OSIRIS) is being developed at the Jet Propulsion Laboratory. OSIRIS uses modern mesh antennas to achieve the large aperture to measure SSS from space (Njoku et al. 1999). The large-aperture antenna is rotated about a vertical axis to obtain a "conical" scan (Figure 9). In a conical scan the ocean surface is observed at constant incidence. In principle, a constant incidence angle makes retrieval of salinity from brightness temperature easier than in across-track scanning, where the incidence angle varies. Also, if the scan angle is sufficiently large, the signals at vertical and horizontal polarization separate and can be used in the retrieval algorithm.

The OSIRIS configuration is comprised of a large mesh reflector with offset feed horns (Figure 9). Measurements are made at L- and S-bands, each with dual-polarization (V and H) and a polarimetric capability. In addition, the system is designed to include an

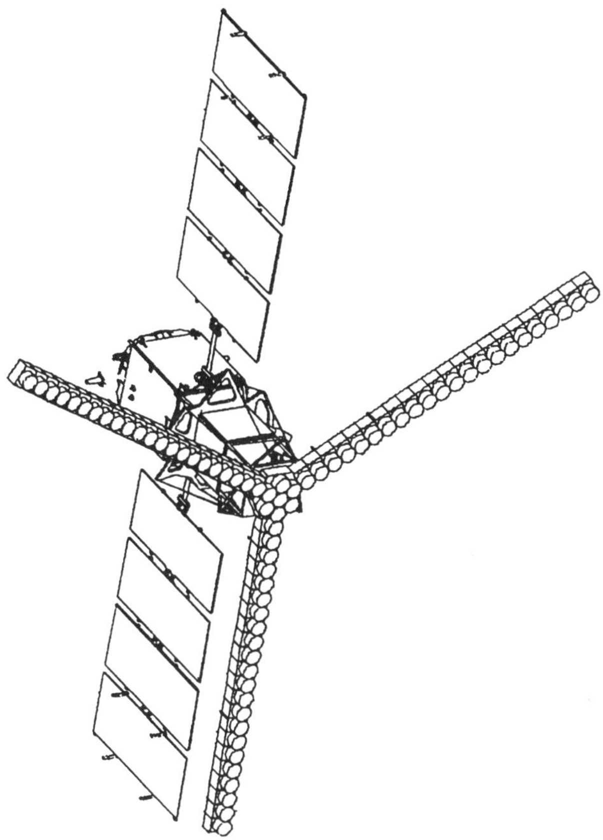

Figure 8. SMOS sensor in its fully deployed configuration.

L-band radar. The multichannel and active-plus-passive sensor approach is intended to provide corrections for SST, wind-induced surface roughness, and Faraday rotation. A contiguous scan (Figure 9) is obtained by adjusting the rotation rate of the antenna and using multiple feeds. The incidence angle of the conical scan is 40° at the surface. An aircraft sensor system with L- and S-band radiometers and L-band radar was developed to evaluate the potential of this technique for retrieval of SSS data. Initial flight tests of the aircraft instrument were made during the summer of 1999.

3.4 Hydrosat

A concept originally proposed about 15 years ago (Swift and MacIntosh 1983), Hydrosat is a viable candidate for remote sensing of SSS from space. In this concept, the antenna rotates about an axis in the direction of motion, which results in an across-track imaging mode (Figure 10). This configuration was originally proposed to correct for

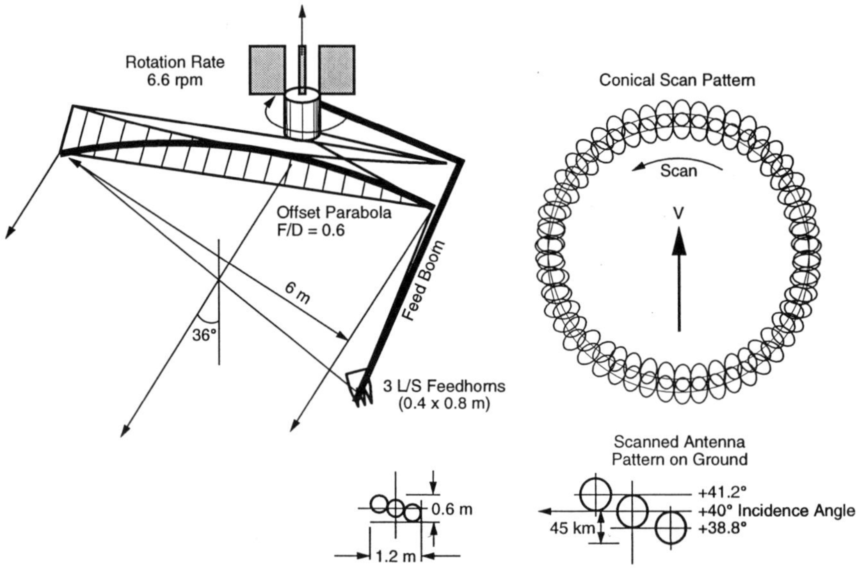

Figure 9. Proposed OSIRIS sensor, with (left) deployed configuration and (right) antenna scan pattern on the ground.

galactic radiation scattered from the ocean into the radiometer and it has other advantages for calibration.

Hydrosat employs an offset parabolic reflector antenna. The antenna spins about the velocity vector of the spacecraft and slip rings connect the antenna to power and a data recorder. The rotation rate of the antenna is adjusted so that a complete scan is obtained in the time required for the spacecraft to move forward by one footprint. Two antennas are required to reduce the rotation rate needed to obtain a contiguous scan (Swift and MacIntosh 1983). This scanning procedure assists in calibration because, during a substantial portion of the rotation cycle of the antenna, the radiometer will be viewing a constant cold-sky background of 2.7 K. The cold sky serves as one of two calibration targets. An option for the second reference is an internal matched load of known temperature employed in a manner similar to that in a Dicke radiometer. Another option is to inject noise while the antenna is viewing cold sky. This procedure is independent of errors introduced by thermal gradients, and Tanner (1998) has shown that the equivalent output temperature of modern noise diodes offers a stable reference source. The noise output of the diode can be monitored by observing targets whose brightness temperature can be readily calculated (e.g., the Great Lakes).

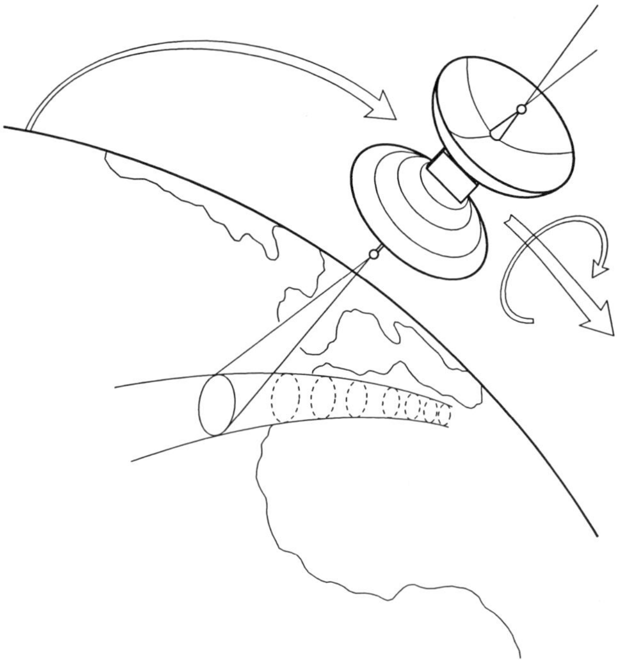

Figure 10. Hydrosat instrument configuration.

4. Conclusions

Significant progress has been made toward creating a sensor for remote sensing of SSS. The basic physics and feasibility of the technology have been demonstrated with aircraft sensors; practical sensor systems to monitor salinity globally from space have been proposed. However, important issues remain before we can move confidently toward remote sensing from space. The issues are different, depending on whether the application is for salinity in the coastal areas or the open ocean.

In coastal areas, the salinity gradient is relatively large. Aircraft research instruments such as ESTAR and SLFMR have demonstrated measurement accuracy on the order of 1 psu, which is adequate for many applications. In order to achieve this accuracy, radiometric stability and precision of 0.5 K is required, which is well within the state of the art. However, the requirements for spatial resolution (1–10 km) present a challenge. Achieving a spatial resolution from space of even 10 km is a major technological challenge at L-band because of the large antennas required. The instruments proposed to date, such as Hydrostar and OSIRIS, have resolutions of about 30 km. Research is under way to extend these technologies to achieve the 10-km goal.

In the open ocean, spatial resolution of about 100 km is probably adequate for many applications. This resolution does not pose a significant problem for antennas in space, even at L-band. However, the salinity gradient is relatively small in the open ocean and additional research is needed to determine if accuracy of 0.1–0.2 psu can be obtained. The technology to build a radiometer with sufficient accuracy (0.05–1.0 K) is probably adequate, but many questions remain about the retrieval algorithm. The biggest unknown is the effect of surface roughness and foam. The extent that roughness contributes to the emissivity of the surface, and the effect of roughness on the corrections needed for the galactic background, the cosmic background, and sun glint are not known. In addition, at this level of accuracy a correction will be needed for Faraday rotation, as will an accurate determination of sea surface temperature. Research is under way to address the algorithm issues.

Acknowledgments. Jim Zaitzeff passed away unexpectedly during the preparation of this manuscript. The manuscript was in large part his idea and is an example of his importance to research on remote sensing of sea surface salinity. Jim was also an important factor in the development of the SLFMR. He will be missed.

References

Blume, H.-J. C., B. M. Kendall, and J. C. Fedors, Measurements of ocean temperature and salinity via microwave radiometry, *Bound. Layer Meteorol.*, *13*, 295–308, 1978.

Goodberlet, M. A., C. T. Swift, K. P. Kiley, J. L. Miller, and J. B. Zaitzeff, Microwave remote sensing of coastal zone salinity, *J. Coastal Res.*, *13*, 363–372, 1997.

Jackson, T. J., and D. M. Le Vine, Mapping surface soil moisture using an aircraft-based passive microwave instrument: Algorithm and example, *J. Hydrology*, *184*, 85–99, 1996.

Klein, L. A., and C. T. Swift, An improved model for the dielectric constant of sea water at microwave frequencies, *IEEE Trans. Antennas Prop.*, *25*, 104–111, 1977.

Lagerloef, G., C. Swift, and D. Le Vine, Sea surface salinity: The next remote sensing challenge, *Oceanogr.*, *8*, 44–50, 1995.

Lerner, R. M., and J. P. Hollinger, Analysis of 1.4-GHz radiometric measurements from Skylab, *Remote Sensing Environ.*, *6*, 251–269, 1977.

Le Vine, D. M., T. T. Wilheit, R. E. Murphy, and C. T. Swift, A multifrequency microwave radiometer of the future, *IEEE Trans. Geosci. Remote Sensing*, *27*, 193–199, 1989.

Le Vine, D. M., M. Kao, A. B. Tanner, C. T. Swift, and A. Griffis, Initial Results in the development of a synthetic aperture microwave radiometer, *IEEE Trans. Geosci. Remote Sensing*, *28*, 614–619, July, 1990.

Le Vine, D. M., A .J. Griffis, C. T. Swift, and T. J. Jackson, ESTAR: A synthetic aperture microwave radiometer for remote sensing applications, *Proc. IEEE*, *82*, 1787–1801, 1994.

Le Vine, D., M. Kao, R. Garvine, and T. Sanders, Remote sensing of ocean salinity: results from the Delaware coastal current experiment, *J. Atmos. Oceanic Tech.*, *15*, 1478–1484, 1998.

Levitus, S., R. Burgett, and T. P. Boyer, World Ocean Atlas 1994, Volume 3: Salinity, NESDIS 3, 99 pp, 1994.

Martin-Neira, M., and J. M. Goutoule, MIRAS—A two-dimensional aperture/synthesis radiometer for soil moisture and ocean salinity observations, *ESA Bull.*, *92*, 95–104, 1997.

Miller, J. L., M. A. Goodberlet, and J. B. Zaitzeff, Airborne salinity mapper makes debut in coastal zone, *Eos*, *Trans. Amer. Geophys. Un.*, *79*, 173 and 176–177, 1998.

Munchow, A., and R. W. Garvine, Dynamical properties of a bouyancy-driven coastal current, *J. Geophys. Res.*, *98*, 20063–20077, 1993.

Njoku, E. G., W. J. Wilson, S. H. Yueh, and Y. Rahmat-Samii, A large-antenna microwave radiometer-scatterometer concept for ocean and land sensing, *IEEE Trans. Geosci. Remote Sensing*, submitted, 1999.

Reynolds, R. W., D. Behringer, M. Ji, A. Leetma, C. Maes, F. Vossepoel, and Y. Xue, Analyzing the 1993–1998 interannual variability of NCEP model ocean simulations: The contribution of TOPEX/Poseidon observations, In *Satellites, Oceanography and Society*, edited by D. Halpern, Elsevier Science Publishers, New York, this book, 2000.

Swift, C. T., and R. E. McIntosh, Considerations for microwave remote sensing of ocean-surface salinity, *IEEE Trans. Geosci. Remote Sensing*, *21*, 480–491, 1983.

Swift, C. T., ESTAR—The Electronically Scanned Thinned Array Radiometer for remote sensing measurement of soil moisture and ocean salinity, NASA Technical Memorandum 4523, 40 pp, 1993.

Tanner, A., Development of a high stability water vapor radiometer, *Radio Sci.*, *33*, 449–462, 1998.

Thomann, G. C., Experimental results of the remote sensing of sea-surface salinity at 21 cm wavelength, *IEEE Trans. Geosci. Electr.*, *14*, 198–214, 1976.

Thompson, A. R., J. M. Moran, and G. W. Swenson, *Interferometry and Synthesis in Radio Astronomy*, John Wiley, New York, 1986.

David Le Vine, Code 975, Goddard Space Flight Center, NASA, Greenbelt, MD 20771, U.S.A. (email, dmlevine@priam.gsfc.nasa.gov; fax, +1-301-614-5558)

Appendix I

List of Acronyms

ADEOS	Advanced Earth Observation Satellite
AERONET	Aerosol Robotic Network
AIM	Atlantic Isopycnic Model
AMI	Active Microwave Instrument
AMUSE	A Mediterranean Undercurrent Seeding Experiment
AOT	aerosol optical thickness
ARIES	Australian Resource Information and Environmental Satellite
ASAR	Advanced SAR
ATOC	Acoustic Thermometry of Ocean Climate
ATSR	Along-Track Scanning Radiometer
AVHRR	Advanced Very-High Resolution Radiometer
AVISO	Archiving, Validation, and Interpretation of Satellite Oceanographic Data
BEST	Benguela Sources and Transport Experiment
BIO	Bedford Institute of Oceanography
CANIGO	Canary Islands Açores Gibraltar Observations
CCSR	Center for Climate System Research
CDOM	colored dissolved organic matter
CLIVAR	Climate Variability and Predictability
CNES	Centre National d'Etudes Spatiales
CPUE	catch per unit effort
CTD	conductivity-temperature-depth
CZCS	Coastal Zone Color Scanner
DLR	German Aerospace Center
DMSP	Defense Meteorological Satellite Program
ECMWF	European Centre for Medium-Range Weather Forecasts
ENSO	El Niño Southern Oscillation
ENVISAT	Environmental Satellite
EOF	empirical orthogonal function
ERS	European Remote-sensing Satellite
ESA	European Space Agency
ESTAR	Electronically Scanned Thinned Array Radiometer
FDP	fast delivery product
FFT	fast Fourier transform

GEOS	Goddard Earth Observing System
Geosat	geodetic satellite
GFDL	Geophysical Fluid Dynamics Laboratory
GFO	Geosat follow-on
GLI	Global Imager
GODAE	Global Ocean Data Assimilation Experiment
GOES	Geostationary Operational Environmental Satellite
GSFC	Goddard Space Flight Center
HNLC	high-nutrient, low-chlorophyll
HRPT	high-resolution picture transmission
ICSOS	International Conference on Satellites, Oceanography and Society
IES	inverted echo sounder
IfM	Institut für Meereskunde
IMARPE	Instituto del Mar de Peru
IMGA	Instituto per lo studio delle Meteodologie Geofisiche Ambientali
IOC	Intergovernmental Oceanographic Commission
IR	infrared
IRS	Indian Remote-sensing Satellite
JD	Julian day
JPL	Jet Propulsion Laboratory
LEWEX	Labrador Extreme Waves Experiment
LIMEX	Labrador Ice Margin Experiment
MARSEN	Marine Remote Sensing project
MERIS	Medium-Resolution Imaging Spectrometer
METEOSAT	Meteorological Satellite
MHI	Marine Hydrophysical Institute
MIPE	Ministerio de Pesca
MIRAS	Microwave Imaging Radiometer with Aperture Synthesis
MODIS	Moderate-Resolution Imaging Spectroradiometer
MORENA	Multidisciplinary Oceanographic Research in the Eastern Boundary of the North Atlantic
MOS	Multispectral Optical Sensor
NAO	North Atlantic Oscillation
NASA	National Aeronautics and Space Administration
NASDA	National Space Development Agency of Japan
NatMIRC	National Marine Information and Research Center
NCEP	National Centers for Environmental Prediction

NOAA	National Oceanic and Atmospheric Administration
NORCSEX	Norwegian Continental Shelf Experiment
NSCAT	NASA Scatterometer
NWP	numerical weather prediction
OCTS	Ocean Color and Temperature Scanner
OI	optimal interpolation
OLS	Operational Linescan System
OSIRIS	Ocean Salinity Soil Moisture Integrated Radiometric Imaging System
POLDER	Polarization and Directionality of the Earth's Reflectance
RAMSES	Radiometrie Appliquée la Measure de la Salinité et de l'Eau du Sol
RAR	real-aperture radar
SAGE	Stratospheric Aerosol and Gas Experiment
SAR	synthetic aperture radar
SAXON-FPN	Synthetic Aperture Radar and X-Band Nonlinearities—Forschungsplatform Nordsee
SCOR	Scientific Committee on Oceanic Research
SeaWiFS	Sea-viewing Wide Field-of-view Sensor
SIR	Shuttle Imaging Radar
SLA	sea level anomaly
SLFMR	Scanning Low Frequency Microwave Radiometer
SMMR	Scanning Multichannel Microwave Radiometer
SMOS	Soil Moisture Ocean Salinity
SOFAR	SOund Fixing And Ranging
SSH	sea surface height
SSIWG	Salinity Sea Ice Working Group
SSMI	Special Sensor Microwave Imager
SSS	sea surface salinity
SST	sea surface temperature
SWADE	Surface Wave Dynamics Experiment
SWM	SAR wave mode
TIROS	Television and Infrared Observation Satellite
TOA	top-of-the-atmosphere
T/P	TOPEX/Poseidon
TOGA	Tropical Oceans Global Atmosphere
TOPEX	Topography Experiment
TOMS	Total Ozone Mapping Spectrometer
TAO	Tropical Atmosphere Ocean

UT	Universal Time
VTPR	Vertical Temperature Profile Radiometer
WAM	Wave Model
WCRP	World Climate Research Programme
WiFS	Wide Field Sensor
WOCE	World Ocean Circulation Experiment
XBT	expendable bathythermograph

Appendix II

Program

International Conference on Satellites, Oceanography and Society

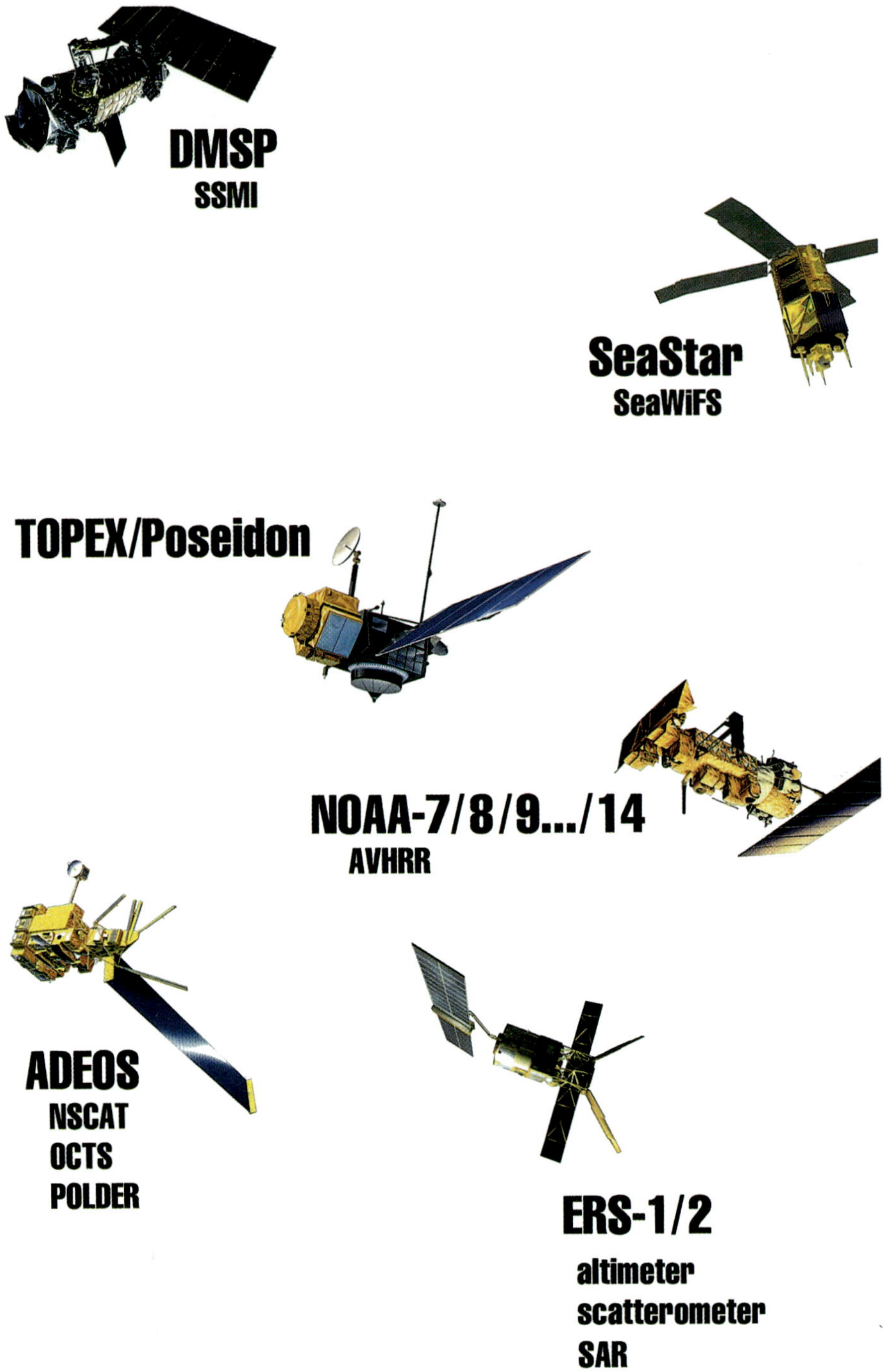
DMSP
SSMI
SeaStar
SeaWiFS
TOPEX/Poseidon
NOAA-7/8/9.../14
AVHRR
ADEOS
NSCAT
OCTS
POLDER
ERS-1/2
altimeter
scatterometer
SAR

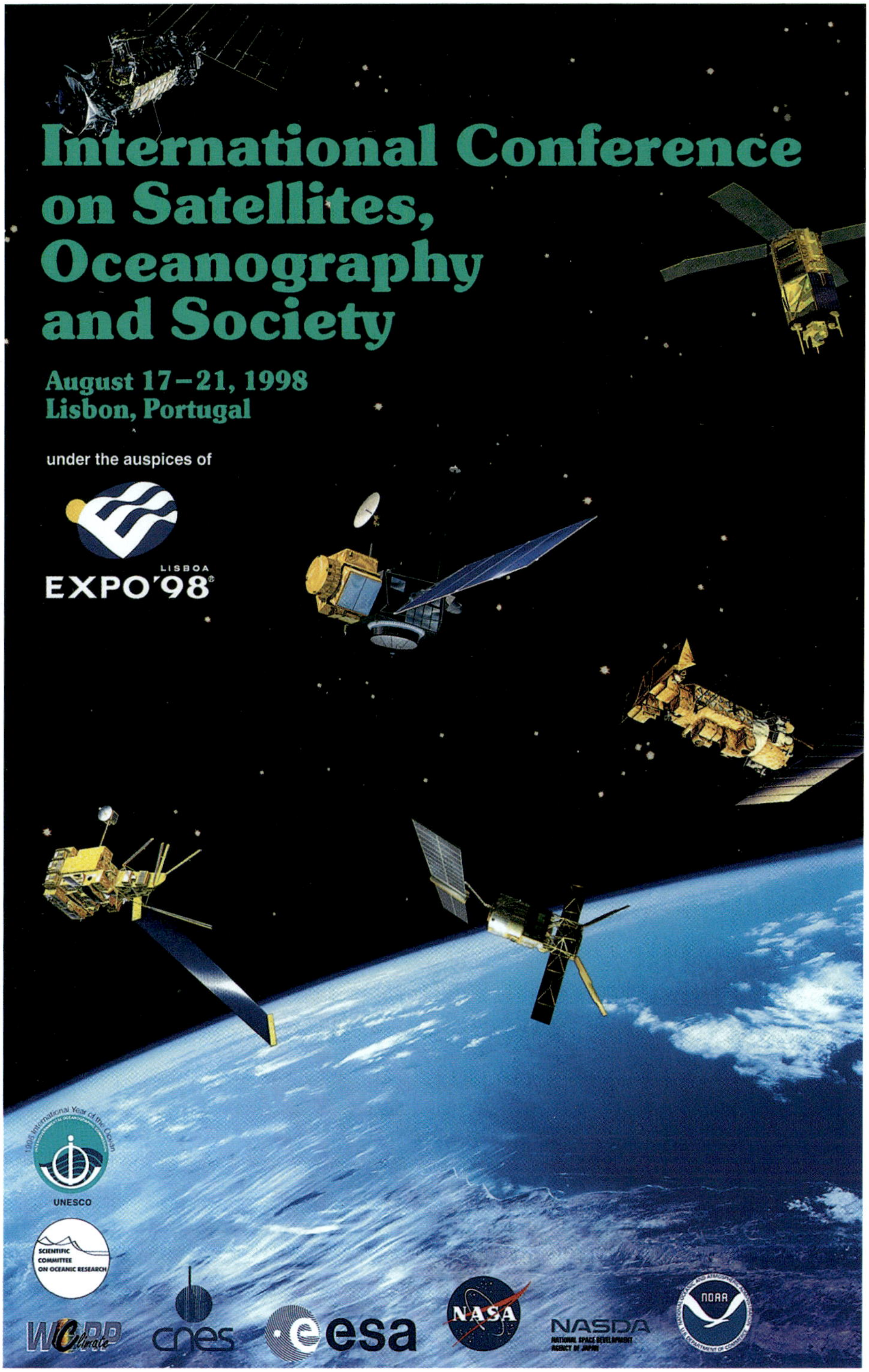
International Conference
on Satellites,
Oceanography
and Society
August 17–21, 1998
Lisbon, Portugal
under the auspices of
LISBOA
EXPO'98
UNESCO
SCIENTIFIC COMMITTEE ON OCEANIC RESEARCH
cnes
esa
NASA
NASDA
NOAA

International Conference on Satellites, Oceanography and Society

August 17–21, 1998—Lisbon, Portugal

Sponsors

Centre National d'Etudes Spatiales (CNES)
European Space Agency (ESA)
EXPO '98
Intergovernmental Oceanographic Commission (IOC)
National Aeronautics and Space Administration (NASA)
National Oceanic and Atmospheric Administration (NOAA)
National Space Development Agency of Japan (NASDA)
Scientific Committee on Oceanic Research (SCOR)
World Climate Research Programme (WCRP)

ICSOS Executive Council

David Halpern, Chair, JPL, Pasadena, USA
Guy Duchossois, ESA, Paris, France
Hartmut Grassl, WCRP, Geneva, Switzerland
Elizabeth Gross, SCOR, Baltimore, USA
Gunnar Kullenberg, IOC, Paris, France
Nancy Maynard, NASA, Washington, USA
Mario Ruivo, EXPO '98, Lisbon, Portugal
Norio Saito, NASDA, Tokyo, Japan
Robert Winokur, NOAA, Washington, USA
Raymond Zaharia, CNES, Paris, France

ICSOS Scientific Organizing Committee

David Halpern, Chair, JPL, Pasadena, USA
Werner Alpers, IfM, Hamburg, Germany
Antonio Busalacchi, GSFC, Greenbelt, USA
Vincent Cardone, Oceanweather, Cos Cob, USA
Peter Janssen, ECMWF, Reading, UK
David McAdoo, NOAA, Washington, USA
Nadia Pinardi, IMGA, Bologna, Italy
Trevor Platt, BIO, Dartmouth, Canada
Akimasa Sumi, CCSR, Tokyo, Japan
Toshio Yamagata, University of Tokyo, Tokyo, Japan
James Yoder, URI, Narragansett, USA

Foreword

The International Conference on Satellites, Oceanography and Society (ICSOS) will be held 17–21 August in Lisbon, Portugal, under the auspices of EXPO '98. Why have an ICSOS and how was ICSOS created? In December 1994 the United Nations General Assembly proclaimed 1998 to be the "International Year of the Ocean." EXPO '98, a world exposition in Lisbon from 22 May to 30 September 1998, will be a meeting point for nations and cultures. The EXPO '98 theme, "Oceans: A Heritage for the Future," will alert the public to the importance of the oceans as critical resources for sustainable development. Approximately 150 countries and organizations will be represented at EXPO '98, which also commemorates the 500th anniversary of the historic voyage to India by Vasco de Gama.

Twenty years ago witnessed the dawn of a new era in ocean sciences. Seasat demonstrated that sea surface height topography and surface wind velocity could be measured from a satellite. Nimbus-7 showed how ocean color measurements provided estimates of phytoplankton. TIROS-N carried the first instrument to adequately record sea surface temperature. The belief was established that satellite observations of radiation at microwave, visible, and infrared frequencies would be beneficial to all. Within 20 years, the internationalization of space missions for oceanography has become de rigueur.

The idea to celebrate the twentieth anniversary of Seasat, Nimbus-7, and TIROS-N with an international conference began in May 1995. Sponsors were found, and two committees, the Executive Council (EC) and Scientific Organizing Committee (SOC), helped organize ICSOS. I am indebted to our Sponsors, whose names are listed on the previous page, for their generous support. I am grateful to members of the EC and SOC, whose names are listed on the previous page, for their humor, kindness, thoughtfulness, and ideas. ICSOS aims to highlight the importance of ocean measurements from satellites, to examine how satellite observations of the ocean contribute to advancement of marine science for the benefit of society, and to encourage the development of valuable marine services made possible by satellite ocean measurements.

ICSOS has an exciting program, both daytime and in the evening. During the day, scientific sessions will be held at the Centro Cultural de Belém (CCB). The CCB is located next to the Tagus River and near the Monument to the Discoveries, the sixteenth century Belém Tower, and the sixteenth century Hieronymite Monastery. To foster further opportunities for the exchange of information between colleagues, old and new, we have programs for four of the five evenings. The Opening and Closing Ceremonies will be held in the internationally-acclaimed Ocean Pavilion at EXPO '98.

In closing, on behalf of the Sponsors, Executive Council, and Scientific Organizing Committee, I welcome you to ICSOS and I trust that you will find opportunities for enrichment. Thank you for your help to make ICSOS a historic occasion.

David Halpern
Chairman, ICSOS Executive Council
and Chairman, ICSOS Scientific Organizing Committee

August 1998

Synopsis of Sessions

Monday 17 August

	Bus Transportation: Lisbon Hotels to Centro Cultural de Belém (CCB)
0945	**Welcome**
1000	**Oral Session 1:** Interannual-To-Decadal Climate Prediction
1100	Refreshments
1130	**Oral Session 1:** Continued
1300	Lunch
1430	**Poster Session 1:** Interannual-To-Decadal Climate Prediction
1545	Refreshments
1545	**Poster Session 1:** Continued
1700	Bus Transportation: CCB to EXPO '98 Sea Entrance
1730	Enter EXPO '98 via Sea Entrance [Tickets Provided to Registrants]
1800	**ICSOS Opening Ceremony:** EXPO '98 Ocean Pavilion — Reception, Viewing Oceanarium
2030	EXPO '98
2200	Bus Transportation: EXPO '98 Sea Entrance to Lisbon Hotels

Tuesday 18 August

	Bus Transportation: Lisbon Hotels to Centro Cultural de Belém (CCB)
0930	**Oral Session 2:** Seasonal-To-Interannual Climate Prediction
1100	Refreshments
1130	**Oral Session 2:** Continued
1300	Lunch
1430	**Poster Session 2:** Seasonal-To-Interannual Climate Prediction
1545	Refreshments
1545	**Poster Session 2:** Continued
1700	Bus Transportation: CCB to Lisbon Hotels
2000	**A Night At EXPO '98** [Tickets Provided To Registrants]

Wednesday 19 August

	Bus Transportation: Lisbon Hotels to Centro Cultural de Belém (CCB)
0930	**Oral Session 3:** Living Resources Assessment And Prediction
1100	Refreshments
1130	**Oral Session 3:** Continued
1300	Lunch
1430	**Poster Session 3:** Living Resources Assessment And Prediction
1545	Refreshments
1545	**Poster Session 3:** Continued
1700	Bus Transportation: CCB to Lisbon Hotels
1930	**ICSOS Banquet,** Ritz Four Seasons Hotel [Tickets Must Be Purchased In Advance]

Thursday 20 August

	Bus Transportation: Lisbon Hotels to Centro Cultural de Belém (CCB)
0930	**Oral Session 4:** Weather and Wave Prediction
1100	Refreshments
1130	**Oral Session 4:** Continued
1300	Lunch
1430	**Poster Session 4:** Weather and Wave Prediction
1430	**Poster Session 5:** New Directions
1545	Refreshments
1545	**Poster Session 4:** Continued
1545	**Poster Session 5:** Continued
1700	Bus Transportation: CCB to Lisbon Hotels

Friday 21 August

	Bus Transportation: Lisbon Hotels to Centro Cultural de Belém (CCB)
0930	**Oral Session 5:** Geophysical Exploration
1100	Refreshments
1120	**Oral Session 6:** New Directions
1300	Lunch
1430	**Panel Discussion:** A Vision Of The Future
1630	Refreshments
1700	Bus Transportation: CCB to EXPO '98 Sea Entrance
1730	Enter EXPO '98 via Sea Entrance [Tickets Provided to Registrants]
1800	**ICSOS Closing Ceremony:** EXPO '98 Ocean Pavilion — Reception
2000	EXPO '98
2200	Bus Transportation: EXPO '98 Sea Entrance to Lisbon Hotels

Program

Monday 17 August

0945 Welcome

Chairperson: David Halpern (Jet Propulsion Laboratory, Pasadena, USA)

1000 Oral Session 1: Interannual-To-Decadal Climate Prediction

Chairpersons: Y. Menard (Centre National d'Etudes Spatiale, Toulouse, France) and F. Schott (Institut für Meereskunde, Kiel, Germany)

1000 Is The Ocean Predictable Over Decades, And How Might One Find Out?
C. Wunsch, Massachusetts Institute of Technology, Cambridge, USA

Monday 17 August (continued)

1030 Acoustic Thermometry Of Ocean Climate (ATOC)
W. Munk, Scripps Institution of Oceanography, La Jolla, USA; and ATOC Group

1100 Refreshments

1130 Studies Of Ocean Circulation Using Satellite Technology As A Method For Tracking Floats And Drifters
R. L. Molinari, NOAA Atlantic Oceanographic and Meteorological Laboratories, Miami, USA; and M. Bushnell, D. Hansen, and M. Swenson

1200 Decadal And Interdecadal Climate Events And Their Impact On The Ocean Circulation In The Indo-Pacific Sector
T. Yamagata, Department of Earth and Planetary Physics, University of Tokyo, Tokyo, Japan; and T. Kagimoto, Y. Masumoto, and P. N. Vinayachandran

1230 Seasonal, Interannual And Decadal Variability Of The Mediterranean Sea Marine Ecosystem And Its Predictability
N. Pinardi, IMGA, Bologna, Italy; and P. Y. Le Traon, N. Ayoub, P. De Mey, and M. Zavatarelli

1300 Lunch

1430 Poster Session 1: Interannual-To-Decadal Climate Prediction
(With Refreshments At 1545)

Chairpersons: Y. Menard (Centre National d'Etudes Spatiale, Toulouse, France) and F. Schott (Institut für Meereskunde, Kiel, Germany)

P1.1 Global Patterns Of Atmospheric Variability Associated With Oceanic Forcing
J. M. Castanheira, Department of Physics, University of Aveiro, Aveiro, Portugal; and C. C. DaCamara

P1.2 A Satellite-Based Study Of Rossby Wave Properties Influencing Climate And Weather Patterns
D. Cromwell, James Rennel Division for Ocean Circulation, Southampton Oceanography Centre, Southampton, UK; and P. Cipollini, P. G. Challenor, and G. D. Quartly

P1.3 Precipitation Over Portugal: Influences Of SST Forcing And The North Atlantic Oscillation
J. A. M. Corte-Real, ICAT, Faculty of Sciences, University of Lisbon, Lisbon, Portugal; and J. P. Diegues and J. G. Pinto

P1.4 Observations Of Indonesian Throughflow Variations And Investigation Of Their Role On ENSO Signals Simulated By An Indian Ocean Model And A Pacific Coupled Ocean-Atmosphere Model
P. Florenchie, Jet Propulsion Laboratory, California Institute of Technology, Pasadena, USA; and C. Perigaud and P. Garnier

Monday 17 August (continued)

P1.5 Combining Altimeter Observations And Oceanographic Data For Climate And Weather
S. L. Garzoli, NOAA Atlantic Oceanographic and Meteorological Laboratories, Miami, USA; and G. Goni

P1.6 Real-Time Altimeter Data Assimilation Experiments In An Eastern North Atlantic Monitoring And Prediction System
S. Giraud, CLS Space Oceanography Division, Toulouse, France; and E. Dombrowsky and P. Bahurel

P1.7 Decadal Climate Change And Satellite Measurement In Korean Waters
S. D. Hahn, Fisheries Oceanography Division, National Fisheries Research and Development Institute, Busan, Korea

P1.8 Using Satellite Remote Sensing Data In The Numerical Modeling Of The Pacific Ocean Circulation
A. Martynov, Institute of Computational Mathematics and Mathematical Geophysics, Russian Academy of Sciences, Novosibirsk, Russia; and V. Moiseyev

P1.9 Study Of Surface Heat Fluxes Over Tropical Indian Ocean And Its Impact on Monsoon Rainfall Over Indian Subcontinent From Satellite Observations
P. M. Muraleedharan, National Institute of Oceanography, Goa, India; and T. Pankajakshan

P1.10 Does The Indonesian Throughflow Affect ENSO Dynamics?
R. Murtugudde, Laboratory for Hydrospheric Processes, NASA Goddard Space Flight Center, Greenbelt, USA; and E. Hackert, A. Busalacchi, and J. Beauchamp

P1.11 Tracking Mediterranean Salt Lenses With In-Situ And Satellite Observations For Monitoring Lagrangian Heat And Salt Sources
P. B. Oliveira, Instituto de Oceanografia & Departmento de Fisica, Universidade de Lisboa, Lisboa, Portugal; and N. Serra, A. F. G. Fiúza, and I. Ambar

P1.12 Interannual And Decadal Variability In The Coupled Ocean-Atmosphere System
A. Polonsky, Marine Hydrophysical Institute, Sevastopol, Ukraine

P1.13 Rainfall Variability Over The Tropical Atlantic From Two Satellite Data Sets
J. Servain, Centre ORSTOM de Brest, Plouzane, France; and A. Faraj

P1.14 The Monitoring Of The Incident Radiation Measurement (1972–1997) As An Indicator Of Air-Sea Interactions In The Oceans
S. Silpipat, Department of Fisheries, Samutprakran, Thailand; and A. Chanpongsang, W. Meecharoen, S. Sornkrut, S. Charoenvong, G. D. Sharp, J. G. Norton, and D. R. McLain

P1.15 Decadal Climate Oscillations In The North Atlantic: Observation And Simulation
M. Watanabe, Center for Climate System Research, University of Tokyo, Tokyo, Japan; and M. Kimoto

Monday 17 August (continued)

P1.16 Long Waves In The Arabian Sea From Geosat Altimeter Observations
C. V. K. Prasad Rao, Naval Physical and Oceanographic Laboratory, Cochin, India

P1.17 A Correlation Between Solar Activity And Sea Surface Temperature In Indonesia
W. Sinambela, Solar and Space Environment Division, National Institute Aeronautics and Space, Bandung, Indonesia; and W. E. Cahyono

1800 ICSOS Opening Ceremony, EXPO '98 Ocean Pavilion
[Tickets Provided For Registrants]

1800 M. Gago (Minister of Science and Technology, Lisbon, Portugal); A.V. Diaz (Director, NASA Goddard Space Flight Center, Greenbelt, USA); G. Duchossois (Head, Earth Observation Office, European Space Agency, Paris, France); W. Munk (Scripps Institution of Oceanography, La Jolla, USA)

1900 Reception And Viewing The Oceanarium

2030 Sessions End For The Day

Tuesday 18 August

0930 Oral Session 2: Seasonal-To-Interannual Climate Prediction

Chairpersons: D. Anderson (European Centre for Medium-Range Weather Forecasts, Reading, UK) and G. Philander (Department of Geosciences, Princeton University, Princeton, USA)

0930 Advances In Seasonal-To-Interannual Climate Prediction: Past, Present, And Future Prospects For Ocean Remote Sensing
A. Busalacchi, NASA Goddard Space Flight Center, Greenbelt, USA

1000 ENSO Mechanisms And Satellite Measurements
J. Picaut, ORSTOM and NASA Goddard Space Flight Center, Greenbelt, USA

1030 Forecasting The 1997–1998 El Niño: The Contribution Of Satellite Observations
R. W. Reynolds, NOAA National Centers for Environmental Prediction, Camp Springs, USA; and D. Behringer, M. Ji, A. Leetmaa, F. Vossepoel, and Y. Xue

1100 Refreshments

1130 Satellite Monitoring For The Season-To-Interannual Climate Fluctuations, Such As 30- To 60-Day Fluctuations And ENSO
A. Sumi, Center for Climate System Research, University of Tokyo, Tokyo, Japan; and T. Nakazawa

1200 Reporting Climate Predictions To The Public
J. M. Nash, TIME Magazine, Chicago, USA

Tuesday 18 August (continued)

1230 Why Care About Season-To-Interannual Forecasts?
M. H. Glantz, National Center for Atmospheric Research, Boulder, USA

1300 Lunch

1430 Poster Session 2: Seasonal-To-Interannual Climate Prediction
(with refreshments at 1545)

Chairpersons: D. Anderson (European Centre for Medium-Range Weather Forecasts, Reading, UK) and G. Philander (Department of Geosciences, Princeton University, Princeton, USA)

P2.1 Monitoring The 1997/1998 El Niño With ERS-2 Radar Altimeter And ATSR
J. Benveniste, European Space Agency - ESRIN, Frascati, Italy; and K. Cardon, P. Goryl, and R. Scharroo

P2.2 High-Resolution Satellite-Derived Sea Surface Temperature Variability Over the Gulf Of Maine And Georges Bank Region, 1993–1996
J. J. Bisagni, Center for Marine Science & Technology, University of Massachusetts, North Dartmouth, USA; and K. W. Seemann

P2.3 Use Of The Satellite Picture At Arabian Sea, Red Sea And Indian Ocean For Weather Prediction On Different Seasons Of Ethiopia
T. Dejene, National Meteorological Service Agency, Addis Ababa, Ethiopia

P2.4 Equatorial Waves And Warm Pool Displacement During The 1992–1997 ENSO Events
T. Delcroix, Groupe SURTROPAC, Centre ORSTOM de Noumea, Noumea, New Caledonia; and F. Masia, J. Picaut, Y. duPenhoat, and B. Dewitte

P2.5 Understanding El Niño 1997–1998 And Its Impact On Society Via Satellites, In-Situ Data, and Dynamical Models
A. Gangopadhyay, Center for Marine Science and Technology, University of Massachusetts, North Dartmouth, USA

P2.6 ENSO Simulation And Prediction With An Intermediate Coupled Model Of The Tropical Pacific: Application To The Major 1997–98 El Niño
D. Gushchina, Meteorological Department, Moscow State University, Moscow, Russia; and B. Dewitte, G. Reverdin, Y. duPenhoat, R. Abarca, and S. Lakeev

P2.7 Multi-Scale, Multi-Resolution Analysis For Coastal Resource Management
J. Helly, San Diego Supercomputer Center, University of California, La Jolla, USA

P2.8 The Tropical Pacific Warm Pool–Cold Tongue As A Dynamical System: Inference From Satellite Observations
W. K. M. Lau, Climate and Radiation Branch, NASA Goddard Space Flight Center, Greenbelt, USA

P2.9 Coupled Data Assimilation And ENSO Prediction
T. Lee, Jet Propulsion Laboratory, California Institute of Technology, Pasadena, USA; and L.-L. Fu, A. Foo, J. P. Boulanger, and R. Giering

Tuesday 18 August (continued)

P2.10 Atmospheric Signals During Onset Period Of 1997 ENSO From ADEOS/NSCAT Data
T. Nakazawa, Meteorological Research Institute, Tsukuba, Japan

P2.11 Simulating And Forecasting El Niño With Intermediate Coupled Ocean-Atmosphere Model And In-Situ/Satellite Data
C. Perigaud, Jet Propulsion Laboratory, California Institute of Technology, Pasadena, USA; and C. Cassou and F. Melin

P2.12 When Do Satellite Measurements Help In Forecasting El Niño?
P. S. Schopf, Institute for Computational Sciences and Infomatics, George Mason University and Center for Ocean-Land-Atmosphere Studies, Calverton, USA; and B. Kirtman

P2.13 The Exceptional Summer 1997 In The Baltic—Warmest August, Oder Flood And Cyanobacteria Bloom
H. Siegel, Baltic Sea Research Institute, Warnemunde, Germany; and M. Gerth

P2.14 Coupling Of New Production And Air-Sea Exchange Of CO_2 In The Western Equatorial Pacific
D. Turk, Department of Oceanography, Dalhousie University, Halifax, Canada; and T. Kawano and I. Asanuma

P2.15 Mechanisms Of Interannual CO_2 Flux In The Equatorial Pacific Ocean
M. A. Verschell, Laboratory for Hydrospheric Processes, NASA Goddard Space Flight Center, Greenbelt, USA; and A. Busalacchi

P2.16 The Role Of Wind Forcing On The Variabilities In Equatorial Pacific Ocean
X. H. Yan, Graduate College Of Marine Studies, University of Delaware, Newark, USA; and Y. He and W. T. Liu

P2.17 Spacebased Observation Of Surface Momentum, Thermal, And Hydrologic Forcing And Their Contribution To The Understanding Oceanic Seasonal To Interannual Variabilities
W. T. Liu, Jet Propulsion Laboratory, California Institute of Technology, Pasadena, USA

1645 Sessions End For The Day

2000 A Night At EXPO '98 [Tickets Provided For Registrants]

Wednesday 19 August

0930 Oral Session 3: Living Resources Assessment And Prediction

Chairpersons: R. Barber (Marine Laboratory, Duke University, Beaufort, USA) and T. Saino (Institute of Hydrospheric-Atmospheric Sciences, Nagoya University, Nagoya, Japan)

Wednesday 19 August (continued)

0930 How Satellite Observation Contributes To The Knowledge Of Oceanic Algal Biomass At Global Scale
A. Morel, Laboratoire de Physique et Chimie Marines, Universite Pierre et Marie Curie, Villefranche-sur-Mer, France

1000 Temporal And Spatial Pattern In Ocean Color Imagery At Regional To Global Scales
J. A. Yoder, Graduate School of Oceanography, University of Rhode Island, Narragansett, USA

1030 Quantifying And Understanding Primary Production Of The Global Ocean
T. Platt, Bedford Institute of Oceanography, Dartmouth, Canada

1100 Refreshments

1130 How Satellite Observation Contribute To Fisheries Management For Conservation and Sustainable Use Of Marine Living Resources
S. Matsumura, National Research Institute of Far Seas Fisheries, Shimizu-shi, Japan

1200 How Satellite Observations Contribute To Coastal Ocean Management
A. F. G. Fiúza, Instituto de Oceanografia & Departmento de Fisica, Universidade de Lisboa, Lisbon, Portugal

1230 Lunch

1400 Poster Session 3: Living Resources Assessment And Prediction
(with refreshments at 1545)

Chairpersons: R. Barber (Marine Laboratory, Duke University, Beaufort, USA) and T. Saino (Institute of Hydrospheric-Atmospheric Sciences, Nagoya University, Nagoya, Japan)

P3.1 Functional Relationships Between Optical Signatures And Phytoplankton Production: Exploitation Of Remotely Sensed Data For Basin-Scale Models
J. Aiken, Plymouth Marine Laboratory, Plymouth, UK; and S. B. Hooker

P3.2 Region-Based Mapping Of Plankton Distribution In The Adriatic Sea With SeaWiFS Data
A. Baraldi; IMGA-CNR, Bologna, Italy; and F. Parmiggiani

P3.3 ASOMIR—Application Of Satellite Oceanographic In Studying The Madeira Island Region, Portugal
R. M. A. Caldeira, Department of Biology, University of California, Los Angeles, USA

P3.4 Remote Salinity and Ocean Color Monitoring And Linkages To Bio-Optical Characteristics: Aspects And Performance In A Coastal Environment
E. J. D'Sa, NOAA National Environmental Satellite, Data and Information Service, Washington, USA; J.B. Zaitzeff and R.G. Steward

Wednesday 19 August (continued)

P3.5 SST Images As Predictor Of Hake Availability In Namibian Waters
A. Gordoa, Centre d'Estudis Avançats de Blanes, Blanes, Spain; and M. Masó, L. Vogues, and O. Chic

P3.6 Using Satellites and Modeling To Monitor And Interpret The Physical Control Of Biological Productivity Off The Coast Of Peru During The 1997–1998 El Niño
M. E. Carr, Jet Propulsion Laboratory, California Institute of Technology, Pasadena, USA; and K. Broad

P3.7 Estimating New Production From Space—A Case Study In The North Pacific
J. I. Goes, National Institute of Oceanography, Dona Paula, India; and J. Ishizaka, D. Longjiang, and T. Saino

P3.8 Routine Monitoring Of Phytoplankton Concentrations For The Prediction Of Oceanic Heat Fluxes
J. P. Huot, Space System Environment Analysis Section, ESA/ESTEC, Noordwijk, The Netherlands; and P. Moreno

P3.9 Remote Sensing Applications For Island Ecosystem Management: Problems And Prospects For Lakshadweep, India
P.K. Dinesh Kumar, National Institute of Oceanography, Cochin, India

P3.10 Satellite Remote Sensing Applications To Fisheries Research
R. M. Laurs, NOAA-National Marine Fisheries Service, Honolulu, USA; and J. Polovina

P3.11 Using Satellite Remote Sensing To Determine Environmental Parameters Of Coastal Ocean Areas Associated With *Vibrio Cholerae* Ecology And Epidemiology
V. Louis, COMB, University of Maryland, Baltimore, USA; and A. Gil, B. Wood, C. Parkinson, A. Huq, and R. Colwell

P3.12 Pigment Variability In The South Atlantic Bight As Derived From Eight Years Of CZCS Data
A. M. Martins, Department of Oceanography and Fisheries, University of Azores, Azores, Portugal; and J. L. Pelegri

P3.13 Remote Sensed Studies Of Seasonal And Interannual Variations Of Surface Chlorophyll Concentration In The Black Sea
N. P. Nezlin, P. P. Shirshov Institute of Oceanology, Russian Academy of Sciences, Moscow, Russia

P3.14 Satellite Remote Sensing Applications In Ocean Resources Utilization And Management
S. Ramachandran, Institute for Ocean Management, Anna University, Madras, India

P3.15 Obtaining Indices To Predict The Presence Of Tunny Fish In The Mediterranean Using NOAA Images
A. Romo, Remote Sensing Laboratory, University of Valladolid, Valladolid, Spain

Wednesday 19 August (continued)

P3.16 The Temporal And Spatial Variability Of Aerosols Over The Ocean As Estimated From Space-Borne Sensors
P. M. Stegmann, Graduate School of Oceanography, University of Rhode Island, Narragansett, USA; and N. W. Tindale

P3.17 Chlorophyllous Tongues Off East Coast Of India As Seen By MOS B Ocean Color Sensor
T. Suresh, National Institute of Oceanography, Dona Paula, India; and E. Desa

P3.18 A Satellite Study Of Current-Generated Bedforms Around The Northeastern Tip Of South America: Oceanography And Management Of A Rich Continental Shelf
M. L. Vianna, Programa Oceanografia, Instituto Nacional de Pesquisas Espaciais, Sao Jose dos Campos, Brazil; and M. S. Santos and A. P. Cabral

P3.19 Environmental Monitoring Of Coastal Oceans For Bio-Optical Properties Using A NOAA Weather Platform
C. E. Woody, NOAA National Data Buoy Center, Stennis Space Center, USA; and G. F. Cota

P3.20 Airborne Remote Sensing Researches Of Marine Environmental Conditions And Biologoies Resources Along Barents Sea Shore
V. B. Zabavnikov, Knipovich Polar Research Institute of Marine Fishery and Oceanography, Murmansk, Russia; and V. I. Chernook

P3.21 Marine Pelagic Fish Stock Assessment And Productivity With Prospects On The Use Of Satellites In The Gulf Of Guinea
C. E. Gabche, IRAD, Research Station for Fisheries and Oceanography, Limbe, Cameroon

1645 Sessions End For The Day

1930 ICSOS Banquet, Ritz Four Seasons Hotel
[Tickets Must Be Purchased In Advance]

Thursday 20 August

0930 Oral Session 4: Weather And Wave Prediction

Chairpersons: S. Imawaki (Research Institute for Applied Mechanics, Kyushu University, Kasuga, Japan) and J. Johannessen (Earth Sciences Division, ESTEC, Noordwijk, The Netherlands)

0930 Satellite Retrievals Of Wave Heights And Wave Spectral Data And Their Impact On The Development And Operation Of Global Wave Models
K. Hasselmann, Max-Planck-Institut für Meteorologie, Hamburg, Germany

1000 Wave Modeling And Altimeter Wave Height Data
P. Janssen, European Centre for Medium-Range Weather Forecasts, Reading, UK

Thursday 20 August (continued)

1030 Societal Benefits Of Improved Surface Wave Prediction
V. J. Cardone, Oceanweather, Cos Cob, USA; and H. C. Chen and G. Z. Forristall

1100 Refreshments

1130 ICEWATCH: Operational Ice Nowcasts And Climate Monitoring By Synergetic Use Of Satellite Data And How Society Can Benefit From Improved Forecast
O. M. Johannessen, Nansen Environmental and Remote Sensing Center, Bergen, Norway

1200 The Use Of Satellite Surface Wind Data To Improve Weather Analysis And Forecasting
R. Atlas, Data Assimilation Office, NASA Goddard Space Flight Center, Greenbelt, USA

1230 Lunch

1400 Poster Session 4: Weather And Wave Prediction
(with refreshments at 1545)

Chairpersons: S. Imawaki (Research Institute for Applied Mechanics, Kyushu University, Kasuga, Japan) and J. Johannessen (Earth Sciences Division, ESTEC, Noordwijk, The Netherlands)

P4.1 Global Estimates Of Significant Wave Height Return Periods From Combined Geosat, ERS-1 And TOPEX/Poseidon Altimeter Measurements
J. H. G. M. Alves, School of Mathematics, University of New South Wales, Kensington, Australia; and I. R. Young and M. L. Banner

P4.2 Towards Medium Range Wave Forecasting
P. D. Cotton, Southampton Oceanography Centre, Southampton, UK; and P. G. Challenor and D. J. T. Carter

P4.3 Interpreting Satellite Imagery Of The Ocean For Australian Yacht Racers
G. Cresswell, Division of Marine Research, CSIRO, Hobart, Australia

P4.4 Sea Surface Temperature Monitoring Over The Atlantic Off The Coast Of Portugal
C. C. DaCamara, ICAT, University of Lisbon, Lisbon, Portugal; and T. J. Calado and J. Corte-Real

P4.5 Use of Landsat Thematic Mapper Data To Predict Oceanographic Situation In Saros Bay
C. Gazioglu, Institute of Marine Sciences and Management, University of Istanbul, Turkey; and E. Dogan and O. Muftuoglu

P4.6 Remote Sensing Of Mesoscale Variability In The Surface Layer Of The Black Sea
A. I. Ginzburg, P. P. Shirshov Institute of Oceanology, Russian Academy of Sciences, Moscow, Russia; and A. G. Kostianoy, D. M. Soloview, and S. V. Stanichny

Thursday 20 August (continued)

P4.7 The Assimilation of Altimeter Data Into An Ocean Wave Model
D. Greenslade, Bureau of Meteorology Research Centre, Melbourne, Australia

P4.8 Monitoring Of Ocean Waves With The ERS-1 SAR—From Global Spectral Wave Climatologies To Individual Wave System Evolution
P. Heimbach, Max-Planck-Institut für Meteorologie, Hamburg, Germany; and S. Hasselmann, K. Hasselmann, and E. Bauer

P4.9 Remote Sensing Of Mesoscale Dynamical Processes In The Coastal Zones
A. G. Kostianoy, P. P. Shirshov Institute of Oceanology, Russian Academy of Sciences, Moscow, Russia; and G. G. Boubnov

P4.10 Satellite-Derived Mesoscale Patterns In The Caspian Sea
H. İ. Sur, Institute of Marine Sciences, Middle East Technical University, Erdemli, Turkey

P4.11 Development Of The New Remote Sensing Methods: Measurement Of The Ocean Surface Tilts And The Wind Speed By Microwave Radar
V. Yu. Karaev, Institute of Applied Physics, Russian Academy Of Sciences, Nizhny Novgorod, Russia

1400 Poster Session 5: New Directions
(with refreshments at 1545)

Chairpersons: E. Lindstrom (Office of Earth Science, NASA, Washington, USA) and A. Moura (International Research Institute for Climate Prediction, Lamont-Doherty Earth Observatory, Palisades, USA)

P5.1 Why Observe Our Oceans From Space?
A. deCharon, Jet Propulsion Laboratory, California Institute Of Technology, Pasadena, USA; and F. Blanc and Y. Menard

P5.2 The ERS Scatterometer At The Service Of Science And Society
P. Lecomte, ERS Mission Coordination and Product Assurance Section, ESRIN/ESA, Frascati, Italy

P5.3 New Era In Oceanography: Not Just A Technical And Scientific Challenge But A Cultural Move That Requires Emergence Of A New Type Of Oceanographer, Or From Williamstown To Lisboa
M. Lefebvre, GOOS/OOPC, Villeneuve Tolosane, France; and P. Bahurel

P5.4 Scientific Experiment Of Ocean Color Imager On ROCSAT-1 Satellite
H. W. Li, Department of Oceanography, National Taiwan Ocean University, Keelung, Taiwan; and C. R. Ho, N. J. Kuo, and W. P. Tsai

P5.5 Marine Geophysics And Gravity From ERS Altimetry Over Polar Seas
D. C. McAdoo, NOAA National Oceanographic Data Center, Silver Spring, USA; and S. Laxon

Thursday 20 August (continued)

P5.6 A Case For Enhanced Dissemination Of Satellite Information To Developing Nations
L. O. Ouko, EdgeWise Communications, Nairobi, Kenya

P5.7 Present Status Of SeaWiFS And Activities Of The SeaWiFS Project
W. D. Robinson, SAIC/General Sciences Corporation, NASA Goddard Space Flight Center, Greenbelt, USA; and M. Darzi, C. R. McClain, and G. C. Feldman

P5.8 The Legacy To Mankind From The ADEOS Ocean Sensors
T. Tanaka, NASDA Earth Observation Research Center, Tokyo, Japan; and I. S. F. Jones

P5.9 Sea Surface Salinity—Toward An Operational Remote Sensing System
J. B. Zaitzeff, NOAA National Environmental Satellite, Data and Information Service, Washington, USA; and C. Swift, M. Goodberlet, D. Le Vine, J. L. Miller, and E. J. D'Sa

P5.10 Oceanography, Satellites, And The Society With Regards To The Problem Of Climate Change: What Outcome After The Kyoto Protocol
C. M. Bomba, CERDE, Yaounde, Cameroon

PS.11 The Role Of Satellite Imagery In Environmental Education For The Coastal Areas Of Mediterranean Europe
F. Papadimitriou, Department of Education, University of Athens, Ilissia, Athens, Greece

1645 Sessions End For The Day

Friday 21 August

0930 Oral Session 5: Geophysical Exploration

Chairpersons: D. McAdoo (NOAA National Oceanographic Data Center, Silver Spring, USA) and L. Mendes-Victor (Institute of Geophysics, University of Lisbon, Lisbon, Portugal)

0930 Geophysical Exploration Of The Ocean Basins With Ships And Satellites
D. T. Sandwell, Scripps Institution Of Oceanography, University of California, La Jolla, USA; and W. H. F. Smith and M. Paton

1000 Global Ocean Floor Topography From Satellite Altimetry
A. Cazenave, CNES Groupe de Recherche de Geodesie Spatiale, Toulouse, France

1030 Mapping Of Bathymetry By Synthetic Aperture Radars
G. J. Wensink, Advisory and Research Group on Geo Observation Systems and Sevices, Vollenhove, The Netherlands

Friday 21 August (continued)

1100 Refreshments

1130 Oral Session 6: New Directions

Chairpersons: E. Lindstrom (Office of Earth Science, NASA, Washington, USA) and A. Moura (International Research Institute for Climate Prediction, Lamont-Doherty Earth Observatory, Palisades, USA)

1130 Recent Progress Towards Satellite Measurements Of The Global Sea Surface Salinity Field
G. S. E. Lagerloef, Earth and Space Research, Seattle, USA

1200 Operational Applications of Global Ocean Satellite Data Assimilation
N. R. Smith, Bureau of Meteorology Research Center, Melbourne, Australia

1230 Measurement Of Ocean Surface Currents By An Along-Track Interferometric Synthetic Aperture Radar
W. Alpers, Institute of Oceanography, University of Hamburg, Hamburg, Germany; and R. Romeiser

1300 Lunch

1430 Panel Discussion: A Vision Of The Future

1430 Panelists: G. Asrar (Associate Administrator, NASA, Washington, USA);
G. Brachet (Director-General, CNES, Paris, France);
A. Rodota (Director-General, ESA, Paris, France);
S. Miura (Executive Director, NASDA, Tokyo, Japan);
R. Winokur (Assistant Administrator, NOAA, Washington, USA)

Moderator: C. Kennel (Director, Scripps Institution of Oceanography, La Jolla, USA)

1630 Refreshments

1800 ICSOS Closing Ceremony, EXPO '98 Ocean Pavilion
[Tickets Provided For Registrants]

1800 R. Zaharia (Centre National d'Etudes Spatiale, Paris, France);
J. Johannessen (ESTEC, European Space Agency, Noordwijk, The Netherlands);
A. Moura (International Research Institute, Palisades, USA);
M. Ruivo (Portuguese Intergovernmental Oceanographic Commission, Lisbon, Portugal)

1900 Reception

2000 Sessions End For The Day

2000 END OF CONFERENCE

Index

C

D

E

F

G

N

O

P

R